Teaching Secondary Mathematics

Technology plays a crucial role in contemporary mathematics education. *Teaching Secondary Mathematics* covers major contemporary issues in mathematics education, as well as how to teach key mathematics concepts from the Australian Curriculum: Mathematics. It integrates digital resources via Cambridge HOTmaths (www.hotmaths.com.au), a popular, award-winning online tool with engaging multimedia that help students and teachers learn and teach mathematical concepts. This book comes with a free 12-month subscription to Cambridge HOTmaths.

Each chapter is written by an expert in the field, and features learning objectives, definitions of key terms, and classroom activities – including HOTmaths activities and reflective questions.

Teaching Secondary Mathematics is a valuable resource for pre-service teachers who wish to integrate contemporary technology into teaching key mathematical concepts and engage students in the learning of mathematics.

Gregory Hine is a Senior Lecturer in Mathematics at the University of Notre Dame.

Robyn Reaburn is a Lecturer in Mathematics education at the University of Tasmania.

Judy Anderson is Associate Professor in Mathematics education, Director of the STEM Teacher Enrichment Academy, and a member of the Academic Board at the University of Sydney. She is currently the President of the Australian Curriculum Studies Association.

Linda Galligan is an Associate Professor with the School of Agricultural, Computational and Environmental Sciences at the University of Southern Queensland.

Colin Carmichael is a Senior Lecturer in Diploma Programs at the University of Southern Queensland.

Michael Cavanagh is the Director of the Teacher Education Program in the School of Education and Senior Lecturer in Mathematics education at Macquarie University.

Bing Ngu is a Lecturer in the Mathematics Education team within the School of Education at the University of New England.

Bruce White is a Lecturer in mathematics and science teacher education at the University of South Australia.

Teaching Secondary Mathematics

GREGORY HINE

ROBYN REABURN

JUDY ANDERSON

LINDA GALLIGAN

COLIN CARMICHAEL

MICHAEL CAVANAGH

BING NGU

BRUCE WHITE

CAMBRIDGE
UNIVERSITY PRESS

CAMBRIDGE
UNIVERSITY PRESS

477 Williamstown Road, Port Melbourne, VIC 3207, Australia

Cambridge University Press is part of the University of Cambridge.

It furthers the University's mission by disseminating knowledge in the pursuit of education, learning and research at the highest international levels of excellence.

www.cambridge.org
Information on this title: www.cambridge.org/9781107578678

First published 2016

Cover designed by Leigh Ashforth, watershed art + design
Typeset by Newgen Publishing and Data Services
Printed in Singapore by C.O.S. Printers Pte Ltd

A catalogue record for this publication is available from the British Library

A Cataloguing-in-Publication entry is available from the catalogue of the National Library of Australia at www.nla.gov.au

ISBN 978-1-107-57867-8 Paperback

Contents

Part 2 Learning and Teaching Key Mathematics Content   212

Preface

Teaching Secondary Mathematics has been written to enrich the professional lives of secondary mathematics educators at any stage of their careers. Throughout the text the audience has access to key theoretical and philosophical perspectives that have been the focus of much empirical research and that underpin best instructional practices within the secondary mathematics classroom. Additionally, various contemporary issues influencing the planning, teaching and evaluation of mathematical learning experiences have been included for reader consideration.

Within each of the 12 chapters, the needs of adolescent learners have been outlined carefully, together with pedagogical approaches educators can use to respond effectively to these needs. At the conclusion of each chapter, readers are provided with various activities to consolidate and extend their learning. First, reflective questions that link closely to the topics presented are offered for individual or collaborative response. Second, a variety of engaging mathematics activities – which are strategically linked to Cambridge HOTmaths – provides teachers of middle and senior school students with resources to use immediately in the classroom. At the time of publication, *Teaching Secondary Mathematics* reflects the thinking and content available in the most current version of the Australian Curriculum.

The authors of the chapters are notable academics hailing from a broad range of Australian universities, and who in their current roles prepare the next generation of mathematics teachers in Australia. Their professional classroom experience and engagement with cutting-edge research combine to produce a clear 'mathematical voice' which resonates in the summaries, insights, teachings, questions and reflections presented. Although the text has been authored primarily for an Australian audience, it is hoped that this voice will inspire and challenge mathematics teachers everywhere for years to come.

Gregory Hine
January 2016

How to use HOTmaths with this book

Once you have registered your HOTmaths access code (see the inside front cover of this book), for subsequent visits the below navigation instructions provide a general overview of the main HOTmaths features used within this textbook.

Log in to your account via at www.hotmaths.com.au.

This will take you to your Dashboard. The different HOTmaths streams can be accessed via the Course dropdown, located on the left-hand side of the toolbar. Here you can change the **Course list** (stream) and **Course** (year level), then select a **Topic** and **Lesson** from the respective lists.

Most lessons contain a number of interactive and printable activities, which can be accessed via the links on the right-hand side of the orange lesson toolbar. These include: **Resources**, **Walkthroughs** and **Questions**. The Resources tab contains the **demonstrations** (videos), **widgets** (animations) and **HOTsheets** (activities) associated with that particular lesson.

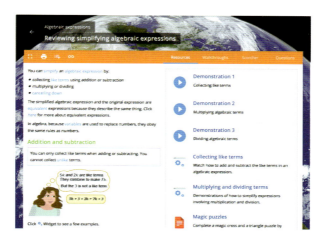

You can also access specific widgets and HOTsheets quickly using the search function, found on the right-hand side of the main toolbar. Upon searching the name of any widget or HOTsheet, the results page will display widgets based on the keywords searched – as denoted by the 'widget' tab being highlighted in blue. If you are looking for a HOTsheet, simply click onto the 'Hotsheets' tab and these results will appear. Using the 'Magic puzzles' HOTsheet as an example, searching this name and clicking on the 'HOTsheets' results tab will provide you a link to the Magic puzzles HOTsheet.

Please note that given its nature HOTmaths is constantly being updated. All pathways and references are up to date as of May 2016 and every effort has been made to provide you with an accurate picture of the functions within HOTmaths.

Acknowledgements

The authors and Cambridge University Press would like to thank the following for permission to reproduce material in this book.

Figure 2.2: Australian Institute of Health and Welfare; **3.7**, **5.2**: Reproduced from K. Stacey (2005). The place of problem solving in contemporary mathematics curriculum documents. *Journal of Mathematical Behaviour*, 24, 341–50; **3.9–3.12**: Reproduced with permission from R. Kincaid (2015). *Weighted dice: A study in applications of probability* (PowerPoint slides). Proceedings of the S.T.E.A.M. & Education Conference. Honolulu, Hawaii; **4.1**: Reproduced by permission of the publisher, © 2012 by tpack.org; **4.2**, **4.3**: Finger, G., Romeo, G., Lloyd, M., Heck, D., Sweeney, T., Albion, P. & Jamieson- Proctor, R. (2015). Developing Graduate TPACK Capabilities in Initial Teacher Education Programs: Insights from the Teaching Teachers for the Future Project. *The Asia-Pacific Education Researcher*, 1–9; **4.5**, **10.6**: TinkerPlots software © Learn Troop 2016; **7.2–4**: Minstrell, J. (2015). Facets of thinking: Diagnostic assessment of speed versus time graph. Retrieved on March 6 2015 from http://www.diagnoser.com; **7.5**: Reproduced by permission of Oxford University Press, Australia from George Booker, 2011, *Building numeracy: Moving from diagnosis to intervention* © Oxford University Press, www.oup.com.au; **10.5** House image © iStock/Getty Images Plus/mstay; **11.13**: PhET Interactive Simulations, University of Colorado, http://phet.colorado.edu. Reproduced under Creative Commons Attribution licence; **12.7**, **12.8**: Created courtesy of International GeoGebra Institute.

ACARA material: © Australian Curriculum, Assessment and Reporting Authority (**ACARA**) 2009 to present, unless otherwise indicated. This material was downloaded from the ACARA website (www.acara.edu.au) (**Website**) (accessed various dates) and was not modified. The material is licensed under CC BY 4.0 (creativecommons.org/licenses/by/4.0/). ACARA does not endorse any product that uses ACARA material or make any representations as to the quality of such products. Any product that uses material published on this website should not be taken to be affiliated with ACARA or have the sponsorship or approval of ACARA. It is up to each person to make their own assessment of the product.

Every effort has been made to trace and acknowledge copyright. The publisher apologises for any accidental infringement and welcomes information that would redress this situation.

Part 1

Contemporary Issues in Learning and Teaching Mathematics

CHAPTER 1

Introduction: the learning and teaching of mathematics

Learning outcomes

After studying this chapter, you should be able to:

- discuss the importance of mathematics in modern society, including fundamental aspects of mathematical literacy and numeracy
- describe the Proficiency Strands of the Australian Curriculum: Mathematics
- summarise the key features of some important theories about student learning
- identify some ways in which teachers can promote student engagement in mathematics lessons.

Introduction

The purpose of this chapter is to consider the nature of mathematics, its importance in the lives of twenty-first century students, and to discuss some key theories about how learning takes place. The chapter begins with a description of some of the key knowledge, skills and understandings that today's students need in order to participate fully in modern society. Next, we discuss some of the significant developments in theories about how students learn mathematics and some implications of these theoretical approaches for your work as a teacher. The chapter concludes with a look at how we might engage students more productively in mathematics lessons in order to cultivate positive attitudes to the subject among students and to improve their learning outcomes.

At various points in the chapter we shall discuss some relevant scholarship and research that may provide guidance for your work of teaching mathematics in secondary classrooms. We will also explore some of the HOTmaths resources (www.hotmaths.com.au) that you could use in your lessons and you will have an opportunity to reflect on their significance.

KEY TERMS

- **Behaviourism:** a theory of learning that is mainly concerned with observable behaviour and assumes that learners respond passively to external stimuli.
- **Constructivism:** a cognitivist theory of learning that views the learner as an active constructor of knowledge.
- **Engagement:** the level of attentiveness, interest and curiosity that students exhibit when they are learning.
- **Financial literacy:** the ability to manage money and financial risks effectively and responsibly in order to achieve one's financial goals.
- **Gestaltism:** a theory that suggests that the mind processes wholes first, before attending to constituent parts.
- **Numeracy:** the ability to understand and work flexibly and efficiently with numbers and other mathematical concepts to solve a variety of problems across a range of contexts.
- **Socioculturalism:** a view of learning that emphasises how learning does not take place solely within the mind of the individual, but also through one's interactions with others.
- **Statistical literacy:** the ability to understand, interpret and evaluate data from a variety of sources.

Mathematics for the twenty-first century

The Melbourne Declaration on Educational Goals for Young Australians (Ministerial Council on Education, Employment, Training and Youth Affairs, MCEETYA, 2008a) is the framework document that will effectively set the national school education agenda for the coming decade. One of the central themes in the Declaration is the need to improve the educational outcomes for all students since this is 'central to the nation's social and economic prosperity and will position young people to live fulfilling, productive and responsible lives' (MCEETYA , 2008a, p. 7).

We live in a world of rapid change and development, especially in terms of advances in technology. As information and communication technologies become more sophisticated, they exert an ever-increasing influence on how we conduct our daily lives. Many of these technological tools, such as data-bases and spreadsheets, are underpinned by mathematics (Siemon et al., 2011). Mathematical understanding is crucial in laying a strong foundation for study beyond secondary school in a range of disciplines, such as engineering, business and finance. In addition, developing mathematical understanding and becoming more numerate helps students to interpret a range of practical situations, allowing them to make more informed decisions in their everyday lives. For example, reading timetables, interpreting graphs, converting frac-tional amounts when cooking, and so on.

The Shape of the Australian Curriculum: Mathematics (National Curriculum Board, 2009) also notes the need for a mathematically literate workforce:

> Successful mathematics learning also provides a workforce that is appropri-ately educated in mathematics to contribute productively in an ever-changing global economy, with both rapid revolutions in technology and global and local social challenges. An economy competing globally requires substantial num-bers of proficient workers able to learn, adapt, create, interpret and analyse mathematical information. (2009, p. 4)

Mathematics education plays a crucial role in driving national productiv-ity and prosperity. School mathematics lays a foundation on which to build the mathematical and scientific literacy every citizen needs to thrive in an increasingly technology-dependent world. But what kinds of mathematical knowledge, skills and understandings do today's students require if they are to be full and active members of our modern, highly technological age? To answer this question we first need to consider the very nature of mathematics and what it means to do mathematics.

The Dutch mathematician Hans Freudenthal developed an approach to teaching known as realistic mathematics education, which he based on learn-ing mathematics by solving well-chosen and carefully sequenced problems taken from daily life. Freudenthal believed that by attempting to solve such problems, students would gradually develop and enrich their mathemati-cal understanding. Although he died in 1990, his work has continued to be very influential in mathematics education reform, not only in Holland, but throughout the world. He once described mathematics as:

> an activity of solving problems, of looking for problems, but it is also an activity of organizing a subject matter. This can be a matter from reality which has to be organized according to mathematical patterns if problems from reality have

to be solved. It can also be a mathematical matter, new or old results, of your
own or others, which have to be organized according to new ideas, to be better
understood, in a broader context, or by an axiomatic approach. (Freudenthal,
1971, pp. 413–14)

As you read this quotation, and perhaps you need to do so more than once,
note how Freudenthal begins with the many different kinds of activities people do in their everyday lives that involve mathematics. He also refers to more
abstract mathematics, but in doing so he emphasises mathematics as a *process*
of inquiry, particularly through his reference to finding and solving problems,
taking a systematic approach in organising one's thinking about these problems, and making generalisations from their solutions to develop mathematical understanding. Another important feature of Freudenthal's description is
the underlying assumption that mathematics is an essentially human endeavour because it is always contextualised as a social and cultural activity. From
Freudenthal's point of view, mathematics education has its beginnings in the
socio-cultural world of each student and the essential aim of school mathematics is to strive to raise students' awareness of the essentially mathematical nature of this everyday human activity (Ryan & Williams, 2007).

If we take Freudenthal's description as our starting point, we can begin to
identify some of the particular mathematical knowledge and skills that are
important for today's students. These are discussed in the following sections.

Statistical literacy

Data are all around us and twenty-first century students need considerable
skill in order to comprehend the large quantity of information they are presented with each day through the media and other sources. Indeed, to be an
active participant in our democratic society requires the ability to deal with
ever increasing amounts of data. Many occupations, such as actuaries, business analysts and meteorologists, also require an ever greater appreciation of
data in far more sophisticated ways than has previously been the case. These
jobs require a sound grasp of statistics – a powerful tool that enables us to
make sense of our data-driven world. We can use statistics to identify trends
over time and make meaningful comparisons among and between data sets.
We can also extrapolate from data in order to hypothesise and predict what
might occur in the future for the purposes of planning and decision-making.

Every day, people use numbers and comparisons of numbers to justify
decisions they make about their lives. Statistics is a way of making sense of
a number of factors or costs when comparing these activities and providing

evidence to support our choices. In **statistical literacy**, students learn how to add rigour to numerical claims and comparisons and how to examine other people's claims using statistical evidence and reasoning. Statistical literacy is the ability to make sense of data; it is one's ability to understand, interpret and evaluate data from a variety of sources. Statistical literacy includes the ability to know where to look and how to find data appropriate to your specific needs and purposes. Being statically literate means understanding and making sense of data when it is presented to you. It involves familiarity with statistical terminology in order to communicate your understanding clearly to others. And statistical literacy is also about adopting a critical stance to deal with data that may be displayed in an ambiguous or misleading fashion (Gal, 2002). As with many other areas of mathematics, we would also want our students to develop positive attitudes and confidence in their statistical abilities so that they can be informed and engaged participants in society.

Some of the activities that could be used to promote statistical literacy include the development of surveys and questionnaires and a discussion about how the wording and ordering of the items can influence responses. Students could consider if they need to survey the entire population (e.g. all of the students in a school) or if a sample would be sufficient. This could lead to discussions about the sample size and the need for the sample to be representative of the population. Raw data often needs to be 'cleaned up' by, for example, disregarding anomalous results. In doing so, students could consider the impact of outliers on measures such as the mean of a set of scores. When the data are collected, decisions need to be made about how it could be represented (in tabular form or graphically). This could lead to discussion about the kinds of graphs that are best suited to representing data that are categorical, numerically discrete or continuous. Then the data need to be analysed so that trends or other patterns can be identified. When dealing with data from other sources, students need to develop a healthy scepticism for claims made by considering how the data were collected and by whom, and for what purpose. This could lead to a discussion of biased or misleading data.

Financial literacy

Another important aspect of mathematics in the twenty-first century is the area of **financial literacy**. Financial literacy is significant because the ability to manage personal finances 'is a core skill in today's world. It affects quality of life, the opportunities individuals and families can pursue, their sense of security and the overall economic health of Australian society' (National Consumer and Financial Literacy Framework, 2011, p. 5). A recent report by the Australian Securities & Investments Commission (ASIC) defines financial literacy as 'being able to understand and negotiate the financial landscape, manage money and financial risks effectively and responsibly, and pursue and attain financial and lifestyle goals' (ASIC, 2014, p. 6). Financial literacy is therefore seen as a core life skill for twenty-first century students so that they can fully participate in modern society. The increasingly complex nature of the world today and the greater number of alternatives for dealing with discretionary income mean that students need to develop the knowledge and skills to take charge of their own financial future. Helping students understand financial issues is particularly important nowadays since they can be expected to deal with more and more sophisticated financial products and services. Levels of financial risk are also rising, especially in areas such as managing savings and investments, purchasing a home, planning for retirement and ensuring the ability to pay for future healthcare needs.

However, surveys and assessment data show that young adults typically have very low levels of financial literacy. For example, in 2012 the Programme for International Student Assessment (PISA) conducted the first financial literacy assessment of approximately 29 000 secondary school students aged 15 years. The survey was conducted in 18 countries and covered financial issues such as understanding a bank statement, calculating the long-term cost of a loan and knowing how insurance works. The results, which were released in July 2014, showed no significant difference between the performance of boys and girls. But they did indicate that students from relatively high socio-economic backgrounds tended to do better in financial literacy than less advantaged students. Importantly, the results also revealed that skills in mathematics and reading were very closely related to financial literacy.

The National Consumer and Financial Literacy Framework (2011, p. 5) sets out some of the key aspects of financial literacy that students should have the opportunities to learn. These include:

- how to manage their finances and plan for needs and wants, now and into the future
- the language of money, how to navigate the ever-changing consumer and financial landscape and where to go to for assistance
- the rights and responsibilities of consumers in modern society and the wider impact of everyday consumer and financial decisions
- the skills needed to develop a range of enterprising behaviours.

The range of skills that students need to develop in order to engage in the kinds of enterprising behaviours envisaged in the Framework document include adaptability, initiative, communication, problem solving, planning and organising, analysing issues and managing identified risks. Many of these skills are closely aligned with some of the central themes of the Australian Curriculum: Mathematics (Australian Curriculum, Assessment and Reporting Authority, 2011). This is especially the case with the working mathematically processes, which we discuss later in the chapter.

Numeracy

Today's students must achieve basic **numeracy** skills such as the ability to count, measure, compare and sequence. Students need to develop competence in many everyday tasks, such as budgeting, shopping, travel and leisure activities, and so on. In some respects, these skills can be learned and practised through simple exercises such as those typically found in school textbooks. However, most mathematics educators argue that to be fully numerate, students must do more than acquire and master basic mathematical routines and algorithms. These skills are necessary and essential, but not sufficient for numeracy (Australian Association of Mathematics Teachers, 1997).

The numeracy demands placed on twenty-first century learners cover much more than just number skills and the ability to recall basic facts such as the multiplication table. Numeracy involves a deep understanding of mathematical concepts and skills from across the discipline, including numerical, spatial, graphical, statistical and algebraic topics. Students need to develop their mathematical thinking strategies as well as more general thinking skills through problem-solving activities which are grounded in realistic contexts. Perso (2011) identifies the following as fundamental characteristics of students who are numerate:

- knowing how to read situations and determine if or what mathematics is needed
- knowing how to choose the methods and tools they will use and explain and justify their reasons based on context
- knowing how to apply the methods, models and tools they have chosen
- knowing how to critique their own (and others') methods and determine whether they make sense in context
- knowing how to communicate their process and results appropriately depending on purpose and audience.

The report of the National Numeracy Review (Commonwealth of Australia, 2008) reinforces the need to view numeracy more broadly than just the acquisition of basic numerical skills:

> Students need to learn mathematics in ways that enable them to recognise when mathematics might help to interpret information or solve practical problems, apply their knowledge appropriately in contexts where they will have to use mathematical reasoning processes, choose mathematics that makes sense in the circumstances, make assumptions, resolve ambiguity and judge what is reasonable. (2008, p. xi)

In other words, students today must not only develop disciplinary knowledge of basic mathematical processes and skills, they also require a deep conceptual understanding as well. That will enable them to engage in problem solving and critical analysis, apply their knowledge creatively in unfamiliar contexts, and communicate their ideas using appropriate mathematical language and symbols (Office of the Chief Scientist, 2014).

ACTIVITY 1.1

An example of a HOTmaths activity that could be used to develop students' numeracy skills is the 'Exploring discounts' HOTsheet. You can access this HOTsheet by logging in to HOTmaths and selecting the 'HOTmaths Global' Course list and the 'Early Secondary' Course. From the Topic list choose 'Using percentages', then the 'Discounts' Lesson. The HOTsheet can be found in the Resources tab.

The activity is designed to help students learn about mental computation methods for calculating percentage discounts and is linked to the Australian Curriculum: Mathematics learning outcome ACMNA187 (Solve problems involving the use of percentages, including percentage increases and decreases, with and without digital technologies).

The advantages of the HOTmaths activity are its clear links to real-life contexts that are familiar to students (calculating discounts on items they wish to purchase) and the emphasis on mental computation (e.g. calculating the amount of the discount using multiples of 10%). Students are also encouraged to make an estimate before calculating and must develop their own strategies to solve a range of problems. At the conclusion of the activity, the teacher can lead a discussion in which the students share their estimation techniques and compare the relative merits of each approach in order to ascertain the most efficient mental computation methods.

EXPLORING DISCOUNTS

TASK 1   **Use 10%**

10% is the same as one tenth. **Use 10% to complete these calculations.**

1 Find 30% discount on $90.
 	
     30% = 3 × 10%
 	
     Discount = 3 × 10% of $90
 	
              = 3 × $9
 	
              =
	2 Find 40% discount on $500.
	
	     40% = 50% – 10%
	
	     50% of $500 =
	
	     10% of $500 =
	
	     Discount =
	
	              =

3 Find 15% discount on $120.
 	
     5% is half of 10%.
 	
     10% of $120 =
 	
     5% of $120 =
 	
     Discount =
	4 Find 60% discount on $8.

Figure 1.1 Exploring discounts HOTsheet

REFLECTIVE QUESTIONS

Read over the 'Exploring discounts' HOTsheet (see Activity 1.1) and then consider the following:

- Why is it important to contextualise the activity using relevant examples and exercises?
- Why is it important to emphasise mental computation and estimation?
- How does the activity promote numeracy?

Mathematical proficiency and the Australian Curriculum

Another way to think of the kinds of mathematical thinking strategies that twenty-first century learners need is found in the work of Kilpatrick, Swafford and Findell (2001). The authors describe five components or *strands* of mathematical proficiency to 'capture what we believe is necessary for any-one to learn mathematics successfully' (Kilpatrick, Swafford & Findell, 2001, p. 116). These include the following:

- *conceptual understanding* – the basic comprehension of fundamental math-ematical concepts, operations and relationships
- *procedural fluency* – the skill to perform mathematical procedures that are completed flexibly, accurately, efficiently and appropriately
- *strategic competence* – the ability to formulate, represent and solve math-ematical problems
- *adaptive reasoning* – the facility for logical thought, reflection, explanation and justification
- *productive disposition* – the predisposition to view mathematics as sensible, useful and worthwhile, combined with a belief in persistence and one's own ability.

Watson and Sullivan (2008) refined the above proficiencies and distilled them into four key areas which became the proficiency strands of the Australian Curriculum: Mathematics (ACARA, 2011a). These strands are Understanding, Fluency, Problem Solving and Reasoning.

Teachers also have a crucial role to play in helping students become more proficient mathematicians through their work in facilitating the mathemati-cal knowledge, skills and understandings needed by twenty-first century learners. Steen and Forman (2000) use the term 'functional mathematics' to describe a range of approaches that teachers could employ to assist students in developing the kinds of attitudes and aptitudes we have discussed. The authors suggest that students learn best when they are active participants in their learning. Lesson activities therefore need to provide regular opportu-nities for students to make choices about the kinds of tools and strategies they will need to solve a range of mathematical problems. Problem-solving tasks will be most effective when students regard them as relevant and interesting, and this encourages students to make connections from the classroom to the mathematics of work and everyday life. We should allow students to work col-laboratively in order to strengthen their communication skills. There is also a need for a progressive formalisation by which students engage with problems first in context, and later with increasing mathematical formality. We should also require that students verify the reasonableness of their solutions in the context of the problem they were attempting to solve.

The learning outcomes associated with statistical literacy, financial liter-acy, numeracy and mathematical proficiency provide a sound basis for stu-dents to take their place in modern society. If they achieve the outcomes, students will possess the requisite knowledge and skills to become full and active participants, contributing to the common wealth and advancement of the nation. In the next section of the chapter we turn our attention to some key theoretical positions on how best to ensure that student learning takes place in mathematics classrooms.

Theories about how students learn mathematics

In general, the common thread running through all learning theories is that they are based on the premise that learning involves some kind of change. However, different theoretical approaches to learning vary in quite fundamental ways in describing the nature of this change and how it might be identified or measured. This section of the chapter focuses on learning theories that have had a significant impact on the practice of mathematics education in recent decades. As a teacher, it is important for you to know and understand these learning theories since, 'For teaching to be effective, it must be grounded in what we know about how students learn' (Goos, Stillman & Vale, 2007, p. 28). We begin our discussion with **behaviourism**.

Behaviourism

Behaviourism was the archetype approach used in educational psychology from the 1920s until the 1950s. For behaviourists, learning is evidenced in changes in a person's behaviour and so behaviourists are concerned with actions that are observable and measurable through empirical data. In general, behaviourist theories suggest that learners are shaped by a range of environmental influences or stimuli to which they respond (Siemon et al., 2011). Behaviourism has its origins in observational studies of animals in laboratory settings. These studies showed that animal behaviour could be influenced through reward and punishment to reinforce desirable actions. For example, rats and pigeons could be trained to perform certain actions, such as pressing levers in order to receive food, or to avoid negative consequences, such as mild electric shocks. Later studies showed that behavioural reinforcement could accelerate learning for adults and children as well.

One of the early proponents of behaviourism was Edward Thorndike. He developed a theory of operant conditioning, which is sometimes referred to as instrumental learning. Thorndike developed his 'law of effect' in which he postulated that an association could be strengthened or weakened as a result of its consequences; behaviours which led to positive outcomes were more likely to be repeated than those which produced disagreeable results. In other words, using positive outcomes to promote desirable behaviours is a much more effective strategy than punishment for unwanted actions. Thorndike's work also examined the bonds that can be formed between a stimulus and its response. He demonstrated that when an association is established between a stimulus and its response and this S–R association is repeated, the response is likely to continue to occur even after the stimulus is reduced or withdrawn.

Thorndike believed that elementary stimulus–response associations were the foundation of learning and applied his theory to the learning and teaching of arithmetic through his use of number bonds. He identified 390 such bonds which were based on the four operations of arithmetic and the numerals 0 to 9. He based this work on three key assumptions. Firstly, that learning is essentially the construction of bonds or connections between situations and responses to them. Secondly, that habitual drill and practice leads to improved memory and better learning outcomes. Thirdly, that such improvement could be best measured by standardised testing.

Later, B. F. Skinner built on Thorndike's work to propose the notion of operant conditioning, in which behaviour is affected and altered by the events that precede and follow it. Skinner also introduced a new idea into the law of effect – the concept of reinforcement. According to Skinner, behaviours which are continually reinforced are more likely to be strengthened and repeated while those which are not reinforced tend to be weakened and diminish over time. In essence, Skinner conceptualised the learning of mathematics as the formation and strengthening of associations between stimuli and responses to them. He focused particularly on observable skills since he believed that learning was evidenced in terms of changes in behaviour.

The work of behaviourists such as Thorndike and Skinner has had a profound impact on mathematics education. Although behaviourist methods are rarely advocated nowadays, it is still possible to see their influence in many mathematics classrooms and syllabuses. The hold of behaviourism is evident in the ways that student learning outcomes are often phrased in syllabus documents, the emphasis on rote learning, the abundance of graded examples and exercises like those often found in school textbooks, a narrow focus on basic skills in assessment tasks, and an approach to teaching firmly grounded in direct instruction (Ryan & Williams, 2007). As Brumbaugh and Rock (2001) note, 'The behaviourist approach treats mathematics as a collection of skills. Learn the skills and learn mathematics' (p. 27).

Another legacy of behaviourism can be seen in the work of Robert Gagné and his hierarchy of learning. Gagné's model classifies various types of learning activities according to the complexity of the mental processes involved in completing them. The higher levels, such as concept learning and problem solving, deal mainly with more cognitive aspects of learning, while the lower-level learning tasks, such as stimulus–response learning, are related to more behavioural features. The hierarchical nature of the model is reflected in Gagné's assertion that higher orders of learning require increasingly greater amounts of prior learning at the lower levels. Gagné proposed that students could achieve learning outcomes by following a predetermined sequence of tasks, each one building on its predecessor. Students could master the tasks at each subordinate level through repetition and practice and then proceed to the next stage in the learning continuum.

Gagné argued that particular kinds of teaching methods were also crucial to ensure students' mastery of prerequisite skills. These included clear objectives for lessons, regular reminders to students of previously learned material and guided instruction. The teacher should then demonstrate worked examples prior to students completing exercises that were very similar to those modelled. Regular feedback on students' progress towards the learning goal is also provided, along with systematic reinforcement of correct procedures (Leder & Forgasz, 1992). Gagné's learning hierarchies were often represented as flow charts which provided teachers and students with a step-by-step program of graded tasks and activities leading to the desired learning outcome. This approach to learning and teaching is reductionist in style since it attempts to reduce learning to a series of discrete tasks and because it focuses primarily on the mathematical content to be learned rather than the processes by which learning takes place.

Gestaltism

Gestaltism takes its name from a German word meaning shape or form. It is based on the belief that the mind tends to perceive the whole form (the gestalt) first and only afterwards are the constituent parts analysed. Gestaltism can therefore be seen as in opposition to the reductionist approach taken by behaviourists such as Gagné, who emphasised a step-by-step approach to learning and teaching. For the gestaltists, learning was far more complex in nature and the sensory whole was always viewed as other than just the sum of its sensory parts. The gestalt theorists were especially critical of the rote learning methods espoused by behaviourists and argued against the behaviourists' insistence on drill and practice as the most efficient method to learn mathematics. The gestaltists based their views on the potentially harmful effects of stimulus–response approaches following experiments which suggested that students who had learned by rote often struggled to solve problems which were presented in unfamiliar settings or varied even slightly from those which they had practised.

An example of one such experiment comes from the work of Max Wertheimer (as cited in Schoenfeld, 1988). Wertheimer asked children in primary school grades to solve arithmetic problems such as

$$\frac{357 + 357 + 357}{3}$$

He noticed that many children promptly began the somewhat laborious task of calculating the sum of the numerator and then performing the division by the denominator. Those children who were competent in the arithmetic operations were able to solve the task correctly; however, they often did so without apparently noticing, even after obtaining the right answer, how the structure of the problem might have been useful in saving all that time and effort. Wertheimer concluded that although these students had mastered the procedures required for addition and division, they had failed to recognise the repeated addition as equivalent to multiplication and division as the inverse of multiplication. They seemed to be so conditioned to focus on the individual parts of the problem that they failed to interpret it as a whole. In other words, the majority of children had clearly not developed a deeper understanding of the underlying arithmetical structure. As Schoenfeld notes,

> This example illustrates that being able to perform the appropriate algorithmic procedures, although important, does not necessarily indicate any depth of understanding. (1988, p. 148)

Schoenfeld then proceeds to discuss a more widely known example from Wertheimer's work which concerns his observations of some mathematics lessons in which students were learning about the area of a parallelogram. As in the previous case from arithmetic, the students had been taught the standard procedure – in this case, how to apply the formula 'Area equals base length multiplied by perpendicular height'. Wertheimer observed the teacher illustrate the formula by cutting away a triangular section from one side of the parallelogram and repositioning it on the other edge to produce a rectangle whose base and altitude were equal to that of the original figure (see Figure 1.2). He noted how the students were able to reproduce the teacher's method and

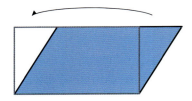

Figure 1.2 Transforming a rectangle into a parallelogram

explain it adequately during the class. However, when Wertheimer later asked some of the students to calculate the areas of parallelograms given in non-standard orientations, the students quickly became confused and were unable to provide the correct answers. Again, Wertheimer deduced that even though the students had memorised the proof and correctly followed the procedures they had learned to complete exercises during the lesson, their lack of appreciation as to how or why their methods worked exposed their lack of conceptual understanding. Instead, their superficial understanding limited what they could do, especially when confronted with tasks that did not closely conform to those they had previously learned to solve.

Richard Skemp's often cited distinction between instrumental and relational understanding in mathematics learning is apt here. Skemp defined the former as 'rules without reasons' and the latter as 'knowing both what to do and why' (Skemp, 1976, p. 20). He also listed some advantages for each type of understanding. For instrumental understanding, Skemp noted that it is usually easier to understand, the rewards are more immediate and correct answers can often be obtained more quickly. As Skemp notes,

> If what is wanted is a page of right answers, instrumental mathematics can provide this more quickly and easily. (1976, p. 23)

However, as Wertheimer's observations remind us, simply being able to obtain the right answer is not always sufficient. Skemp identified the advantages of relational understanding as its adaptability to new tasks: relational understanding is easier to remember, it can provide intrinsic motivation for learners, and it is organic in nature (by which Skemp meant that the development of relational understanding in one area of mathematics encourages a search for similar levels of understanding in others). On balance, Skemp believed that the longer-term advantages of relational learning far outweigh any immediate benefits associated with instrumental understanding.

Skemp also acknowledged that despite the obvious advantages of a relational approach to learning and teaching mathematics, instrumental learning remains prevalent in the majority of mathematics classrooms and he provides some important reasons why this might be the case. These include: firstly, the powerful influence of high stakes external examinations (which tend to focus on testing understanding at a more instrumental level of recall of basic rules and procedures); secondly, overcrowded mathematics curricula (which discourage teachers from allocating the time that is often necessary, at least in the initial stages, to teach for relational understanding); thirdly, the difficulties associated with assessing students' relational understanding (which, to be valid and reliable, typically requires interviewing individual pupils about their work); and fourthly, the challenges associated with encouraging teachers to discard long-held views about the nature of mathematics and mathematics instruction (which can be difficult to shift).

Constructivism

Skemp's advice on the long-term benefits of a relational understanding of mathematical concepts fits nicely within a cognitivist outlook on learning. Cognitivism is the paradigm in psychology and learning sciences that overtook behaviourism in the 1960s during the so-called Cognitive Revolution (for example, see Miller, 2003). One of the pillars of cognitivism is the notion that knowledge is not simply received by learners but rather that it is actively constructed by them. While there are now many forms of **constructivism**, they all share this fundamental premise (Bobis, Mulligan & Lowrie, 2013).

Constructivism regards learning as a cognitive process and is derived principally from the work of Piaget. He viewed learners as individuals who interact with their environment in order to make sense of it (Jaworski, 1994). Indeed, constructivism sees the learner as the central agent in the process of knowledge construction, through assimilating one's experience in the world (Ryan & Williams, 2007). Assimilation is a process by which we interpret input from the world around us through our senses. This information is interpreted in terms of our existing cognitive structures or schemata (Siemon et al., 2011). For constructivists, to understand a concept is essentially to assimilate it into an appropriate schema.

At times, individuals may have an experience which cannot be adequately explained by their existing knowledge schemata. In these situations, a conflict or perturbation may arise which is unable to be resolved simply through assimilation of the new knowledge or experience (Glasersfeld, 1995). Constructivists believe that the desire to maintain an equilibrium – by which we continue to make sense of our experiences – leads to an accommodation or structural reorganisation of existing schemas and the construction of new ones as well (Peled & Suzan, 2011). As Davis and Tall (2002) remark, the assimilation and accommodation processes are very much dependent on the experience of each individual:

> Essential to an understanding of schemes is the focus of attention of a learner. What it is a person focuses on in an action scheme determines the consequent structure of that scheme for them. (2002, p. 133)

Ben-Hur (2006) suggests that there are some fundamental characteristics of the structural changes in schemas that arise as a result of accommodation. The schematic restructuring is *pervasive*, since profoundly new experiences that require accommodation lead to rearrangements of the entire cognitive structure. Accommodation promotes *centrality* because when new learning brings about a structural change in cognitive structures it places the learner in the position to learn something more advanced in the same area. And the reorganisation is regarded as *permanent* because the kind of structural changes that result from the need to accommodate new experiences are neither reversible nor can they be forgotten.

Given that new learning is seen to arise as a consequence of the individual's desire to maintain equilibrium (Lerman, 2001), constructivism proposes that teachers deliberately create situations that are likely to engender a cognitive conflict between the student's current way of thinking and the new knowledge to be learned. This approach is often referred to as conflict teaching and it can produce significant moments in the learning process, as it is designed 'both as a source for conflict and instability, and at the same time as an opportunity to reflect, reorganize one's knowledge and construct new knowledge' (Peled & Suzan, 2011, p. 75). For example, it has been observed that students who learn instrumentally are likely to over-generalise rules, formulas or procedures and use them in contexts where they do not apply. On such occasions, the use of non-examples can be a profoundly important means of generating disequilibrium. If the teacher can encourage students to make sense of the unexpected situation for themselves, exploring those non-examples may lead to deeper insights into the underlying structure of the problem. Wertheimer's use of parallelograms in non-standard positions referred to previously is a case in point. Students who did not recognise that the area formula requires that the height of the parallelogram is perpendicular to its base could be helped to understand the formula by attempting to find the areas of non-standard figures and explaining their results.

Constructivists view the learner as an active participant in the learning process because they hold that knowledge is not passively received but instead is actively constructed. The role of the teacher also changes from that of transmitting knowledge to students (who are viewed as 'blank slates' or 'tabula rasa') to facilitating knowledge construction by the students themselves. Therefore, the teacher's task in a constructivist lesson is to select meaningful learning experiences, to encourage students to engage productively in the learning activities and to guide students to develop their understanding of fundamental concepts.

Another important aspect of constructivist approaches to learning is the need for students to reflect on what they have done. Reflection allows students to make sense of their learning and abstract the underlying properties of a mathematical concept. Hence reflection is regarded as an essential step in triggering the process of accommodation (Boaler, 2000). Piaget makes an important distinction in his work between what he refers to as *empirical abstraction* and *reflective abstraction*. The former is based on our experiences and observations which allow us to derive the common properties of objects and make generalisations from specific instances or occurrences. For example, a child is likely to have often experienced fair sharing, where each person receives an equal part of the whole. From many such experiences in a variety of different contexts the child is able to empirically abstract the concept of 'half' and then generalise its meaning in new and unfamiliar situations.

Reflective abstraction, on the other hand, is internalised and leads to different kinds of generalisations. These generalisations are more concerned with interrelationships among actions and allow new meanings to be constructed. For example, when learning the commutative property of addition (that the total is independent of the order in which the numbers are added together), a child might work with concrete materials to count and rearrange a group of objects several times. Each of these physical actions is represented internally in such a way that the child can reflect on them and realise that they give the same result each time. These reflections then lead to an abstraction of the concept of commutativity.

Piaget (1972) refers to the *encapsulation* of a process as a mental object in such a way that

> the whole of mathematics may therefore be thought of in terms of the construction of structures, ... mathematical entities move from one level to another; an operation on such 'entities' becomes in its turn an object of the theory, and this process is repeated until we reach structures that are alternately structuring or being structured by 'stronger' structures. (1972, p. 70)

Reflective abstraction, therefore, is activating the restructuring of one's knowledge schema/models such that processes that were implicit at one level become the objects on which the learner acts in the next (Boaler, 2000). In a similar vein, Sfard (1991) uses the term *reification* to describe 'a sudden ability to see something familiar in a totally new light ... an instantaneous quantum leap: a process solidifies into object, into a static structure' (p. 20).

Why, then, is this restructuring of processes as objects so important? The answer lies in the fact that when mathematical ideas and processes are reified as objects they can be stored in long-term memory in a compressed form and promptly reactivated when required. The results of research by Gray (1991) may help to illustrate this important point. In the study, Gray asked teachers to select children whom they identified as above average, average and below average in terms of their ability to perform simple arithmetic. He interviewed 72 children between the ages of seven and 12 from two schools and asked the children to solve simple addition and subtraction questions while he recorded the methods they used. He identified four main strategies which the children used: counting all (where all of the numbers are counted), counting on (where the child starts counting from one of the given numbers), using known facts (results which the child has committed to memory) and using derived facts (a flexibility in using the numbers to make the process of calculating easier).

The results of Gray's study showed clearly that lower performing children used complicated methods such as counting all of the numbers from one or counting on from the smaller number. These methods involved additional cognitive load and were more likely to result in errors. Higher performing children, on the other hand, made greater use of derived facts and employed more efficient strategies. Gray concluded that these two groups were, in fact, doing a different kind of mathematics. High performers were able to retrieve a range of facts and procedures with ease and adapt them to derive new facts, which could then be accumulated into their memory and added to their stored repertoire. Low performers struggled with memory-intensive procedures that constrained their ability to move beyond the most laborious and error-prone of methods. In other words, the process of reflective abstraction enabled the higher performing students to develop a relational understanding of the mathematics.

Socioculturalism

If constructivists view learning as a process of knowledge construction in the mind of the individual, sociocultural theorists regard learning as 'a collective process of enculturation into the practices of the mathematical community' (Goos et al., 2007, p. 28). Learning therefore leads to changes in participation and identity (Lave & Wenger, 1991) as students grow in confidence and improve their ability to make sense of mathematical ideas. The social constructivist view of knowledge is as

> not fixed, rather it is socially negotiated, and is sought and expressed through language. In other words, students do not learn only by listening but also by engaging in experiences that contribute to their learning, and having the opportunity to share approaches to tasks with others.
>
> (Sullivan, Clarke & Clarke, 2013, p. 9)

Much of the sociocultural approach to learning builds on the work of Vygotsky. He proposed that knowledge construction develops from social interactions and that the context in which these interactions take place acts as a mediator for the learning that results. His belief that social interaction preceded development is in contrast to the constructivist view proposed by Piaget, who regarded development as a necessary precursor for learning. The importance of mediated learning is summarised in Vygotsky's concept of *internalisation,* which he described as 'internal reconstruction of an external operation' (Vygotsky, 1978, p. 56). That is, learning first occurs on the social plane between individuals, and then within individuals on the psychological plane.

Crucial to Vygotsky's insistence on the primacy of social interaction is what he termed the *Zone of Proximal Development,* or ZPD, which he described as

> the distance between the actual developmental level as determined by independent problem solving and the level of potential development as determined through problem solving under adult guidance or in collaboration with more capable peers. (Vygotsky, 1978, p. 86)

The ZPD therefore describes the region between what students can do independently and that which can be done only with the assistance of others. Hence the ZPD is the realm of knowledge and skills which have not yet become apparent but could emerge as the student interacts with more knowledgeable peers, parents, texts or teachers (Lerman, 2000). In this respect, Vygotsky believed that teachers should pitch their instruction within the ZPD rather than at the level of the student's current level of independent performance. By carefully selecting activities beyond what the student can do on his/her own, the teacher would therefore assist in promoting knowledge growth to produce developmental gains.

The ZPD has other implications for mathematics teaching, particularly in terms of the provision of *scaffolding* by which social support from the teacher or the peer group is used to promote new learning. Scaffolding enables students to explore new concepts in a more sophisticated manner and more efficiently than if they were attempting to learn them on their own. It is especially important in the initial stages of learning about a new concept or process and can be gradually withdrawn as the student becomes more skilled in the new activity (Ryan & Williams, 2007). Scaffolding can take many forms in mathematics lessons, such as:

- using assessment *for* learning to check prerequisite knowledge and understanding
- allowing students to work collaboratively in small groups
- illustrating mathematical concepts using examples and non-examples
- presenting new material incrementally in small, manageable steps
- ensuring that new knowledge and skills progress from the simple to the more complex
- providing regular feedback on students' work
- teacher questioning
- visual aids
- highlighting key information and using procedural prompts.

In recent years, there has been a growing emphasis on the social and cultural aspects of mathematics education and how these can potentially influence the learning and teaching of mathematics in the classroom. In particular, researchers have increasingly considered issues related to equity in mathematics education and focused their attention on the relative under-performance of disadvantaged groups (Jorgensen, 2010). A sociocultural perspective on learning views the classroom as an environment in which students and teachers from a range of cultural and social settings come together for the purpose of learning and teaching mathematics. Each individual brings their own considerable knowledge and prior experiences shaped by personal social and cultural backgrounds. This is significant for the teacher, who must consider how the classroom norms and practices may impact on students from a diverse range of circumstances.

Student engagement in learning mathematics

Engagement typically refers to how actively involved students are in their learning. Engagement is an important consideration for mathematics teachers since the quality and the intensity of student involvement in classroom activities can have a profound impact on the rate at which they learn, the depth of understanding they develop and their ability to generalise new concepts so that they can apply them in unfamiliar contexts (Helme & Clarke, 2001). Stipek (1996) suggests that students who are actively engaged in their learning will be excited and enthusiastic when undertaking challenging tasks and be more willing to make a concerted effort to persist, even in the face of difficulties.

Poor student engagement is a significant issue in Australian secondary mathematics classrooms (McPhan et al., 2008) and research suggests that many students are disengaging from mathematics (Sullivan, Tobias & McDonough, 2006). They are not opting to study higher levels of mathematics in the final years of high school (Kennedy, Lyons & Quinn, 2014), nor are they choosing to pursue their study of mathematics at the tertiary level (Office of the Chief Scientist, 2014). Boaler (2009) suggests that a major contributing factor leading to disengagement in mathematics classrooms is the mistaken belief that success in mathematics is a sign of general intelligence and that some people can do mathematics while others simply cannot. Boaler regards this belief as so pervasive and widely held that it is rarely mentioned and she refers to it as the 'elephant in the classroom'.

The incorrect notion that some students just cannot do mathematics is reminiscent of the work of Dweck (2000), who distinguished between two quite different perspectives on ability and intelligence. The first, which she called an *entity view*, treats intelligence as something which is fixed and stable. People who hold an entity view are likely to believe that their intelligence was predetermined at birth and will remain so throughout their lives. Dweck proposed that students who believe in the entity view require easy success in order to maintain motivation and they typically regard challenging work as a threat. These students are unlikely to remain engaged in mathematical problem-solving tasks, preferring instead a more repetitive, drill and practice style of instrumental learning where they can obtain correct answers quickly with a minimum of effort. The *incremental view*, on the other hand, regards intelligence as malleable and people who view intelligence in this way believe that they have some control to change their level of intelligence or achievement. Students with incremental beliefs gain satisfaction from learning something new and are not bothered that they might appear foolish in their initial attempts. These students enjoy a challenge and will persevere in a problem-solving activity as long as they feel they are gaining some valuable new insights or skills from doing so.

How students establish their personal learning goals can be linked to the ideas about intelligence. Since an entity view is concerned with looking smart, students with this orientation are concerned primarily with performance goals. They can easily lose confidence in themselves as soon as any obstacles arise and are susceptible to a kind of learned helplessness because they believe that they have little control over their work. So, they tend to avoid situations that they perceive to be challenging, perhaps by procrastinating or even denigrating the activity. In contrast, the incremental view is more concerned with learning new things and so these students are mainly interested in mastery or task goals. When faced with a challenge they tend not to give up but rather to seek out alternative approaches and are likely to redouble their efforts until they can progress or achieve success.

The distinction between entity and incremental approaches can therefore lead to quite different levels of student engagement, so it is worth considering some of the implications of Dweck's ideas for teaching. For instance, performance goals might well be motivated in students whose teachers have tended to place too great an emphasis on their past achievements. These teachers may focus on performance but protect the student from negative feedback, perhaps by not discussing errors or simply reducing the requirements of the lesson activities. Dweck claimed that these teachers are likely to promote a belief in students that they can achieve success without effort and, though well intentioned, the practice of reducing the demands of tasks may simply serve to create a situation where difficult challenges are avoided rather than confronted.

Attard (2011) conducted a longitudinal case study of the engagement levels of 20 students over three years. She tracked the students from their final year of primary school into their first two years of secondary school. Her aim was to identify factors which encouraged or inhibited student engagement in mathematics lessons. Data for the study were drawn from student interviews, focus group discussions and classroom observations. A major finding from the study concerns the role of what Attard referred to as 'positive pedagogical relationships' between teachers and their students. Attard found that these relationships must be developed as a foundation for sustained engagement. Some of the essential ingredients in positive pedagogical relationships include implementation of a range of teaching styles on a regular basis to ensure sufficient variety of learning tasks while still maintaining a sense of order and structure within lessons. Teachers should encourage active student participation in lessons by providing appropriate academic challenges. There also need to be opportunities for social interaction through activities such as group work. In addition, the relevance of the mathematics being studied must be emphasised so that students can appreciate how the work they are doing is applicable to their current and future lives. On the other hand, pedagogical practices can also serve to decrease student engagement. For example, if they are largely based around individual tasks that do not promote interaction and dialogue, or where tasks are perceived to have little or no relevance to students' lives. Attard also noted that while students are generally happy to work from textbooks, an over-reliance on these resources is counter-productive.

The Australian Association of Mathematics Teachers revised and republished its *Standards for Excellence in Teaching Mathematics in Australian Schools* (Australian Association of Mathematics Teachers, 2006). The document describes some of the key knowledge, skills and dispositions required for good teaching of mathematics. These are arranged in three domains: professional knowledge, professional attributes and professional practice. In terms of student engagement, the domain of professional practice is most relevant. The Standards suggest that teachers purposefully establish a classroom environment that encourages active student participation to maximise opportunities for student learning. Students are 'empowered to become independent learners ... motivated to improve their understanding ... and develop enthusiasm for, enjoyment of, and interest in mathematics'. When planning lessons, teachers should be mindful to include opportunities for self-directed learning of substantive mathematical content based on a variety of resources and teaching styles, including the use of a range of technological tools and devices where appropriate. Teachers should also draw on students' backgrounds and prior experiences to develop activities in which students can explore and apply mathematics in a range of contexts. According to the Standards, good teaching 'promotes, expects and supports creative thinking [and] mathematical risk-taking'.

ACTIVITY 1.2

Online learning tools such as HOTmaths provide a rich source of activities that could be used to promote student engagement and active participation in mathematics lessons. An example is the HOTmaths activity 'Designing spinners'. The activity assists students in learning about theoretical probability and is linked to the Australian Curriculum: Mathematics learning outcome ACMSP168 (Assign probabilities to the outcomes of events and determine probabilities for events). You can access this HOTsheet by logging in to HOTmaths and selecting the 'HOTmaths Global' Course list and the 'Early Secondary' Course. From the Topic list choose 'Looking at probability', then the 'Theoretical probability' Lesson. The HOTsheet can be found in the Resources tab.

DESIGNING SPINNERS

You could design a triangle spinner with probabilities of $\frac{1}{3}$ or $\frac{2}{3}$.

You can create a square spinner showing probabilities of $\frac{1}{4}$, $\frac{1}{2}$ or $\frac{3}{4}$.

You can create many different spinners using circles. Here are two examples:

TASK 1   Create fair spinners

Design two different spinners using two *different* shapes from the template sheet on page 2. Design each of your two spinners so that there is a $\frac{1}{3}$ chance of getting blue, a $\frac{1}{3}$ chance of red and a $\frac{1}{3}$ chance of green.

Figure 1.3 Designing spinners HOTsheet

REFLECTIVE QUESTIONS

Read over the 'Designing spinners' HOTsheet (see Activity 1.2) and then consider the following:

- How does the activity promote student engagement?
- What mathematical concepts and skills could students learn by completing the activity?
- Is there anything else that students might learn by completing the activity?
- How might you use this activity in the classroom?

ACTIVITY 1.3

Now that you have read the chapter, consider the following discussion questions:

- Look back over the definition of mathematics by Freudenthal at the start of this chapter. How would you describe the nature of mathematics?

- Do you see yourself as a teacher of numeracy as well as a mathematics teacher? How can you ensure that your students develop sound personal numeracy skills?
- What aspects of behaviourist learning theories could you usefully adopt in your teaching?
- Constructivists believe that learning occurs when students are active participants in their lessons. What are some practical strategies that you can use in your teaching to ensure that this occurs?
- How can you promote students' relational understanding of mathematical concepts?
- What are some advantages of collaborative group work activities in mathematics lessons?
- How would you deal with students who have become disengaged from learning mathematics?
- How can you support creative thinking and mathematical risk-taking by your students?

Summary

Discuss the importance of mathematics in modern society, including fundamental aspects of mathematical literacy and numeracy

The chapter began with an overview of some fundamental aspects of mathematical learning which are central to the lives of today's students and will allow them to fully participate in the nation's social and economic prosperity. In particular, we discussed the prevalence of information and communication technologies and noted how their increasing sophistication requires new patterns of mathematical understanding from students. We also looked at the role of schooling in providing students with a firm foundation of mathematical and scientific literacy so that they can contribute in our increasingly technology-dependent world.

Statistical literacy is the ability to understand, interpret and evaluate data from a variety of sources. It involves the ability to find data appropriate to specific needs and purposes, to make sense of data when they are presented, and a familiarity with statistical terminology in order to communicate clearly to others. Financial literacy means being able to understand and negotiate the complex financial landscape of our modern world in order to manage money and take appropriate financial risks in order to reach one's financial and lifestyle goals. Numeracy involves understanding the full range of mathematical concepts and skills, including numerical, spatial, graphical, statistical and algebraic. It incorporates mathematical thinking strategies and general skills such as problem solving and reasoning.

Describe the Proficiency Strands of the Australian Curriculum: Mathematics

The four Proficiency Strands which we discussed in the chapter are: **understanding**, where students develop a deep, relational understanding of key concepts and processes, knowing what to do and why it works; **fluency**, where students develop skill in efficiently and correctly carrying out procedures and can readily recall basic facts and fundamental concepts; **problem solving**, where students develop their ability to make choices about the most appropriate methods to employ in a given situation and can work in unfamiliar problem contexts by applying their knowledge, communicating their ideas, and verifying the reasonableness of the solutions they obtain; and **reasoning**, where students can explain their thought processes and develop their capacity for logical thinking through analysis, evaluation, explanation, justification, inference and generalisation.

Summarise the key features of some important theories about student learning

In this chapter we discussed some of the most significant theories about how learning takes place. We contrasted the behaviourist view of learning, as evidenced by changes in behaviour that are observable and measurable, with the cognitivists, who view learning as taking place in the mind of the individual, and socio-cultural theorists, who emphasise social interactions and the context for learning. In considering these different theoretical perspectives we related them to your work in helping students learn some fundamental mathematical concepts and skills. We discussed the crucial distinction between instrumental and relational understanding and identified some potential pitfalls in an over-reliance on a transmissionist teaching style.

Identify some ways in which teachers can promote student engagement in mathematics lessons

We explored the notion of student engagement as it applies in mathematics lessons and discussed two different perspectives on ability and intelligence: the *entity view*, which sees intelligence as fixed and stable, and as predetermined from birth, and the *incremental view*, where intelligence is regarded as changeable, and where we have some ability to control and change intelligence or achievement thorough effort. We also looked at some of the ways in which teachers can promote student engagement through their positive pedagogical relationships, which include implementing a range of teaching styles to provide sufficient variety of learning tasks for students while maintaining a degree of structure within lessons.

CHAPTER 2

Language and mathematics

Learning outcomes

After studying this chapter, you should be able to:

- recognise the importance of language in the teaching and learning of mathematics
- identify the elements of the mathematics register in English and other languages and discuss the impact on learning
- reflect on the appropriate use of language by the student and the teacher in the mathematics classroom
- write mathematics correctly (technical communication).

Introduction

In a typical mathematics classroom, students read, write, listen and speak mathematics. In a typical secondary mathematics classroom there is a lot of listening to the teacher speaking the mathematics, but is this the best way for students to learn? In addition, in these four language modes there are particular issues in the context of mathematics. Consider the following:

1   There are just four numbers, after unity, which are the sums of the cubes of their digits. The word 'diameter' and the word 'radius' are closely related. In Chinese, diameter is 直径 [zhíjìng], literally meaning 'straight path', with radius being 半径 [bànjìng], meaning half path.

2 Recently a pre-service teacher commented: 'I think at this stage I'd make a poor maths teacher because I've never spoken this language. I read it at school, but never "said it aloud" BIG difference' (Galligan & Hobohm, 2013, p. 328).

3 A student says: 'I can see this bit here has to be timesed by 3 outside the brackets, and then take away the bits in the brackets …'

4 In solving the simultaneous equations,

$$x + y = 4 \qquad\qquad (1)$$

$$x + 3y = 10 \qquad\qquad (2)$$

a student wrote as an answer:

$$x = 1; y = 3$$
$$1 + 3 \times 3 = 10$$
$$10 = 10$$

Each of these scenarios highlights some of the language issues in teaching and learning mathematics. These issues fall into a number of categories:

1 The mathematics register (i.e. the words, phrases and associated meanings used to express mathematical ideas). This includes the etymological view of the words of mathematics as well as the **syntax**, **semantics**, **orthography** and **phonology** of the language itself and its impact on understanding mathematics.

2 Language in the classroom: the use of language by teachers to communicate ideas and the dialogue used by students to communicate and learn mathematics.

3 Technical communication: i.e. the accepted standard use of language and symbols to communicate mathematical ideas, both orally and in the written form.

In this chapter we will highlight these issues, ask you to reflect on particular problems and offer activities that can be used in the secondary classroom. Do you have the mathematical language fluency to communicate to your students? Are you able to create a communicative classroom so students do more of the speaking and the teacher more of the listening?

As this chapter is on language (oral and written), we suggest that you have access to a recording device. The best would be a tablet device with a stylus for writing, but using a video recording device and a whiteboard would work as well. We want you to get used to listening to your own voice and being critical of what you say.

This is what a student said of the experience:

I first must say that I was absolutely terrified at doing this [recording and writing], not the math writing part but the talking part. For a couple of days I would sit down to do it, but lost all my nerve. A few days went by and I decided that maybe if I practiced that would help, … so I sat down today, … and started to talk into the microphone and just kept going blank after a couple of seconds, that went on for a couple of hours, ok it was more like five hours. Then all of a sudden, I just started talking and didn't stop. I must have forgotten all about the microphone until about the last minute or so when I thought about it again and got a bit panicky, but was able to finish it. Besides all that I have actually enjoyed doing this and getting result. (Galligan & Hobohm, 2013, p. 329)

KEY TERMS

- **Etymology:** the study of the origins of mathematical words.
- **Literacy in mathematics:** refers to students being able to access the mathematics in words and to make sense of the context and clarify what is required (Council of Australian Government, 2008, p. 33)
- **Mathematical literacy:** defined by Programme for International Student Assessment (PISA) as 'an individual's capacity to identify and understand the role that mathematics plays in the world, to make well-founded judgements and to use and engage with mathematics in ways that meet the needs of that individual's life as a constructive, concerned and reflective citizen' (Thompson et al. 2013, p. 7).
- **Mathematics register:** the set of meanings that is appropriate to a particular function of language *together with* the words and structures which express these meanings.
- **Morpheme:** the most elemental unit of meaning; it is the minimal linguistic sign in which there is an arbitrary union of a sound and a meaning that cannot be further analysed.
- **Morphological complexity:** the proportion of the lexicon's total description length that is due to the description lengths of the affixes and signatures. There is little morphological complexity in Chinese, but English is more complex and Hungarian more complex still.
- **Orthography:** the representation of the sounds of language by written symbols.
- **Phonology:** the system of sound patterns.
- **Semantics:** the study of meaning in language.
- **Syntax:** the sentence pattern in language.

Why language is important

Both National Reports on Education and the Australian Curriculum recognise the importance of language in mathematics. If you read the Australian Curriculum: General Capabilities, you will see that it highlights literacy as an important tool in the teaching and learning of mathematics, from interpreting word problems to discussing mathematics in the classroom. A nationally commissioned report (Council of Australian Governments, 2008) recommended that the language and literacies of mathematics be explicitly taught since language can be a significant barrier to understanding mathematics. As teachers routinely assess students' understanding of mathematics through literacy (often through reading and writing), an important question to ask yourself when setting or marking assessment tasks is: Are your students struggling to understand the mathematics or are there specific language difficulties associated with assessment tasks you have set?

In addition to **literacy in mathematics**, **mathematical literacy** has an important role in understanding mathematics. Mathematical literacy is related to the role mathematics plays in understanding life, whether it is school, work or everyday activities. In addition to this awareness, mathematical literacy is also related to a person's confidence and competence in using the mathematics in the appropriate context.

ACTIVITY 2.1

1   Read Section 2.4 of the National Numeracy Review Report on 'The role of language in mathematics learning' (www.coag.gov.au/sites/default/files/national_numeracy_review.pdf).
    Do you understand the difference between:
    • mathematical language as distinct from mathematical literacy and literacy in mathematics?
    • use of everyday English terms and those same terms in mathematics classrooms?
2   Read the section on General Capabilities – Literacy in the Australian Curriculum www.acara.edu.au/verve/_resources/Mathematics_-_GC_learning_area.pdf. It discusses students using literacy in many aspects of the mathematics curriculum. Literacy also involves learning vocabulary and using a variety of texts typical of mathematics.

    List the different aspects of literacy in mathematics, and next to this list give an example of each that would be relevant in high school. The Australian Curriculum Document has a few examples, so think of different ones.

REFLECTIVE QUESTIONS

Can you see that there is a strong relationship between mathematics achievement and literacy for high school students? Perhaps you have seen this in your own experience either as a student, a pre-service teacher or in other settings. What sort of activities would you develop in your mathematics classroom to improve literacy? What could you do to ensure that students carefully read mathematical word problems?

The mathematics register

In everyday English numbers are usually adjectives (i.e. one book, three days), but in mathematics 'one' and 'three' are nouns. In everyday English we speak of the 'function of government' or 'factors that influence climate change'. The use of 'function' and 'factor' is quite different in mathematics. In a classroom, teachers and students talk about mathematics differently from a written text. In a class, the language is often informal and natural, but more formal mathematical language 'makes varied use of complex, rule-governed writing system mainly separate from that of the natural language into which it can be read' (Pimm 1991, p. 20). This collection of terms is often called the **mathematics register**. The mathematics register is more than the words and phrases in mathematics, it is the 'set of meanings that is appropriate to a particular function of language *together with* the words and structures which express these meanings' (Halliday, 1978, p. 195). As teachers, you need to be aware of the subtle differences in language. In some instances words used in a mathematics classroom may have altered meaning and grammatical function. Adams, Thangata and King (2005) suggest that teachers need to support students to use technical language when talking about concepts. You should encourage students to make the connections between everyday meanings and the mathematical ones, such as function, derive, mean, rational or root. Ensure you provide time for students to 'talk about mathematics as they solve problems, encouraging them to articulate patterns and generalisations' (p. 446). So there is something about the words and how they are linked together that can make understanding mathematics easier (or harder). First, let's have a look at the words more closely.

Words used

Discussion of the origins of mathematical words (their **etymology**) can be of interest to students, particularly if you have students whose first language is not English (see Chapter 6, which has a further discussion of this issue). The beginning of this chapter mentioned diameter and radius, which in Chinese are words that are quite similar. Another is the word 'parabola' in English, which comes from the Greek *para* 'alongside, nearby, right up to' and *–bola*, from the verb *ballein* 'to cast, to throw' (Schwartzman, 1994). In order to get access to the meaning, students need a lesson in ancient Greek. These types of words have been referred to as 'dead metaphors'. In contrast, the corresponding word in Chinese is 抛 物 线, which literally means 'throw object line' (no intermediary language required). This comparison is not just interesting information. There has been research (e.g. Ellerton & Clements, 1996) to suggest that characteristics of the English language have a negative effect on students' performance in mathematics, particularly the processing of mathematical text. Words are the gateway to understanding and processing concepts in mathematics and we continually move back and forth from thought to word (Vygotsky, 1962). Take the word 'diameter': while the word may only supply the reader with access to the concept definition, it may be helpful to access the total cognitive structure (i.e. all the words which associate with the word diameter, such as 'radius', for example). In Chinese, diameter 直径 [zhíjìng], which literally means 'straight path', and radius 半径 [bànjìng], which means 'half path', allow more direct access to the meaning of both words. Therefore, the way in which those words are assembled and presented, and the way they sound, may contribute to the ease with which we can get to the concept. If that gateway is more difficult, then a role of the teacher is to open the gateway to better understanding.

The names of number words may help or hinder early learning. For example, twelve and thirteen are not part of a regular named-value system. By contrast, in Asian languages (e.g. Burmese, Japanese, Korean and Thai) number words are said and then the value of that number is named (5726 – five thousand seven hundred two ten six) (Fuson & Kwon, 1991). In addition, the **orthography** of the language may assist in the reading and processing of text. For example, the word thirteen in Chinese is 十三 (ten three) and the word twenty is 二十 (two ten); the word for March is 三月 and for triangle it is 三角形. (Galligan, 2001). Again, if a teacher can highlight these structures to students who are struggling in high school, this may assist their understanding of the subject, particularly around the concept of place value.

Invention of words

The words that are chosen in mathematics are constantly being invented. An interesting project in New Zealand in the 1990s developed Māori words in mathematics which did not exist at the time (Barton et al. 1998). As Māori is the first language and the language of instruction in some schools, the construction of more meaningful terminology was thought to have motivational and bilingual benefits for students who identify with the Māori culture. For example, the word 'hōkai' is a brace or stay used to keep an eelpot open and it was used for the word 'diagonal' (Barton et al., p. 5).

But it is not only words in another language that need to be invented or adapted; new words need to be continuously invented as we create more mathematics. Again, returning to New Zealand, consider this discussion between three topologists and a mathematician at a dinner (no, it is not a joke) as they are discussing the origin of the phrase 'opens sets' in topology:

"Ah," says the first topologist, "that is easy. Actually any word could have been used, so long as it had an opposite, since it is the relationship between an open set and a closed set that is what is important. Open/closed. Yin/yang. Black/white. It could have been any of these. It is the sense of complementarity that is being expressed." "What?" queried the second. "I don't think so. The meaning of open in this context is the one used of an international border: anything can pass through, there is no well-defined restriction on what makes the border." "Oh," mutters the third quietly (he was the junior member of the group), "I always thought that what was meant was the idea of without any boundary at all – like we refer to an open field, the open sea, or an open question." Fortunately, not being a topologist, my view was neither expected nor important. Which was just as well, because I had imagined that the sense of open being referred to was that of a door. It can be open or shut, it depends what you want to do with it. (Barton, 2008, pp. 60–61)

ACTIVITY 2.2

Look at the statement at the introduction of this chapter: 'There are just four numbers, after unity, which are the sums of the cubes of their digits' (Parker 2014). Try turning this statement into something more symbolic. I recently read this last statement in a book by Matt Parker *Things to Make and Do in the Fourth Dimension*. This book is worth reading if you have time. The phrase 'sums of the cubes of their digits' is a very dense statement (dense because all the words are necessary to extract meaning). The interesting part of the statement 'there are just four numbers' gets lost in the attempt to turn the 'word problem' into a mathematical one.

ACTIVITY 2.3

Log in to HOTmaths and select the 'HOTmaths Global' Course list and then choose the 'Middle Secondary' Course. From the Topic list choose 'Algebraic expressions', then the 'Using algebra' Lesson. From the Resources tab, select the 'Algebra from statements' HOTsheet.

Create a set of cards as shown in the HOTsheet for students to use. Expand the set to make it useful for different year levels. There are other such activities in different topics that you could explore.

HOTsheet   **Using algebra**

ALGEBRA FROM STATEMENTS

Cut out the cards and match each rule with the algebraic statement.

The product of two square numbers is a square number.	$c + d = d + c$
The sum of five consecutive numbers is divisible by 5.	$A = \pi r^2$
Adding two odd numbers together results in an even number.	$h^2 - h + 11$ is always prime.
In Pythagorean triads, the square of the largest number is equal to the sum of the squares of the two smaller numbers.	$n - 2 + n - 1 + n + n + 1 + n + 2 = 5n$
Square any number. Subtract the number from the answer then add eleven. The answer is always prime.	$(uv) \times w = u \times (vw)$
Commutative law — addition *The order in which numbers are added does not matter.*	$P = 2l + 2b$

Figure 2.1 Algebra from statements HOTsheet

ACTIVITY 2.4

Students may be surprised about how 'young' some words in mathematics are. Investigate the meaning of the following words and find out their origin. You may need to use a dictionary or an online site that has etymological details:

- negative number
- vulgar fraction
- septendecillion

- googol
- algebra
- Cartesian plane.

For the senior students:

- calculus
- Chaos theory
- fractal
- cycloid
- repdigit
- Gaussian method of elimination
- kissing circles.

REFLECTIVE QUESTIONS

1   Do you know of mathematical words or phrases in other languages that may be of interest to others? How could you incorporate this discussion into a lesson, project or homework activity?

2   Look up the words 'square' and 'trapezium' in various dictionaries (or online) and compare them. You will see that it is not a unified definition. Are you surprised by this? Perhaps we are used to the idea that mathematics seems to be a very black and white subject, but it is good to suggest to students that this is not always the case. Consider everyday words, such as 'restaurant'. Does a restaurant have to be indoors or must it provide table service? The answer to this question helps to characterise the 'restaurantness' of an establishment. Restaurant is a somewhat fuzzy concept. People are usually comfortable with this uncertainty. So for a word to be understood and used it isn't necessary to know its tightly defined meaning (Leung, 2005, p. 129). As a teacher you can explore the fuzziness of word meanings, even generalising and extending meaning from one instance to another. Learning vocabulary, particularly in terms of its associated concepts and linguistic properties, is an ongoing activity that can be fostered in the classroom.

Definitions

Morgan (2005) discussed the ways definitions are presented at different school levels and the roles that they play in different mathematical practices.

She compared the language used in two textbooks, one for intermediate Year 10/11s and one for higher Year 10/11s. Read both and note any differences. If you are familiar with functional grammar you will notice differences in agency and modality, but if not, you should still see the different ways the textbook authors are talking to and engaging the student.

General Certificate in Secondary Education (GCSE) intermediate textbook

In investigation 15:1, you found that the ratio $\dfrac{\text{shortest side}}{\text{longest side}}$ i.e. $\dfrac{\text{opposite}}{\text{hypotenuse}}$ is the same for each of these triangles. This ratio is given a special name; it is called the sine of 40° or sine 40°.

The ratio $\dfrac{\text{adjacent}}{\text{hypotenuse}}$ is called the cosine 40°. The ratio $\dfrac{\text{opposite}}{\text{adjacent}}$ is called tangent 40°.

The abbreviations sin A, cos A and tan A are called trigonometrical ratios, or trig. ratios.

General Certificate in Secondary Education (GCSCE) higher textbook

The ratios sin θ and cos θ may be defined in relation to the lengths of the sides of a right-angled triangle.

$$\sin θ \text{ is defined as } \dfrac{\text{length of opposite side}}{\text{length of hypotenuse}}$$

$$\cos θ \text{ is defined as } \dfrac{\text{length of adjacent side}}{\text{length of hypotenuse}}$$

Since θ < 90, sin θ and cos θ defined in this way have meaning for angles less than 90°. We will now look at an alternative definition for sin θ and cos θ which has meaning for angles of any size. This gives the following alternative definition for the ratios cos θ and sin θ.

The ratios cos and sin may be defined as the coordinates of a point P on the Unit Circle where OP makes an angle of θ with the positive x-axis and is of length l. Defined in this way, the ratios cos θ and sin θ have meaning for angles of any size.

Morgan concludes that while the first text reads like a set of unquestionable facts ('the ratio is …', the second text provides ambiguity or choice ('the ratio cos may be defined …'), suggesting the mathematics definitions can be negotiated.

More than words

It is not just the nouns and adjectives that are of importance in mathematics. There is a correlation between reading and mathematics performance for students to understand mathematical language, as there is a complex interaction between the **semantics** and the **syntax**. Consider, for example, the importance of the word 'before' in problems posed in the question below. Why is there a difference in the wording 'down to a stop' and 'eventually stopping'?:

- The car is moving with a constant speed before it slows down to a stop.
- The car starts from rest before moving and then eventually stopping.

The following question is also mentioned in Chapter 7 (Anderson, 2009): consider the importance of the words 'decrease in', and '50% of':

- How much is the decrease in electricity bill?
- Is the decrease less or greater than 50% of the first electricity bill?

Chapter 8 discusses some further issues around solving such word problems.

Table 2.1 summarises some of the possible language issues in mathematics (Galligan, 2001):

Table 2.1 Possible language issues in mathematics

Syntax	Orthography	Phonology	Semantics
Subordinate clauses, topic prominence.	Number of words, morphological complexity, boundaries.	Acquisition.	Content words, connecting words, context.
Word order.	Character complexity.	Word rules.	
Question structure, passive voice, markers.	Text density.	Syllable structure and recall span, sentences.	
Redundancy.	Types of characters, script layout, access to meaning.		
Sentence type. Cues. Conditional statements.			

Some examples of language issues are listed in Table 2.2. Try to think of similar examples in both the Junior and Senior Curriculum.

Table 2.2 Examples of language issues in mathematics

Issue	Example	Discussion
Subordinate clauses and text density	Find the side length of a square whose diagonal is of length 10 cm	Simplify this by breaking it into two sentences: A square has a diagonal of 10 cm. Find the side length.
Topic prominence	Instead of: How long is this line?, Look at this line. How long is it?	Getting students to pay attention to the topic of the sentence, again breaking it into two sentences.
Word order	What number is two less than five? Find this square's side length.	The first example is difficult in English. Rewrite it: Consider the number 5. Pick a number two less than this. Perhaps change it to: Find the side length of this square.
Passive voice	When 15 is added to a number the result is 21.	Change to an active statement: Add 15 to a number you get 21.
Redundancy	Ben is taller than George; Dave is shorter than George – Who is the tallest?	Compare this to: Ben is taller than George; Dave is shorter than George – Of the three, who is the tallest? This removes ambiguity.
Conditional statements	How much string *would have* remained if 3 cm *had been* cut from it.	Simplify this to: I cut 3 cm from the string. What remains?
Morphological complexity	Parallelogram, quadrilateral.	Words in mathematical English tend to be complex. Assist students by breaking it down into **morphemes** and explaining each. Parallel//ogram, quadri//lateral.

(Continued)

Table 2.2 (*cont.*)

Issue	Example	Discussion
Content words	Diameter and radius are linked conceptually but not semantically.	This is a feature of English. Explain to students the origin of these words.
Connecting words	If a number is greater than 10 or divisible by three, what could the number be?	A feature of mathematical English is this 'inclusive or'. This is different from 'I am going to the shop' or 'going to the movies'.

Problem solving: a problem of language?

Word problems in mathematics are often cited as an area of difficulty for students, and in English the syntax of English can contribute to this difficulty. There is a classic problem, called the 'student–professor problem' that is also cited in Chapter 8 (Clement, 1982). Try it:

> 'There are six times as many students as professors. Write this in a mathematical sentence.'

Typically students write $S = 6P$ instead of $P = 6S$. Researchers in mathematics suggest that positioning information and the unknown in sentences may have an impact on the ease of processing the sentence (MacGregor, 1993), and this is particularly true for comparison problems in English, such as the example above. Many comparison problems in Chinese have a relational proposition that is explicitly shown. A Chinese multiplicative example is translated syntactically as *men are women six times* (there are six times as many men as women). Here the relational proposition is clear. In a similar additive comparison problem *men compare women more five* (there are five more men than women). This again suggests a clear relational proposition, but there is a subtle change in syntax. This creates a semantic difference and there is evidence to suggest a difference in the processing in Chinese (Galligan, 1997; Lopez-Real, 1997).

When students encounter a mathematical word problem they need to read and understand the words, sentences and then solve the problem. In a well-cited research on students' errors in solving mathematical word problems (Ellerton & Clements, 1996) the major error highlighted was students' inability to read the problem and then translate it into correct mathematical form. They drew on research by Newman (1983) that challenged teachers to redefine what is 'basic' in school mathematics and highlighted errors in *reading, comprehension, transformation, process* and *encoding*. The questions you can ask the students include:

1   Please read the question to me (reading).
2   Tell me what the question is asking you to do (comprehension).
3   Tell me a method you can use to find an answer to this question (transformation).
4   Show me how you worked out the answer to the question. Explain to me what you are doing as you do it (process skills).
5   Now write down your answer to the question (encoding).

Newman's research has generated a large amount of evidence pointing to the conclusion that far more children experience difficulty with the semantics, vocabulary and symbolism of mathematics than with standard algorithms. As a result of Newman's research it has been found that many 'remedial' mathematics programs pay little attention to the first three questions above.

When students 'read' mathematics it is not just the words and symbols, but also diagrams. Graphs and charts, like word problems, are very dense. Kemp and Kissaine (2010) suggest that table reading and interpreting skills can be taught and learned and that it should not be assumed that students should be able to perform these tasks without instruction. For example, in the graph below from the Australian Institute of Health and Welfare (www.aihw.gov.au/asthma/deaths), the point highlighted could be described in words: '*In the year 2000 in the Australian population, there were approximately 2.75 female deaths per 100 000 population*', with the information extracted from both axes, the legend and the notes.

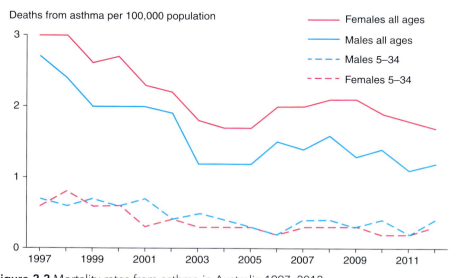

Figure 2.2 Mortality rates from asthma in Australia 1997–2012
Source: ACAM and AIHW analysis of AIHW National Mortality Database.

ACTIVITY 2.5

The following table is based on a worksheet by MacGregor and Moore (1991, p. 134). Can you create your own worksheet similar to this for use in your classroom (at different levels)?

If we	multiply	$20	from	......	we get	......
	increase	4kg	by			
	share		to			
	halve		into			
	reduce		between			
	divide					
	add					

ACTIVITY 2.6

Log in to HOTmaths and select the 'HOTmaths Global' Course list and then choose the 'Middle Secondary' Course. From the Topic list choose 'Algebraic expressions', then the 'Reviewing algebraic expressions' Lesson. Select the 'Questions' tab and then the subsequent 'Challenge' tab. Look at the problems in the Challenge questions. Record yourself explaining to

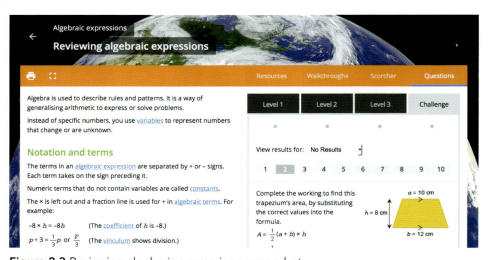

Figure 2.3 Reviewing algebraic expressions screenshot

a fellow student (who pretends to be a school student) what you would do to solve one of the problems. In particular, use the Newman's Error Analysis framework.

REFLECTIVE QUESTIONS

There are many more graphs such as the one on Australia's health shown in Figure 2.2 produced by the Australian Institute of Health and Welfare that could be used for class discussion. Can you think of ways that graphs such as these could be used? And in what secion of the curriculum would this be used?

Second language learners

The **mathematics register** in English is of particular concern to teachers when there are second language learners in the classroom. There has been much debate and research on the importance of learning and doing mathematics in students' native tongues, rather than in a second language such as English (e.g. Spanos et al., 1988; Nathan et al., 1993). While most researchers agree that the native tongue allows for more efficient conceptual processing, there may be differences in the nature of that conceptual processing. Cognitive effort required to process certain mathematical word problems is high in English (Clement et al., 1981; Cooper & Sweller, 1987; MacGregor, 1991). Similar research in other languages on mathematical word problems has also been done in Europe (e.g. Malle, 1993), as well as in Asian–English language comparison (e.g. Galligan, 1997; Li & Nuttall, 2001). As a teacher, it is important that you are aware of the difficulties students may have with reading mathematics and ensure you make both written and oral communication as clear and simple as possible, but appropriate to the level of the learner. A New Zealand study investigating the relationship between English language and mathematics learning for non-native speakers (Barton & Neville-Barton, 2005) found, for one senior high school, that there was on average a 15 per cent disadvantage in the test because it was in English, and it appeared that it was the syntax more than the vocabulary that caused these problems. In many cases teachers were unaware of this disadvantage.

Furthermore, for all four of the five institutions in the study, the students were not aware of the extent of their difficulties, and 'all studies reported that students in general performed worse than the teachers/lecturers anticipated' (Barton & Neville-Barton, 2005, p. 10). The main problems appeared to be with prepositions, word order, logical structures such as implication, conditionals and negation, and mathematics couched in everyday contexts. Chapter 6 discusses issues around teaching mathematics to English language learners.

ACTIVITY 2.7

Log in to HOTmaths and select the 'HOTmaths Global' Course list and then choose the 'Middle Secondary' Course. From the Topic list choose 'Logarithms', then the 'What are logarithms?' Lesson. Here you will find a range of walkthroughs. Take a few of these exercises and record yourself solving them as if you were explaining them to a class.

Figure 2.4 What are logarithms? screenshot

REFLECTIVE QUESTIONS

Think of the meanings of the words unknown vs. variable. $3x + 7 = 10$ is an equation with an unknown (i.e. we can find the value of x). $3x + 7 = y$ is an equation with two variables (i.e. the x and the y vary). Look up the meaning of these words and you will see that these statements could be contested. For example, in the Collins Dictionary of Mathematics it states that: 'An unknown is a variable, or the quantity that it represents, whose value is to be discovered by solving an equation' (Borowiski & Borwein, 1989, p. 616).

Goos, Stillman and Vale (2007) state that 'the notion of variable is fundamentally different from the concept of unknown'. What is your view of the difference between these two concepts? Have you noticed students having trouble with these concepts? Ask your students: what does the symbol n in the expression $2n + 3$ stand for?

Language in the classroom

In a recent review of research on teaching mathematics, Sullivan (2011) identified six key principles of effective teaching of mathematics. Principle five, on structuring lessons, is: Adopt pedagogies that foster communication and both individual and group responsibilities, use students' reports to the class as learning opportunities, with teacher summaries of key mathematical ideas. Sullivan goes on to argue that students can benefit from either giving or listening to explanations of strategies or results, and that this can best be done along with the rest of the class with the teacher participating, especially facilitating and emphasising mathematical communication and justification. A key element of this style of teaching and learning is students having the opportunity to see the variability in responses, and confirming that this variability can indicate underlying concepts for students (Watson & Sullivan, 2008). The communication can occur at any segment of a lesson, including the end.

Cheeseman (2003) argued that a lesson review must contain reflective activities that ask students to talk about the concept and their thinking. It is beyond the scope of this chapter to investigate in detail the communication strategies that can be used. However, to create a positive classroom culture, Sullivan (2011) recommends that communication needs to be actively encouraged and planned. As a teacher you need to create a learning community by including tasks that ensure that all students participate in both group and whole of class discussions. Research suggests that students learn not only from teachers and the tasks, but from other students as well.

The approach used in teaching primary students is also worth investigating. Students and teachers move from everyday language, to materials language, to mathematical language, to symbolic language as they develop conceptual understanding (Larkin et al., 2012). The first three are usually oral and the latter written. At the same time, when teaching the concept, a teacher moves from using familiar objects, substituted objects, photographs and diagrams to non-word symbols. While this can be seen in young students' learning (Irons, 2014), it can also be seen in the secondary classroom. For example, in teaching the concept of function, teachers need to recognise the organising power of the concept of functions from middle school mathematics through more advanced topics in high school and beyond (Leinhardt et al., 1990). Teachers also need to see the potential to discuss the connections between graphical and algebraic representation of functions and how that understanding increases over time. Too often 'functions' appear in Year 11 as something alien or new, yet the concept may have been introduced in primary school using everyday language; later, teachers need to 'wean pupils off the use of informal everyday language and to privilege the use of formal technical vocabulary'(Leung, 2005). One method of weaning is to reframe the way students talk in mathematically acceptable language. This provides teachers with the opportunity to enhance connections between language and conceptual understanding. In reporting on their work, O'Connor and Michaels (1996) used the term 'revoicing' to mean the repeating, rephrasing or expansion of student talk in order to clarify or highlight content, extend reasoning, include new ideas or move discussion in another direction. Walshaw and Anthony suggest that in classrooms where revoicing is used, 'there is a greater tendency for students to provide the explanations ... and for the teacher to repeat, expand, recast, or translate student explanations for the speaker and the rest of the class' (Walshaw & Anthony, 2008, p. 119).

Typical errors in language

When you speak in a classroom, it is easy to make small mistakes. For example a pre-service teacher recently reflected on a recording she did on explaining the phrase 'expanding an expression'. In one short (less than five-minute screencast) she picked up on five errors in explanation:

> ... enjoy my 'warts and all', spur of the moment screencast I recorded. The main issue I pick up on my screencast is the misuse of terminology. Firstly I use the term 'equation' where I should be using 'expression' and I think I said solve rather than expand. I also mention the mnemonic F.O.I.L which I don't think is quite right for this circumstance ... Then I fall into the dreadful use of English with the word 'timesing' rather than multiplying. Finally there is the tricky negative versus minus ...

Below are five language 'bloopers' that teachers make in the everyday language of the classroom:

1 **Negative or minus**: Students will often say minus 10 when they mean negative 10. The first is an operation of subtraction; the second is a number. The expression $x - 10$, 'x minus 10', can be changed to an equivalent expression $x + -10$, 'x plus negative 10'.

2 **Timesing and plussing**: These are not English words. Use correct words such as 'multiplying' and 'adding'.

3 **Mathematical tricks**: There are various mnemonics that are a memory aid, such as

   - FOIL (first ... last)
   - BIMDAS (brackets ... subtraction)
   - SOHCAHTOA (sin ... adjacent).

   Make sure students don't use these mnemonics as an explanation of how mathematics works. The explanation of the expansion of the product of two binomials (i.e. $(x + a)(y + b) = xy + xb + ay + ab$ should include the phrase 'distributive law').

4 **Equation or expression**: $3x + 7$ is an expression. $3x + 7 = 10$ is an equation (it has an equals sign in it). $3x + 7$ is an expression of two terms (i.e. they are separated by a + symbol), and the first term has two factors (x and 3). See Chapter 8 for more discussion on these two terms.

5 **Unknown vs variable** $3x + 7 = 10$ is an equation with an unknown (i.e. we can find the value of x). $3x + 7 = y$ is an equation with two variables (i.e. the x and the y vary). See Chapter 8 for more discussion on these two terms.

Talking about errors

Have you tried to ask students to talk about errors they have made? Think back to when you did not understand some mathematics. Did you talk about those errors to others or your teacher/lecturer? Students seem to find it difficult to express themselves when they make an error. They may find they do not have the language to talk about the issues clearly, they may have been humiliated in the past or the atmosphere of the classroom may not be conducive to such discussions. However, a teacher should encourage students to change their mindset, saying their mind grows when they make errors and then act on those errors. For a discussion on the issue of growth mindset, you can read discussions by Jo Boaler (2013) or a book by Julie Ryan and Julian Williams which has an appendix with common errors and a section on discussion prompt sheets (Ryan & Williams, 2007).

Balance between talking and listening

An effective classroom teacher needs to have an eye on the 'mathematical horizon', using listening skills and attending carefully to what students say, what they mean and what they do not say. In addition, while teachers need to communicate effectively, they also need to be careful about giving too much feedback or 'teacher lust' (Maddern & Court, 1989). Maddern and Court cited a teacher who had created a positive learning environment and had shown a keen desire for talk to occur in the classroom, but the actual mathematical talk was minimal. Walshaw and Anthony (2008) went on to explain that there were limited opportunities to speak the mathematics. On a number of occasions no pauses were made for students' thinking time and students were sometimes 'talked over'.

Enabling opportunities for talking is especially important in high school, where students are expected to use the more mathematical and symbolic language. As mentioned earlier in this chapter, students need fluency in the mathematics register in order to be successful mathematics learners and they need to be taught explicitly to translate between the everyday and the mathematics registers. In particular, Marton and Tsui (2004) suggest that: 'the teacher who makes a difference … is focused on shaping the development of novice mathematicians who speak the precise and generalizable language of mathematics' (p. 532). This development could take the form of argumentation. Students can take and defend particular positions. O'Connor and Michaels (1996) pointed out that this process depends on a skilful teacher pre-empting mathematical conversations. The skill 'provides a site for aligning students with each other and with the content of the academic work while simultaneously socializing them into particular ways of speaking and thinking' (p. 65).

A study by Goos (2004) based on a senior secondary school class in Queensland showed how a teacher can initiate a culture of inquiry in the classroom. While the students tended to use informal language in discussions among themselves, the teacher insisted that students use mathematical language when explaining their solutions to the whole class. The teacher's approach can be reflected in an interview extract with this teacher:

> I do think it's important that they're able to communicate with other people and their peers. They will learn at least as much from each other as they will with me. To be able to do that they have to talk to each other. It's also a part, one of the reasons I often use force them to say things because they need to be able to use the language because language itself carries very specific meanings; and unless they have the language to be able to, obviously communicate, but I think it also has something to do with their understanding as well. (Goos, 2004, p. 271)

The student's voice can also be heard in this same class. Here, two students (Dean and Adam) are exploring a novel problem. Dean comments:

> Adam helps me [...] see things in different ways. Because, like, if you have two people who think differently and you both work on the same problem you both see different areas of it, and so helps a lot more. More than having twice the brain, it's like having ten times the brain ... (Goos, 2004, p. 278)

Questioning

Many books have discussed the importance of good questioning in the classroom. Often they are categorised as closed (what is the value of x in this equation?) or open (how do you know the value of x is 10?), with the open-ended questions usually promoting active learning. However, closed questions are sometimes useful and open questioning is not enough. Wiggins and McTighe (1998) suggested six types of questions connected to six facets of understanding. Table 2.3 shows these six facets with some examples.

Table 2.3 Facets of understanding and examples

Facet	Sample question
Understanding	We know that −3 times −5 is 15. But why is this so? How can you explain this to someone else?
Interpretation	What happens when we add a large number to a series of numbers where we want to find the mean? What happens to the median?
Application	Why would a real estate agent use the median instead of the mean to help sell a house?
Perspective	In what circumstances would the median be used instead of the mean?
Empathy	We know that −3 times −5 is 15. Why do you think people find this a difficult concept to understand and remember?
Self-knowledge	We know that −3 times −5 is 15. Do you remember when you found this difficult to understand and what was it that helped you realise this was true?

Questioning may differ depending on whether the questions are to the whole class, a small group or an individual student. Basic principles for whole class questioning include (Huetinck & Munshin, 2008):

- Do not ask a question unless you want a thoughtful response.
- Try not to embarrass a student if they cannot answer a question (know your students) but ensure that all students have a chance to answer at least one question during the week.
- Ensure that you address the problem of a student saying 'I don't know' or 'I have no idea', and if this type of answer persists, express your concern to the student (privately) that you want the student to increase their mathematical understanding.
- Write higher-order thinking questions in your lesson plans.
- Ensure that you have 'wait time' (more than three seconds).

Further reading can be found in the book by David R. Johnson *Every Minute Counts: Making Your Math Class Work* (1982). For example, Johnson suggests leaving a question unanswered at the end of the lesson, but ensuring that you come back to it at a later date. He also suggests not to overuse the questions 'Do you have any questions?', 'Does everyone understand?'

Another suggestion is to take notice of the series of questions that are asked. Herbel-Eisenmann and Breyfogle (2005) suggest a series of 'focusing' questions. These could include questions that could focus on translating, clarifying, interpreting, applying, synthesising or evaluating (reflecting Bloom's Taxonomy).

If you have access to mini whiteboards you can gauge students' level of understanding without embarrassing individual students. A more high-tech version of this is learner–response devices (using clickers or mobile phones) that can provide teachers with instant feedback on a range of questions. Higher-tech again is a collaborative classroom management software program where all students have a tablet wirelessly connected to a teacher's learning management system, so a teacher can individually gauge a student's performance and ask students to answer questions displaying the results on a screen, without the student standing in front of the class.

ACTIVITY 2.8

The questions below can be improved. Write alternatives to elicit deeper understanding:

- What is Pythagoras' theorem?
- What is the derivative of?

- How about this angle?
- Would you say x is 10?
- Does everyone understand?
- Has anyone worked out the answer to question 4?

REFLECTIVE QUESTIONS

1   Many teachers in their explanations of procedures use the word 'simply'. What appears simple and straightforward to you may be confusing to a student. Similarly, in discussing their own errors, students sometimes use the phrase 'it was a silly mistake'. Often these mistakes are common and easy to make. They are not trivial, and need further investigation. Have you found yourself using these words?

2   A student writes: $21^3 \times 21^4 = 21^{12}$. Record yourself (and a partner) explaining the error to the student (your partner). Critically review your recording.

Technical communication

Communicating mathematical material is a skill that comes with practice and noticing. This section focuses on the presentation of mathematical and numerical data in textbooks and reports, and on the way you, as a teacher, present the mathematics to your students. If you consciously implement these suggestions in your writing and lesson planning, you will find that after a while you will automatically implement these ideas into your teaching and pass on your knowledge to your students. There are Australian Standards on how to write much of the quantitative information in scientific and technical reports, and also in trade and industry (www.australia.gov.au/topics/science-and-technology/standards).

There are a number of formatting software programs that can assist you to type mathematics correctly. If you are a professional mathematician you would probably use LATEX. However, if you are used to using Word, then there may be an Equation editor (in the 2013 version select the 'insert' tab and choose the 'equation' option). MathType® is an alternative. It can easily format your mathematics. If you have one of these alternatives, try to replicate some of the mathematics that is presented in this chapter.

Standard form of writing numbers

Numbers below 10 are written in words. Most scientific writing guides recommend that 0 should be written as 'zero'. For example: five cars, one egg, zero students.

Exceptions to these are when numbers are used with units of measure, age, time, dates, page numbers, percentages, money, ratios and proportions. For example: two metres; $6; nine-second intervals.

Write all numbers as numerals when two or more numbers are used in the same section. For example: 'A full irrigation system contains 10 metres of pipe, two water pumps and 12 water storage tanks'. An exception to this would be if none of the numbers is greater than nine: 'A full irrigation system contains nine metres of pipe, two water pumps and six water storage tanks.'

Formatting mathematics

Numbers larger than 999 should be written with a space to indicate thousands (or thousandths) of units. For example you should write: 10 001, 100 000 001, 0.043 219. (There are exceptions – for example, in nursing the current recommendation is to use commas to ensure there are no errors in drug delivery by a power of 10.)

The format you choose should be:

- consistent throughout the document
- in a format easily understood by your audience
- suitable for any comparisons you wish to make between the numbers.

Most non-scientific audiences find it easier to think in terms of 100 million or 25 000 instead of 10^8 or 2.5×10^4. However, if you want to compare orders of magnitude of numbers it is easier to compare 5.1×10^3 and 5.1×10^8, rather than having to count zeros.

Use the singular, rather than the plural to describe a fraction less than one. For example, 0.5 *kilometres* may seem correct, but since 0.5 is less than one we have the singular, not plural. You would not say 'half a kilometres'. The correct term is 0.5 kilometre. For example: 0.9 kg, 0.25 cup of flour.

In Australia a full stop (not a raised point) should be used to represent the decimal point (e.g. 34.76). In Europe it is common to use a comma as a decimal marker (e.g. 34,56), but this is not favoured in English-speaking countries, and therefore should not be used.

A zero should always be placed before the decimal point in numbers less than one. That is, .34 should always be written 0.34. This ensures clarity and consistency, particularly when such numbers are mixed with numbers greater than one. Show students that decimals are expressed this way on the calculator.

In Australia, we use the Standard International (SI) units: for example, length is measured in metres and the symbol is 'm'. There are many units derived from the basic units. Area is measured in square metres and the symbol is m² and luminance is measured in candela per square and the symbol is cd/m².

Formatting to note:

- Names of units and prefixes, when spelt out in full are expressed in lower-case letters (except at the beginning of a sentence). The exception to this is the capital C in degree Celsius. Example: 10 kilometres per hour and 10 degrees Celsius.
- Unit symbols are usually expressed in lower case. There are some exceptions: e.g. millilitre (mL), units named after people, i.e. hertz (Hz), pascal (Pa), newton (N), watt (W), etc. and symbols containing exa, peta, giga and mega, i.e. gigawatt (GW), gigametre (Gm).
- The term 'per' should only be used when the unit is expressed in words, whereas '/' should only be used with symbols. For example kilometres per hour or km/h *not* kilometres per h or km/hour.
- Symbols for units should not be set in italics. The only exceptions are ohms which is represented by the Greek letter Omega (Ω) and the prefix 'micro', which is the Greek letter mu (μ).
- A space (thin if possible) should be included between numerical value and unit names and symbols (e.g. 27 m, not 27m). Exceptions to this are the symbols for degrees of arc (°), minute ('), and second (") (e.g. note the difference between 10°, and 30°C).

Mathematical equations

Letters used as variables should be typed in italics (A, B, C, x, y, z) **unless they represented vector/matrix quantities, in which case they should be typeset in bold** ($\mathbf{A}, \mathbf{B}, \mathbf{C}, \mathbf{x}, \mathbf{y}, \mathbf{z}$). It is essential that the same font be used for the same variable throughout the document.

Equations should be centred on a separate line in documents unless they are short and simple. For example:

$$ax^2 + bx + c = 0 \tag{A.1}$$

Equations appearing on a separate line should be numbered in the order that they appear in the document. This makes it easy to reference a given equation later in the document. Numbering all equations provides the document with a look of consistency, which makes it easier for the reader to follow and find specific references. For example:

$$y = ax^2 + bx + c \tag{A.2}$$

The general form of a quadratic is given in Eqn. A.2.

Short equations may be placed on a separate line or inline, whichever you prefer.

The current in the wire was calculated using $E = I\,R$, where E is the current, I is the electric potential, and R is the resistance.

Or:

The current in the wire was calculated using

$$E = I\,R \tag{A.3}$$

where E is the current, I is the electric potential, and R is the resistance.

All equal signs, fraction lines, multiplication, addition and minus signs should be horizontally aligned.

Examples:

$$\Delta T = \frac{4}{\pi} K \frac{T}{T_0} \tag{A.4}$$

$$\frac{y - y_0}{x - x_0} = \frac{y_1 - y_0}{x_1 - x_0} \tag{A.5}$$

The equal signs should be aligned for a series of connected equations.

Example:

$$z = r\,(\cos\theta + i\sin\theta) \tag{A.6}$$

$$= r\,e^{i\theta} \tag{A.7}$$

Note that in this case it is acceptable to not number all the equations in the block. For example it is perfectly acceptable to follow the following example, which just references the key equation in an equation block.

$$z = r\,(\cos\theta + i\sin\theta) = r\,e^{i\theta} \tag{A.8}$$

As you would not write $2 = 3$**, you should not write a mathematic expression or equation that is not correct.** For example *do not write* something like the following:

$$z = \frac{12 - 6}{6}$$
$$= 1$$
$$= 0.16 \qquad\qquad (A.9)$$

The above statement cannot be true as 1 is not equal to 0.16. In this case, the author meant the following.

The standard z score is

$$z = \frac{12 - 6}{6}$$
$$= 1 \qquad\qquad (A.10)$$

This gives the area in the tail of the standard normal of 0.16.

The hierarchy of brackets is normally parentheses, square brackets, and then braces {[()]}. According to the *Style Manual for Authors, Editors and Printers* (Commonwealth of Australia, 2002) this is stylistic rather than obligatory. However, you should be aware that in certain mathematical contexts brackets have special meaning. For example [1, 2] denotes a line segment from 1 to 2 inclusive, while (1, 2) denotes an ordered pair.

Bill Barton (2008) talks about the misuse of the terminology by teachers:

In the graph [below] the variable x and the function $f(x)$ are each used in two ways ... The 'x' in the expression $f(x) = 2 \times 2 + 1$ is any value of x at all – x is a variable. But the meaning of 'x' in the label $P(x, f(x))$ and the label on the horizontal axis is a particular, but unspecified, value of x. In this situation it is more correct to label the particular value as x_1, but often teachers do not do this, and slip between particular and general uses of a variable without thinking – to the confusion of their students. One more implication. The need to communicate, the need to play, the need to explore, and the need to learn about mathematics means that those charged with teaching the subject must themselves be more mathematically literate than ever before. If a teacher is to recognise, follow, and utilise the diverse mathematical thinking of children, then the more links, experiences and applications on which to draw, the better. They must know other ways of approaching the same idea; they must sense different directions in which the idea can be taken; they must be able to make use of cognitive conflicts that arise and new situations the children imagine. (pp. 156–57)

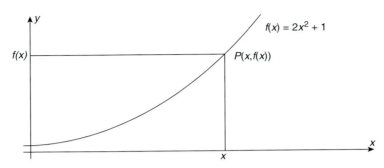

Figure 2.5 Multiple meanings for x

ACTIVITY 2.9

What is the difference between a square metre and a metre squared? Google this topic and you may be surprised by the amount of confusion in the general community. Again, think of the structure of the English language. We write m^2, so in reading from left to right the 'metre' is first encountered, but we say square metres.

ACTIVITY 2.10

Go to any mathematics textbook that you have used and look at the way the mathematics is written. You will see that they follow these conventions. For example, log in to HOTmaths and select 'HOTmaths Global' Course list and then choose the 'Middle Secondary' Course. From the Topic list choose 'Simultaneous linear equations', then the 'Using the substitution method' Lesson. From the Resources tab, select the 'Solving by substitution SOLUTIONS' HOTsheet.

Solve a set of simultaneous equations and check you have the same conventions. Record yourself solving these equations as if you were teaching students and check that your language is correct. Listen particularly to the words and phrases you use. For example:

- Did you use the word 'simply' at any stage or 'just do'? (Remember, students are learning this, so it is not simple.)
- Did you use the correct terms such as 'equations' and 'expressions' in the right context?

- Did you use the terms 'unknown' and 'variable' in the right way?
- Did you put '=' signs under '=' signs?
- Did you explain from one line to the next what to do and write it down, e.g. 'substitute $c = 2$ into (i)'?
- Did you have the final statement in proving your solution is correct as something like '$10 = 10$'?

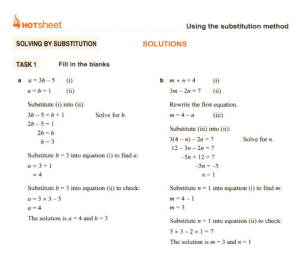

Figure 2.6 Solving by substitution SOLUTIONS HOTsheet

Summary

Recognise the importance of language teaching and learning of mathematics

Language plays a pivotal role in the teaching and learning of mathematics. National and international documents on mathematics education attest to this. As a teacher, you need to be fluent in the everyday language of the classroom and the particular language of mathematics. In addition, you need to recognise the important role language plays in the ease with which students understand mathematical concepts. A good extension of the issues found in this chapter can be read in the Australian Council for Educational Research (ACER) 2010 Research Digest publication issue which focused on research studies about language in the mathematics classroom (Meiers, 2010).

Identify the elements of the mathematics register in English and other languages and discuss the impact on learning

The mathematics register is the words, phrases and associated meanings used to express mathematical ideas. As a teacher you need to be aware of the impact on learning: that the words used in mathematics may have a different meaning in mathematics, that 'definitions' in mathematics may have different meanings depending on the level of the student, that the syntactic and semantic structure of word problems can make interpretation of a word problem much more difficult and that second language learners have particular difficulties in the mathematics classroom, but that they may also bring teachable moments through rich discussion on the mathematics register in other languages.

Reflect on the appropriate use of language by the student and the teacher in the mathematics classroom

A key to effective teaching of mathematics is the appropriate use of language in the classroom, both oral and written. In particular, a teacher should encourage meaningful, directed and fluent student talk at the appropriate level and at the appropriate times. This involves modelling the correct language, encouraging speaking and framing questions to promote active learning.

Write mathematics correctly (technical communication)

Closely linked to the objectives already mentioned is the correct use of the particular symbolic mathematical language often seen as the 'grammar' of mathematics. A prepared teacher will always write mathematics in the correct way and explain why the mathematics is written this way. These days, writing mathematics is relatively easy, with programs such as MathType®. More powerful are tablet and recording devices, where teachers can model how mathematics is written correctly (as they are saying it) and ask students to record their answers as well.

CHAPTER 3

Making mathematical connections

Learning outcomes

After studying this chapter, you should be able to:

- articulate the importance of helping students make mathematical connections in their learning
- define the term 'mathematical connections' in a secondary context
- understand teachers' practices in promoting mathematical connections
- outline key features, instructional approaches, and challenges associated with STEM education
- outline key instructional guidelines and tasks that can help students connect with mathematics.

Introduction

The Australian Curriculum and Assessment Reporting Authority (ACARA) recognises the importance of making mathematical connections. Specifically, ACARA (2015a) articulates this importance in its curriculum *Aims*, which are to ensure that students:

- are confident, creative users and communicators of mathematics, able to investigate, represent and interpret situations in their personal and work lives and as active citizens
- develop an increasingly sophisticated understanding of mathematical concepts and fluency with processes, and are able to pose and solve problems and reason in Number and Algebra, Measurement and Geometry, and Statistics and Probability

- recognise connections between the areas of mathematics and other disciplines and appreciate mathematics as an accessible and enjoyable discipline to study.

The importance of helping students of mathematics make connections both within the discipline and to the real world is a claim that receives consistent international reinforcement (Boaler, 2002a; Gainsburg, 2008; Sullivan, 2011). Although learner might make connections spontaneously, teachers cannot 'assume that the connection will be made without some intervention' (Weinberg, 2001, p. 26). Implied within this statement is the exhortation for teachers to take an active role in making mathematical connections clear to students. For instance, Ma (1999) contended that to help students make connections, teachers must understand mathematics as an interrelated web of ideas, and possess the appropriate **pedagogical content knowledge** to know which strategies and activities best facilitate student learning. Moreover, the National Council of Teachers of Mathematics (NCTM) asserts that teachers must be instrumental in developing a deep conceptual understanding of mathematics within students, and assist in departing from a perception of mathematics as a 'set of isolated facts and procedures' (NCTM, 2009, p. 3). Moreover, the Australian Curriculum: Scope and Sequence documents can be viewed as a conceptual 'road map' of content which can be used strategically by educators in planning units of work replete with meaningful connections.

In this chapter we shall commence with an examination of the importance of making mathematical connections in the secondary classroom. Next, various conceptualisations of the term 'mathematical connections' will be presented in an attempt to define the term itself. Then, the work of practitioners and researchers is offered in the way of useful and appropriate instructional practices and guidelines. Science, Technology, Engineering and Mathematics (STEM) education will also be presented as a way of making mathematical connections clear and relevant to secondary students. Finally, a variety of activities are included for teachers to help students make mathematical connections both within the discipline and to real-world settings.

KEY TERMS

- **Instrumental understanding:** the acquisition of rules without the promotion of thinking about how familiar concepts and procedures can help in new situations.
- **Pedagogical content knowledge:** the teacher's interpretations and transformations of content knowledge in the context of facilitating student learning.
- **Procedural fluency:** the knowledge of mathematical procedures (knowing both when and how to use them appropriately), and the demonstrable skill in performing these procedures flexibly, accurately and efficiently.

- **Relational understanding:** learning to make connections between mathematical concepts and advancing understanding by linking those concepts to prior knowledge.
- **Zone of proximal development:** the region that exists between what students can do independently and that which can only be completed with the assistance of others.

The importance of making mathematical connections

The current literature underscores the importance of making mathematical connections across mathematical concepts (Businskas, 2008) and to the real world (Boaler, 1997a; Gainsburg, 2008). Leikin and Levav-Waynberg (2007, p. 350) assert that 'connecting mathematical ideas means linking new ideas to related ones and solving challenging mathematical tasks by seeking familiar concepts and procedures that may help in new situations'. This assertion supports the seminal work of Skemp (1987), who discussed two types of mathematical understanding critical for progress in mathematics learning: relational and instrumental. While **instrumental understanding** implies habitual learning, or acquiring 'rules without reasons', it does not promote thinking about how familiar concepts and procedures can help in new situations. By contrast, **relational understanding** requires the learner to make connections between different mathematical concepts and to make progress by linking these concepts to prior knowledge (Skemp, 1987). Mathematical connections are thought to benefit student learning (Gainsburg, 2008) by enhancing students' capacity to remember, appreciate and use mathematics (Businskas, 2008) and to develop conceptual understanding (Anthony & Walshaw, 2009a). Additionally, the use of real-world connections can motivate mathematics learning (National Academy of Sciences, 2003) and help students apply mathematics to real problems – particularly those in the workplace (NRC, 1998). To make connections in the secondary classroom, authors argue that teachers must constructively use students' prior knowledge (Hattie, 2009; Swan, 2005) so that a 'deeper and more lasting understanding' can occur (NCTM, 2000, p. 64). After discerning what students can and cannot do, it is recommended that teachers build connections from previous lessons, experiences and thinking (Anthony & Walshaw, 2009b; Clarke & Clarke, 2004).

Mathematics connections in Australia: status quo

In addition to the emphasis ACARA places on mathematical connections, various documents point to this importance. For instance, the Australian Education Report (Sullivan, 2011) drew upon research findings and other sets of recommendations for teaching actions, to present a set of six principles to guide teaching practice. Of these six principles, *Principle 2: Making connections* is elaborated for teachers as such:

> Build on what students know, mathematically and experientially, including creating and connecting students with stories that both contextualise and establish a rationale for the learning. (Sullivan, 2011, p. 26)

In the most recent Trends in International Mathematics and Science Study (TIMSS), Australian students continued to rank in the top third of all participating countries. Interestingly, 48 per cent of Singaporean Year 8 students demonstrated performance at the *Advanced Benchmark*, while only nine per cent of Australian Year 8 students reached this same benchmark (Thomson et al., 2013). A decade ago, the *Advanced Benchmark* was attained by 44 per cent of Singaporean students and only seven per cent of Australian students (Thompson & Fleming, 2004). The Advanced International Benchmark is included in Table 3.1.

In an examination of the Singaporean mathematics curriculum, Kaur (2001) noted an emphasis on 'the development of mathematical concepts and skills, and

Table 3.1 TIMMS Advanced Benchmark Descriptor

Advanced Benchmark	
	Students can organise and draw conclusions from information, make generalisations and solve non-routine problems. Students can solve a variety of fraction, proportion and per cent problems and justify their conclusions. Students can express generalisations algebraically and model situations. They can solve a variety of problems involving equations, formulas and functions. Students can reason with geometric figures to solve problems. Students can reason with data from several sources or unfamiliar representations to solve multi-step problems.

Source: Thompson et al., 2013, p. 21.

the ability to apply them to solve problems' (p. 141). During lessons, she observed mathematics teachers placing emphasis on students solving non-routine problems. In an Australian context, research conducted into teaching approaches for Year 8 students suggested that few complex problem-solving opportunities are being provided (Hollingsworth, 2003). Furthermore, and consistent with Skemp's notion of instrumental understanding, Stacey (2003, p. 119) appraised the average Australian mathematics lesson as one exhibiting 'a syndrome of shallow teaching, where students are asked to follow procedures without reasons'. Stigler and Hiebert (1999) proposed that one factor influencing the lack of adoption of problem-solving strategies concerns primarily the teachers' knowledge and beliefs about mathematics teaching and learning.

The most recent Programme for International Student Assessment (PISA) report (De Bortoli & Macaskill, 2014) highlights that along with England and the United States, Australian students demonstrated excellent performance in problem solving. According to De Bortoli and Macaskill (2014), this may suggest that 'in these countries, top performers in mathematics have access to – and take advantage of – the kinds of learning opportunities that are also useful for improving their problem-solving skills' (p. 46). Additionally, when compared to other countries 'Australian students are comparatively stronger on both the *Exploring and Understanding* and the *Representing and Formulating* processes, and are relatively weaker on the *Planning and Executing* process' (p. 46). Despite this acclaim, there has been a significant decline in the performance of Australian students who are classified as top performers, average performers and low performers. In other words, there have been significant declines for Australian students at the 10th, 25th, 75th and 90th percentiles between PISA 2003 and PISA 2012 (Thomson et al., 2012, p. 44). Following the earlier comparison between Singaporean and Australian students, the proportion of Singaporean students (40 per cent) who achieved Proficiency Level of 5 or 6 far exceeded that of Australian students (15 per cent). These Proficiency Levels have been included in Table 3.2.

Table 3.2 PISA Proficiency Levels 5 and 6

Level 6	[Students] conceptualise, generalise and use information. They are capable of advanced mathematical thinking and reasoning; have a mastery of symbolic and formal mathematical operations and relationships; and can formulate and precisely communicate their findings, interpretations and arguments.

(Continued)

Table 3.2 (*cont.*)

Level 5	[Students] develop and work with models for complex situations; select, compare and evaluate appropriate problem-solving strategies for dealing with complex problems; work strategically using broad, well-developed thinking and reasoning skills; and reflect on their work and formulate and communicate their interpretations and reasoning.

Source: Thompsonet al., 2012, p. 13.

Given the evidence that Year 8 students experience little complex problem solving in mathematics classes, as well as the results from PISA and TIMSS, there is considerable room for change (Anderson, 2005). To this point, Stacey (2003, p. 122) recommends that 'there needs to be a greater emphasis on explicit mathematical reasoning, deduction, connections and higher-order thinking in lessons'. Drawing on the work of scholars within this section, one can posit that Australian students require access to learning activities that intentionally promote mathematical connections.

ACTIVITY 3.1

Log in to HOTmaths and select the 'HOTmaths Global' Course list and then choose the 'Early Secondary' Course. From the Topic list choose 'Decimals & money', then the 'Changing fractions to decimals' Lesson.

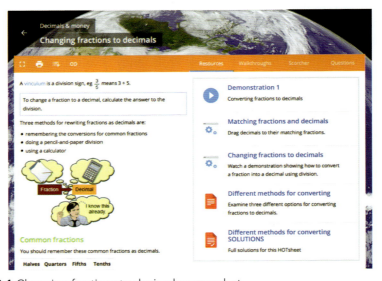

Figure 3.1 Changing fractions to decimals screenshot

Looking at the Resources, outline the key mathematical connections you expect students to make. What will your strategies be to ensure that these connections are made? Assume that you are teaching a co-educational Year 7 class.

ACTIVITY 3.2

Log in to HOTmaths and select the 'HOTmaths Global' Course list and then choose the 'Middle Secondary' Course. From the Topic list choose 'Logarithms', then the 'Approximating logarithms' Lesson.

Looking at the Resources, outline the key mathematical connections you expect students to make. What will your strategies be to ensure that these connections are made? Assume that you are teaching a co-educational Year 11 class.

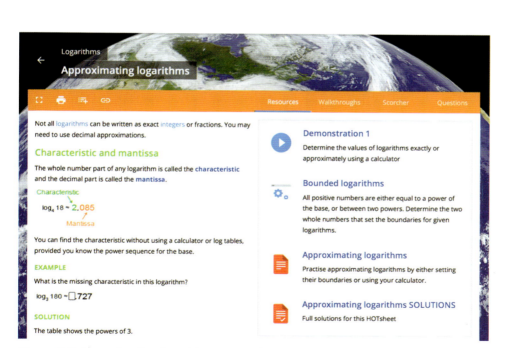

Figure 3.2 Approximating logarithms screenshot

REFLECTIVE QUESTIONS

In your own words, define the term 'mathematical connections'. In your definition, what key words did you use? How do you plan to help your secondary students make these connections during the lessons and units you will teach?

What are mathematical connections?

So what is meant by the term 'mathematics connection'? The Oxford English Dictionary offers eleven definitions for 'connection', with perhaps the most appropriate definition listed as: 'The condition of being related to something else by a bond of interdependence, causality, logical sequence, coherence, or the like; relation between things one of which is bound up with, or involved in, another'. If we commence with the idea that mathematical connections are 'causal, logical or coherent relationships' between ideas and concepts, further questions must be asked. For instance, Businskas (2008, p. 7) proposed that a connection is a feature of the content matter (e.g. a relationship between ideas), a relationship that is constructed by the learner (e.g. a mental construction in the learner's mind), or a 'process that is part of the activity of doing mathematics'. In the United States, the National Council of Teachers of Mathematics (NCTM) outlines six principles which are 'are statements reflecting basic precepts that are fundamental to a high-quality mathematics education' (2015, p. 2). The second principle is Curriculum, which is articulated as:

In a coherent curriculum, mathematical ideas are linked to and build on one another so that students' understanding and knowledge deepen and their ability to apply mathematics expands. An effective mathematics curriculum focuses on important mathematics that will prepare students for continued study and for solving problems in a variety of school, home, and work settings. A well-articulated curriculum challenges students to learn increasingly more sophisticated mathematical ideas as they continue their studies.

This principle focuses explicitly on both those connections which link mathematical ideas together, as well as the appreciation and application of mathematical concepts to contexts outside the classroom.

Connections: feature, relationship or process?

Much of the literature presents the notion of mathematical connections as a process (Boaler, 2002a; Evitts, 2004; Ma, 1999). For instance, Ma (1999) contended that having a profound understanding of mathematics required an ability to connect ideas within a topic and to central concepts of the discipline. In a similar vein, Boaler (2002a, p. 11) highlighted that 'the act of observing relationships and drawing connections, whether between different functional representations or mathematical areas, is a key aspect of mathematical work, in itself, and should not only be thought of as a route to other knowledge'. Moreover, Evitts (2004) examined the problem-solving activities of pre-service teachers as a process of making mathematical connections. He observed that the pre-service teachers were engaged in making a variety of connections, which were categorised as:

- Modelling: Attempting to find some aspect of prior mathematical knowledge that could be used to portray some real-world component of the problem in a mathematical way.
- Representational: Using two or more representations to talk about the same mathematical idea.
- Structural: Discussing and using similarities found between a real or mathematical component of the problem and another real or mathematical situation.
- Procedure-Concept: Describing or using procedures via a rule or formula-based approach. Indicating a conceptual basis for utilising a procedure.
- Between Strands of Mathematics: 'Crossing over' from one strand of mathematics to another when analysing he problem. Using references to other areas of mathematics (Evitts, 2004, p. 56).

Coxford (1995) conceptualised connections as features, or very broad ideas linking different topics in mathematics. In this conceptualisation, Coxford identified three categories of mathematical connections: (a) unifying themes, (b) mathematical processes and (c) mathematical connectors. *Unifying themes* are themes (e.g. change, data, shape) that may be used to draw learners' attention to the connected nature of mathematics. For instance, the theme of change may connect algebra, polynomials, differential calculus and geometry. Key *mathematical processes* include representation, application, problem solving and reasoning. To illustrate, lower secondary students should develop competency in moving fluidly through the concrete–representational–abstract notions of fractions, decimals and percentages. Coxford (1995, p. 7) maintained that these connections 'are vital if students are to make sense out of later operations on numbers'. Third, *connectors* are mathematical ideas (e.g. graph, algorithm, variable, proportion, function) which arise in relation to studying a wide spectrum of topics. According to Coxford, these ideas 'permit the student to see the use of one idea in many different and, perhaps, seemingly unrelated situations' (Coxford, 1995, p. 10). While this section has dealt with the notion of connections as features and processes, one should be careful not to dismiss the extent to which mathematical connections exist in the minds of learners. On this point, Hodgson (1995, pp. 14–15) commented that 'if students are unable to establish connections, then the connections cannot be used in problem situations regardless of whether they exist or not'. In other words, the province of mathematical efficacy remains the enterprise of the learner, and the importance of teachers in making connections known explicitly to learners is underscored.

Teachers' conceptualisation of connections

Teachers view mathematical connections as a function of instructional practice (Thompson, 1992), and a component of their mathematical content knowledge (Gainsburg, 2008). Businskas (2008, pp. 150–51) found that secondary teachers perceived mathematics as an interconnected web of concepts, and such connections comprised mathematical knowledge that became a strategic element of their teaching. Additionally, while some teachers saw mathematical connections as integral to the way they taught, others were conflicted, assuming that an emphasis on connections would be time consuming and may detract from their responsibilities to 'cover the curriculum' and to prepare their students for external assessments. Fennema and Franke (1992) concluded that when teacher content knowledge has been defined in a manner consistent with the nature of mathematics (or when a conceptual organisation of knowledge was considered) there existed a positive relationship between teacher content knowledge and their instruction. Building on this idea, Ball and Bass (2000, p. 89) viewed this relationship as the foundation for developing students' understanding, or as either a 'pedagogically useful understanding' or pedagogical content knowledge. Concerning the latter term, Shulman (1986, p. 9) asserted that 'to think properly about content knowledge requires going beyond knowledge of the facts or concepts of a domain. It requires understanding of the structures of the subject' (1986, p. 9).

Constraints

Teachers' attempts at engaging students in real-world problem solving can be constrained by many factors. One factor includes meeting requirements for standardised testing and externally mandated curricula, although some teachers believe real-world connections can prepare students for tests (Gainsburg, 2008). Sullivan (2011) notes that teachers experience students who avoid risk taking and do not persist when challenges arise. After discerning the avoidance strategies used by students, teachers can also become complicit in adopting avoidance strategies of their own (Sullivan et al., 2006). For instance, teachers sometimes modify tasks at the planning stage if they expect that students will not engage with the tasks without considerable assistance (Tzur, 2008). Other teachers avoid the challenge of dealing with students who have given up by either reducing the potential demand of the task (Desforges & Cockburn, 1987), or those who do not respond to activities as anticipated (Charalambous, 2008). Gainsburg (2008) found that although some teachers valued tasks that require critical thinking or promote literacy development, more feared that complex, ill-structured or language-intensive tasks would overwhelm students. As such, the key finding supported the notion that teachers were concerned more about over-challenging their students than under-challenging them.

ACTIVITY 3.3

Log in to HOTmaths and select the 'HOTmaths Global' Course list and then choose the 'Middle Secondary' Course. From the Topic list choose 'Index laws', then the 'Extending index laws' Lesson.

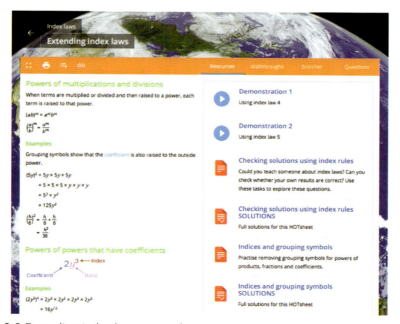

Figure 3.3 Extending index laws screenshot

Looking at the Resources, hypothesise the conceptual difficulties you expect middle secondary students to make. What strategies will you use to prevent or overcome these difficulties?

ACTIVITY 3.4

Log in to HOTmaths and select the 'HOTmaths Global' Course list and then choose the 'Middle Secondary' Course. From the Topic list choose 'Polynomials', then the 'Dividing polynomials' Lesson.

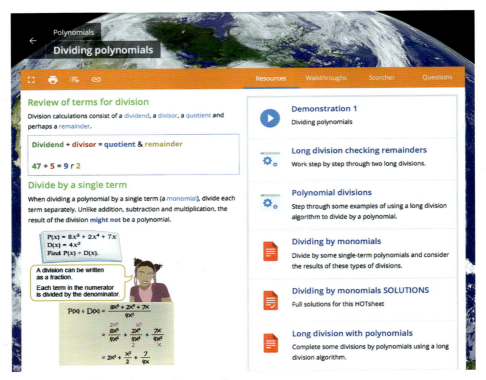

Figure 3.4 Dividing polynomials screenshot

Looking at the Resources, hypothesise the conceptual difficulties you expect senior secondary students to make. What strategies will you use to prevent or overcome these difficulties?

REFLECTIVE QUESTION

Reread the sub-section on how teachers conceptualise mathematical connections. Then, after selecting a year level and mathematical strand from the Australian Curriculum, outline specifically how you plan to help students connect mathematics ideas.

Teachers' practices in promoting mathematical connections

The literature is replete with practical ideas for teachers to make mathematical connections (Gainsburg, 2008; Sullivan, 2011). Commonly, such ideas include careful scaffolding of learning (Stephens, 2009) and using problem solving as an instructional strategy (Anderson, 2009). Given the recent implementation of an Australian Curriculum, teachers have many opportunities to teach problem solving and use real-life problems as a focus of learning in mathematics lessons. Drawing upon earlier comments from Stacey (2003), there are concerns about the extent to which Australian students solve problems other than those of low procedural complexity. While some teachers may use non-routine and problem-centred tasks in lessons (Anderson & Bobis, 2005), research suggests that many do not (Anderson, Sullivan & White, 2004). Instead, there remains a preference for teachers to rely on presenting lessons focused on the types of questions found in examinations and in textbooks (Doorman et al., 2007; Kaur & Yeap, 2009; Vincent & Stacey, 2008). In this section, we shall look at some of what the current literature offers in a way of general instructional guidelines for making connections, the role of scaffolding within mathematical learning and various problem-solving strategies teachers can use.

General guidelines

In a summary of mathematics education literature, Gainsburg (2008, p. 200) posited a range of instructional practices to assist teachers in creating connections for students:

- simple analogies (e.g. relating negative numbers to subzero temperatures)
- classic 'word problems' (e.g. 'Two trains leave the same station ...')
- the analysis of real data (e.g. finding the mean and median heights of classmates)
- discussions of mathematics in society (e.g. media misuses of statistics to sway public opinion)
- 'hands-on' representations of mathematics concepts (e.g. models of regular solids, dice)
- mathematically modelling real phenomena (e.g. writing a formula to express temperature as an approximate function of the day of the year).

Following a similar line of thought, Sullivan (2011) listed the following recommendations for teachers wishing to engage students in learning mathematics. These recommendations include:

- examining the development of the 'big ideas' that underpin the main strands of the mathematics curriculum, and being able to use the content descriptions of the new Australian mathematics curriculum to inform long-term and daily planning
- ways of appropriately emphasising numeracy and practical mathematics in teaching and assessment in the compulsory years
- approaches to engaging all students through increasing opportunities for decision-making, connecting learning to their experience, and illustrating the usefulness of the learning
- selecting and using a range of tasks that engage students in meaningful mathematics and numeracy and building these tasks into lessons
- exploring the specialised content knowledge involved in mathematical tasks and developing strategies for identifying aspects of common content knowledge that may be needed, including strategies for learning that knowledge when it is required

- examining pedagogies that are appropriate with heterogeneous classes, including specific actions to support students experiencing difficulty and to extend those who are ready.

In addition to these recommendations, finding relevance for what teachers do in their instructional lessons, Sullivan (2011) contends that these same items are important mathematics education components for the professional development of prospective primary and secondary teachers.

Scaffolding

Bearing in mind that scaffolding enables workers to operate at their current level with assistance – and to make progress in their work – we look at how this term can be applied to an educational context. To commence, the work of Vygotsky (1978) suggested that the **zone of proximal development** describes the region that exists between what students can do independently and that which can only be completed with the assistance of others. For students to learn, Vygotsky believed that activities which 'bridged' or 'scaffolded' this zone were required – as opposed to tasks pitched at students' current level of learning. Mathematics teachers who possess high-level knowledge work with students who clearly have more basic forms of that knowledge (Stephens, 2009). To understand and build on what the students already know, Fennema and Romberg (1999) suggested that teachers use an approach of 'cognitively guided instruction'. Such an approach places a greater demand on teachers' mathematical knowledge for teaching (MKT), as students' responses and strategies can steer the lesson in various directions. According to Stephens (2009, p. 30), the teacher's role is then 'to draw together those different directions with a clear focus on enhancing students' understanding'.

In addition, Stephens (2009) identified a range of interaction patterns or scaffolding practices undertaken by teachers during mathematics lessons. Although these patterns and practices were drawn from a primary context, these instructional practices can support secondary teachers to make informed decisions about how they can meet the learning needs of all students in the most appropriate manner possible. The twelve scaffolding practices that contribute to improved student learning outcomes are listed and described in Table 3.3. According to Stephens (2009, p. 30) 'when teachers used these scaffolding practices it had an effect on their own perceptions of what to teach and how to help students learn – that is, on their MPCK'. Consequently, there was a significant shift in what teachers perceived to be associated with effective mathematics teaching during the project.

Table 3.3 Scaffolding practices

Excavating: drawing out, digging, uncovering what is known, making it transparent	Teacher systematically questions to find out what students know or to make the known explicit. Teacher explores children's understanding in a systematic way.
Modelling: demonstrating, directing, instructing, showing, telling, funnelling, naming, labelling, explaining	Teacher shows students what to do and/or how to do it. Teacher instructs, explains, demonstrates, tells, offers behaviour for imitation.
Collaborating: acting as an accomplice, co-learner/problem-solver, co-conspirator, negotiating	Teacher works interactively with students in-the-moment on a task to jointly achieve a solution. Teacher contributes ideas, tries things out, responds to suggestions of others, invites comments/opinions on what she/he is doing, accepts critique.
Guiding: cuing, prompting, hinting, navigating, shepherding, encouraging, nudging	Teacher observes, listens, monitors students as they work, asks questions designed to help them see connections, and/or articulate generalisations.
Convince me: seeking explanation, justification, evidence; proving	Teacher actively seeks evidence, encourages students to be more specific. Teacher may act as if he/she doesn't understand what students are saying, encourages students to explain, to provide/obtain data.
Noticing: highlighting, drawing attention to, valuing, pointing to	Teacher draws students' attention to particular feature without telling students what to see/notice (i.e. by careful questioning, rephrasing or gestures), encourages students to question their sensory experience.
Focusing: coaching, tutoring, mentoring, flagging, redirecting, re-voicing, filtering	Teacher focuses on a specific gap (i.e. a concept, skill or strategy) that students need to progress. Teacher maintains a joint collective focus and provides an opportunity for students to bridge the gap themselves.
Probing: clarifying, monitoring, checking	Teacher evaluates students' understanding using a specific question/task designed to elicit a range of strategies, presses for clarification, identifies possible areas of need.

Orienting: setting the scene, contextualising, reminding, alerting, recalling	Teacher sets the scene, poses a problem, establishes a context, invokes relevant prior knowledge and experience, provides a rationale (not necessarily at the beginning of the lesson, but at the beginning of a new task/idea).
Reflecting/reviewing: sharing, reflecting, recounting, summarising, capturing, reinforcing, reflecting, rehearsing	Teacher orchestrates a recount of what was learnt, a sharing of ideas and strategies. This typically occurs during whole class share time at the end of a lesson where learning is made explicit, key strategies are articulated, valued and recorded.
Extending: challenging, spring boarding, linking, connecting	Teacher sets significant challenge, uses open-ended questions to explore extent of children's understanding, facilitate generalisations, provide a context for further learning.
Apprenticing: inviting peer assistance, peer teaching, peer mentoring	Teacher provides opportunities for more learned peers to operate in a student-as-teacher capacity, endorses student/student interaction.

Source: Stephens, 2009, p. 3

ACTIVITY 3.5

Log in to HOTmaths and select the 'HOTmaths Global' Course list and then choose the 'Early Secondary' Course. From the Topic list choose

Figure 3.5 Exploring trapeziums & kites screenshot

'Exploring quadrilaterals', then the 'Exploring trapeziums & kites' Lesson.

Looking at the Resources, describe how you could adapt them to create a rich and investigative task for students to engage with. Assume that you are teaching a co-educational Year 8 class.

ACTIVITY 3.6

Log in to HOTmaths and select the 'HOTmaths Global' Course list and then choose 'Middle Secondary' from the Course. From the Topic list choose 'Transforming non-linear graphs & graphing regions', then the 'Regions on the Cartesian plane' Lesson.

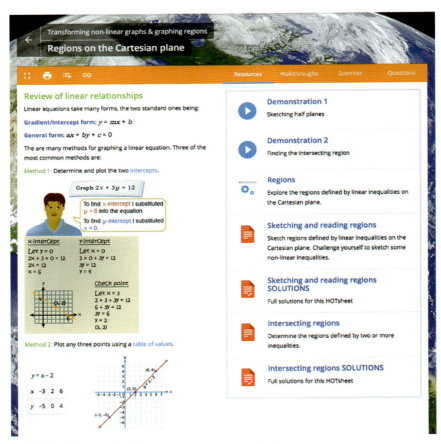

Figure 3.6 Regions on the Cartesian plane screenshot

Looking at the Resources, describe how you could adapt them to create a rich and investigative task for students to engage with. Assume that you are teaching a co-educational Year 11 class.

REFLECTIVE QUESTIONS

Which type of learning activities do you plan to use in your secondary classes? As an exercise, create a list of those activities you feel would be useful in helping Middle School and Senior School students make mathematical connections. Next to each activity, estimate what percentage of instructional time you plan spending (ensure that these numbers equal 100%) on these activities. Finally, justify the percentage you have allocated to each of the learning activities.

Science, technology, engineering and mathematics (STEM)

Science, Technology, Engineering and Mathematics (STEM) education has maintained significant interest from educators, politicians and industry personnel worldwide for decades (Fan & Ritz, 2014). The Australian Academy of Sciences (2006) has identified high-end mathematical skills as crucial to Australia's success in a wide range of fields, including: mining and resources, manufacturing and trade, biotechnology, statistics and finance, and environmental risk assessment. Around the same time as this identification, Tytler and Symington (2007) underscored the value of having a scientifically informed and oriented public to support research and development in emerging technological and science-driven areas. At a school level, both mathematics and science are critically important subjects in preparing students for future roles interacting with STEM or working as STEM professionals – as well as engaging them both conceptually and aesthetically (Darby, 2008; Zembylas, 2005). Despite this importance, science and mathematics are viewed differently by educators and curriculum designers. To illustrate, mathematics is considered to be much more sequential and structured than science (Siskin, 1994), whereby students must master one concept before commencing the next. However, science is perceived as a more topic-based subject (and less strictly sequential) with progression through topics governed less through conceptual mastery. Nevertheless, Tytler et al. (2008, p. 116) note that for both subjects 'ultimately the interest resides in an appreciation of the conceptual explanations and structures of the discipline and their power in making sense of the world'.

There remains debate over the best method to deliver STEM education. For instance, Bybee (2010) advocates an *integrated approach* based on contemporary STEM issues. An integrated approach to STEM education 'removes the walls' placed between each of the STEM content areas and teaches them as one subject (Breiner et al., 2012; Morrison & Bartlett, 2009). The integrated approach begins with an age-appropriate challenge or problem that engages students. As students explore options and gain an understanding of the problem, they must access the respective STEM disciplines and apply knowledge and skills to the problem. The *silo approach* to STEM education is characterised by a teacher-driven classroom where each individual STEM subject is taught in isolation (Dugger, 2010). The concentrated study of each individual subject allows students to learn course content at greater depth, with little opportunity to 'learn by doing' (Morrison, 2006). A third approach is the *embedded approach*, which requires students to acquire content knowledge through an exploration of real-world situations and problem-solving techniques within social, cultural and functional contexts (Chen, 2001). According to Roberts and Cantu (2015), an embedded approach aids learning as it reinforces and complements materials students learn in other classes. To illustrate, a technology education teacher uses embedding to emphasise technological content (as it would be done via the silo approach) and this maintains the integrity of the subject matter. In contrast to the silo approach, however, embedding promotes learning through a variety of contexts (Rossouw, Hacker & de Vries, 2010).

Features of STEM programs

Within the literature, successful STEM programs have a common set of features. Typically – and from a mathematical perspective – STEM programs engage students in mathematical problem-solving situations that stimulate intellectual activity (Tytler et al., 2008), and offer opportunities for students to develop deep, connected understandings of mathematics that will enable them to solve unfamiliar, non-routine problems (Batterham & Miles, 2000; Kilpatrick, 2002). Following an evaluation of STEM programs within secondary schools, Brody (2006) classified key features of STEM programs as: (a) exposure to strong content knowledge in mathematics and science based on academic instruction and hands-on demonstration, (b) an appreciation of the utility of STEM subjects in the workplace, (c) access to role models working in STEM fields and (d) collaboration with peers who share interests in STEM. In the United States, Sanders (2009) notes that the 'flavour' of present-day, integrative STEM education efforts resembles several accreditation standards that were developed from past engineering education reform efforts.

To advance STEM education within schools, a fundamental clarification of STEM literacy is required (Tytler et al., 2008). Accordingly, these authors created a modified, working definition of STEM literacy from the PISA 2006 Science framework (OECD, 2006). According to Tytler et al. (2008), this working definition of STEM literacy referred to the following:

- acquiring scientific, technological, engineering and mathematical knowledge and using that knowledge to identify issues, acquire new knowledge and apply the knowledge to STEM-related issues
- understanding the characteristic features of STEM disciplines as forms of human endeavours that include the processes of inquiry, design and analysis
- recognising how STEM disciplines shape our material, intellectual and cultural world
- engaging in STEM-related issues and with the ideas of science, technology, engineering and mathematics as concerned, effective and constructive citizens.

To conclude, Tytler et al. (2008) contend that translating this definition of STEM literacy into school programs and instructional practices requires a way of organising education so that the respective disciplines can be integrated and instructional materials designed, developed and implemented.

STEM in action

The work of Tytler et al. (2008) reviewed extant literature concerning supports of and barriers to STEM education in Australia during the primary–secondary transition. In particular, this work highlighted various enrichment initiatives where students participated in contemporary STEM occupations as part of their school STEM curriculum. Such enrichment and enhancement initiatives can be

> embedded in curriculum materials or achieved by links being made to STEM professionals through excursions or incursions, web-based explorations of new developments, curriculum modules designed to embed this sort of material, or school-community linked units of work. (Tytler et al. 2008, p. 126)

According to Stagg (2007, p. 12), 'direct contact between students and people working in scientific jobs tends to be identified by the students themselves as the most effective way to learn about careers'. Other researchers note the enthusiastic reactions STEM-based initiatives generate within students (Cripps Clark, 2006), as well as school teachers deriving significant professional learning from working with scientists. One project highlighted by Tytler et al. (2008) is the Australian School Innovation in Science, Technology and Mathematics (ASISTM) project. Through strategic partnerships between clusters of schools, scientific and industrial organisations, universities and government organisations, this project developed innovative curriculum projects for students. One of these projects is entitled 'Bloodstain Pattern Analysis', run through the University of Western Australia. In this project, Year 10 students participate in a range of activities that require them to make observations, collect, analyse and interpret data before forming a conclusion. For the project – which aims to engage students in real-life forensic science – teacher and student support is made available through learning modules, background information and copter-based resources. In a study of the ASISTM project, Tytler et al. (2007) designed an innovation framework to interpret the experiences provided. Specifically, these scholars concluded that two key benefits were the technology projects offered schools access to expensive and contemporary technologies, and a number of projects focused specifically on alerting students to STEM career opportunities in their local regions.

Challenges for STEM education

Advancing STEM education presents several significant challenges for educators (Bybee, 2010; Hoachlander & Yanofsky, 2011). As mentioned earlier, there is no consensus on what is the best approach for STEM instruction. For instance, when considering the benefits of an integrated STEM approach one must weigh up the associated shortfalls; for example, teaching through integrative approaches requires considerable pedagogical training, and a mastery of all four disciplines. Equally, one should be reminded of the benefits of others approaches – such as the interactive approach, which develops interaction between STEM subjects (Williams, 2011). According to Williams (2011, p. 32), this interactive approach can be achieved 'by fostering cross-curricular links in a context where the integrity of each subject remains respected'. A second challenge involves actively planning for technology and engineering learning opportunities in school programs (Bybee, 2010). While 'scaling up' these courses in schools may appear to be concomitant with a 'silo approach', the appropriate inclusion of technology and engineering in science and mathematics education appears a reasonable way to meet this challenge (Bybee, 2010). Third, while Australia implements a national mathematics curriculum from Kindergarten through to Year 12 in 2016, there is no attention directly focused on STEM education by way of syllabus documents (e.g. scope and sequence), general capabilities or proficiency standards. Instead, it has been left up to educators and curriculum designers to develop units of STEM work carefully and creatively to prepare students for future roles in STEM-related careers.

Putting activities into practice

In the final section of this chapter, it seems appropriate to examine some activities that help students make mathematical connections. By their nature, these activities are problem based and therefore require a problem-solving approach. According to Anderson (2009, p. 342), problem solving is 'recognised as an important life skill involving a range of processes including analysing, interpreting, reasoning, predicting, evaluating and reflecting'. Developing students who are competent problem solvers is a complex task that requires a range of knowledge, skills and dispositions (Stacey, 2005). Moreover, and consistent with Figure 3.7, Stacey (2005) contends that students require good communication skills, the ability to work cooperatively, as well as a teacher who possesses the appropriate personal attributes for organising and directing their efforts.

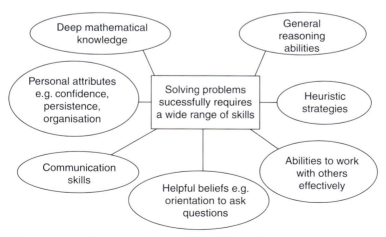

Figure 3.7 Factors contributing to successful problem solving
Source: Stacey, 2005, p. 342.

The *General Capabilities* of the Australian Curriculum outline the importance of problem solving in learning mathematics. Clearly, problem solving is an important component of doing mathematics, and consequently students should be given regular opportunities to solve complex, real-world problems. For instance, the Literacy and Numeracy capabilities are included in Table 3.4.

Table 3.4 General Capabilities (Literacy and Numeracy)

Literacy	Students use literacy to understand and interpret word problems and instructions that contain the particular language features of mathematics. They use literacy to pose and answer questions, engage in mathematical problem solving, and to discuss, produce and explain solutions.
Numeracy	It is important that the mathematics curriculum provides the opportunity to apply mathematical understanding and skills in context, both in other learning areas and in real world contexts.

Source: ACARA, 2015e.

The famine problem

One question that can promote mathematical connections in the secondary classroom is the famine problem (Stillman & Galbraith, 1998). The question is presented as such:

Every day during the month of July, Relief Aid Abroad trucked supplies of food into the famine stricken areas of Nacirema. On the first day, 1000 tonnes were shifted; on the second day 1100 tonnes were shifted; on the third day, 1200 tonnes were shifted and so on until a maximum amount was reached. The supply of food then declined by 100 tonnes per day until the end of the month. If the total food supplied for the month was 59 300 tonnes, on which day of the month was the maximum amount trucked out? (1998, pp. 159–60)

Stillman and Galbraith (1998) suggest that a student's solution to the problem contains the following essential aspects in some form.

- Obtain $T_n = 900 + 100n$ for amount trucked on days $n = 1, 2, 3 \ldots m$ where $T_m = 900 + 100m$ gives maximum on day m.
- Obtain $T_m + r = 900 + 100(m - r)$ for amount trucked on day $n = m + r$ so $T_{31} = 200m - 2200$ for amount trucked on day $n = m + r = 31$.

- Obtain $S_m = m(950 + 50m)$ for total trucked from $n = 1$ to m and $S_{31-m} = (31 - m)(150m - 700)$ for total trucked from $n = m + 1$ to 31.
- Note that $S_m + S_{31} - m = 59\,300$ for total trucked during month leading to $m^2 - 63m + 810 = 0$ with solutions $m = 18, 45$.
- Note that $m < 31$ (days in July) so $m = 18$ (maximum amount is trucked on July 18).

Looking at the verbal nature of the problem it is necessary that students:

- recognise the Arithmetic Progression (AP) pattern in the data
- form T_n and S_n expressions in terms of correct variables identified from the context
- interpret key words such as:

  o 'maximum' T_n stops increasing when $n = m$
  o 'decline per day' common difference is negative in a second AP
  o 'total food' addition of sums of two different APs (1998, p. 160).

Looking at the real-world aspect of the context, it is necessary that students additionally import a numerical datum (31), not explicitly provided, into the solution process

- 31 days in July to give the link $m + r = 31$ between the APs
- 31 days in July means that $m < 31$ so $m = 45$ is rejected (1998, p. 160).

Stillman and Galbraith (1998) intentionally designed this problem 'to entail a large complex dataset for initial perception, representation and analysis'. In looking at the nature of the problem, it was expected that students would make mathematical connections with a number of variables and relationships in generating a solution based on abstract reasoning. For this to occur, students would need to augment their working memory with an external representation (e.g. pen and paper).

The Licorice Factory

The Licorice Factory is an activity for middle school students that has been drawn from the Maths300 website (Lovitt & Williams, 2015), which is a subscription-based online repository for rich, investigative mathematical tasks. To commence, introduce the story to the class:

> Once upon a time in Lolly Land, there was a grand Licorice Factory. In Lolly Land, customers order their licorice in any lengths from 1 to 100. For example, if a customer wants some 36-length licorice, a Factory Employee goes to the 36 machine, feeds in the unit length pieces, and the machine stretches it out to

length 36. In the Factory there is room to walk between the machines, and soon you will be given a Floor Plan.

All was going well until one day someone ordered a Number 6 length licorice and … they found the Number 6 machine was broken! The boss was very worried until one of the workers figured out how to make Number 6s without that machine. What do you think she proposed that they should do? She said: We could feed the unit pieces into Number 2 machine and get length 2 licorice and then feed those into the Number 3 machine and they will be stretched three times to become length 6.

So, they tried it and it worked! They didn't really need the Number 6 machine. The boss gave a year's supply of licorice to the worker who thought up this idea … Then a week later the Number 10 machine broke … When the boss realised that more than one machine was not needed, he offered a lifetime supply of licorice to the worker who could tell him all the machines he could shut down. (Lovitt & Williams, 2015)

Although this is the complete situation, teachers may find that it is useful to stop the class at various points throughout the story to model the 'stretching' of licorice (both with real licorice, and with unifix cubes), and to allow students to try it for themselves. Producing large, laminated cards with numbers 1–36 and placing these on the classroom floor can also assist in creating the factory floor where students can walk around and stand on numbers (machines). At this stage, distribute the Factory Floor Plan to the students (this also works nicely if they work in partners with one large, paper-based Floor Plan). Their task is to 'cross off' as many machines as possible, but at the same time, they must still have access to machines that can produce licorice to any length from 1–100.

Some key questions to ask the students here are:

- What strategy are you using to cross off machine numbers?
- How does this activity require knowledge of numbers?
- Are there any patterns in numbers that make the process of crossing off easier?
- Are you able to justify *every* decision to cross off the machines on your Floor Plan?

Students may recall and use some number facts from their prior learning, e.g. *prime numbers, factors, multiples* and *composite numbers*. Encourage them to use these terms when asserting a position to 'cross off' or 'not cross off' a machine.

After creating a Master List of Licorice Machines that will remain working after as many as possible have been shut down, conclude as a class that the machines still working are the *prime numbers* from 1–100. Using Number 24 Machine as an example, demonstrate on board how 24 can be written as $2^3 \times 3$. This can be done by introducing the 'Factor tree' approach to break the machines down into their prime factors (use 24 and 36) – see Figure 3.8.

Figure 3.8 Writing 24 as a product of its prime factors, i.e. $2 \times 2 \times 2 \times 3 = 2^3 \times 3$

Ask students to choose 10 Machines that have been shut down, and then to write each of the numbers as a product of its prime factors.

To conclude, asks the students what they have learnt as a result of the activity. This activity emphasises a shift from additive thinking to multiplicative thinking, introduces the ideas of prime numbers and composite numbers, factors, multiples and number relationships. To extend the task, a teacher can draw students' attention to divisibility tests (e.g. the sieve of Eratosthenes), number properties (e.g. amicable numbers, deficient and abundant numbers), and various famous number puzzles (Riemann hypothesis, Goldbach conjecture).

The weighted dice

This activity commences with the teacher posing a question concerned with simple probability: Can we alter the outcome of rolling a die if one of the sides has been altered or 'loaded'? As a corollary, students are asked if they think an odd number is twice as likely to occur with a loaded die. To make loaded dice, Kincaid (2015) suggests that the materials needed are: regular playing dice, small fishing sinkers and a $\frac{1}{8}$ inch drill bit. With these materials, drill a hole from the '4' side to the '3' side along the corner where the '5' and '6' sides meet. After pressing sinkers into the holes created by drilling, the result will be a loaded die (and the mass has been increased by approximately 1.8 g, or 39 per cent of the original mass). In Figure 3.9 an original die is compared with a loaded die.

After creating a class set of loaded dice, Kincaid engaged students in a dice rolling activity for single die rolls ($n = 1440$), two dice sum rolls ($n = 1439$) and three dice sum rolls ($n = 1120$). The results for each roll are displayed in Figure 3.10, Figure 3.11 and Figure 3.12.

Figure 3.9 An original die and a loaded die
Source: Kincaid, 2015.

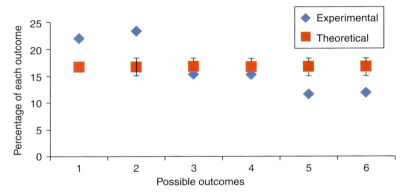

Figure 3.10 Results: single die rolls
Source: Kincaid, 2015.

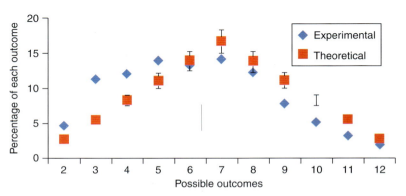

Figure 3.11 Results: two sum dice rolls
Source: Kincaid, 2015.

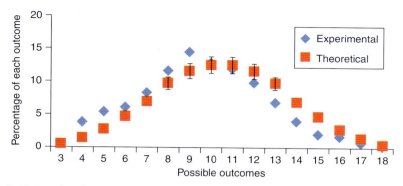

Figure 3.12 Results: three sum dice rolls
Source: Kincaid, 2015.

As can be noted from Figure 3.10, Figure 3.11 and Figure 3.12, Kincaid was able to engage his students in an activity where theoretical and experimental probabilities could be compared. To illustrate, from the single die roll activity the theoretical probability of rolling a 1, 2, 3, 4, 5, or 6 is $\frac{1}{6}$ or 16.6%. Using a loaded die, the probability of rolling a 1 or a 2 has increased by approximately 10%. Obtaining a 3 or 4 has approximately the same probability, while P(5) and P(6) are approximately 10% lower than their respective theoretical probabilities.

With the two dice rolls, Kincaid asked his class to determine the probability of obtaining a 9 with regular dice. This calculation can be expressed:

$$P(9) = P(3) \times P(6) + P(4) \times P(5) + P(6) \times P(3)$$

$$= \frac{1}{6} \times \frac{1}{6} + \frac{1}{6} \times \frac{1}{6} + \frac{1}{6} \times \frac{1}{6} + \frac{1}{6} \times \frac{1}{6}$$

$$\cong 0.11 \text{ or } 11\%$$

Comparing this theoretical probability with the experimental probability, students saw that P(9) with loaded dice decreased from the 'expected' to approximately 7.783%. In addition to providing a fun and interesting real-world application of probability theory, this activity can be used to make mathematical connections between Australian Curriculum strands Number and Algebra and Statistics and Probability. Moreover, these connections can be made by extending the understanding of predictability of single or multi-dice sums, and through an examination of data with calculations of fractions, decimals and percentages.

ACTIVITY 3.7

Log in to HOTmaths and select the 'HOTmaths Global' Course list and then choose the 'Early Secondary' Course. From the Topic list choose 'Graphs on the Cartesian plane', then the 'Linear graphs from equations' Lesson.

Looking at the Resources, describe how you could adapt them to create a challenging and meaningful problem for students to solve. Assume that you are teaching a co-educational Year 7 class.

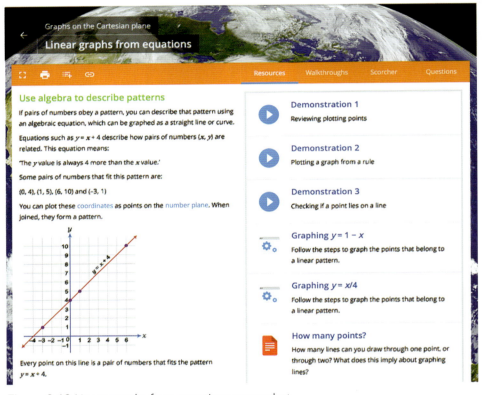

Figure 3.13 Linear graphs from equations screenshot

ACTIVITY 3.8

Log in to HOTmaths and select the 'HOTmaths Global' Course list and then choose the 'Middle Secondary' Course. From the Topic list choose 'Circle geometry', then the 'Tangents, radii & chords' Lesson.

Looking at the Resources, describe how you could adapt them to create a challenging and meaningful problem for students to solve. Assume that you are teaching a co-educational Year 12 class.

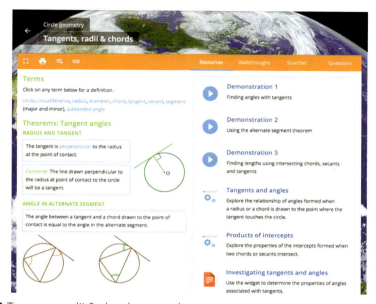

Fig. 3.14 Tangents, radii & chords screenshot

REFLECTIVE QUESTIONS

In your experience as a learner, what were some lessons or activities that you regard as useful for making mathematical connections? List the key features of these lessons and recall how the teacher facilitated the learning process. How do you plan to help your students make these connections in your classroom?

Summary

Articulate the importance of helping students make mathematical connections in their learning

Mathematical connections benefit student learning by enhancing students' capacity to remember, appreciate and use mathematics and to develop conceptual understanding. Additionally, the use of real-world connections can motivate mathematics learning and help students apply mathematics to real problems – particularly those in the workplace.

Define the term 'mathematical connections' in a secondary context

Mathematical connections are 'causal, logical or coherent relationships' between ideas and concepts. A connection can be a feature of the content matter (e.g. a relationship between ideas), a relationship that is constructed by the learner (e.g. a mental construction in the learner's mind), or a 'process that is part of the activity of doing mathematics'.

Understand teachers' practices in promoting mathematical connections

The literature is replete with practical ideas for teachers to make mathematical connections. Commonly, such ideas include careful scaffolding of learning and using problem solving as an instructional strategy. To make mathematical connections, teachers can use simple analogies and classic 'word problems', use and analyse real data, discuss the use of mathematics in society and mathematically model real phenomena.

Outline key features, instructional approaches and challenges associated with STEM education

Science, Technology, Engineering and Mathematics (STEM) education initiatives are identified typically by the following key features: (a) exposure to strong content knowledge in mathematics and science based on academic instruction and hands-on demonstration, (b) an appreciation for the utility of STEM subjects in the workplace, (c) access to role models working in STEM fields and (d) collaboration with peers who share interests in STEM. Popular STEM instructional approaches include the *integrated approach* (all STEM areas taught as one subject), the *silo approach* (each STEM area taught in isolation) and the *embedded approach* (content delivered via engagement with real-life situations and problem-solving techniques). Three key challenges associated with STEM education are offered as: determining best practice for STEM instructional practice, teachers actively planning for technology and engineering learning opportunities within secondary schools and a lack of nationally mandated curriculum documents for STEM.

Outline key instructional guidelines and tasks that can help students connect with mathematics

The most common practical ideas teachers use to make mathematical connections within lessons include contextual word problems and investigations. When planning for these activities, teachers should consider some general guidelines (e.g. simple analogies, classic word problems, real data analysis, mathematical discussions, hands-on representation of concepts, modelling real phenomena) as well as the roles that scaffolding and problem solving will play in students' learning.

CHAPTER 4

Using technology in mathematics education

Learning outcomes

After studying this chapter, you should be able to:

- understand the reasons why it is important to incorporate technology into the teaching and learning of mathematics
- articulate your strengths and areas of development in terms of the TPACK model
- be able to choose the appropriate technology for the teaching and learning of mathematics and have a range of strategies to implement the technologies
- find and adapt resources from the internet and know how to use them in class and online.

Introduction

There are a number of terms that are used when describing the use of technology in mathematics education. Information Communication Technology (**ICT**), Learning Technology (LT), Digital Technology (DT), Information Technology (IT), Educational Technology and e-learning are some of the more common ones. Often these terms are used interchangeably. However, for the purposes of this chapter technology will be used to describe digital devices such as calculators, laptops, tablets, interactive whiteboards, online communications such as email, social media and blended learning; in essence, all forms of technology that can be used in mathematics education.

This chapter will look at the use of technology in a general sense. Research into the general use of technology in mathematics education will be used to mount a case for the use of technology. The Technological Pedagogical Content Knowledge (TPACK) framework (Koehler & Mishra, 2009) that identifies the knowledge teachers need to teach effectively with technology will be outlined and will give you one way of looking at your readiness to implement technology into your teaching. This will be followed by an overview of the key technologies used in mathematics education, and a brief discussion of how technology is being used for teacher professional learning.

One of the features of the use of technology in teaching and learning is that it is constantly evolving and so this chapter will look at ways of using technology rather than specific devices. As such it is important for you as a pre-service teacher to be aware of the current technological tools, learn how to use them and explore their use in mathematics education. It is not possible in a single chapter to give detailed instructions on how to use a variety of devices and so the focus will be on how to use selected devices with particular functionality in a variety of ways for teaching. One example of this would be devices that allow you to sketch and write mathematics, such as tablet computers and interactive whiteboards; these can be used to give visual representations that are important for the learning of mathematics and to solve mathematical problems.

KEY TERMS

- **Blended learning:** combining face-to-face with virtual learning.
- **CAS:** Computer Algebra Systems.
- **DGS:** Dynamic Geometry software that allows the user to create and manipulate geometrical objects.
- **GC:** Graphing/Graphics Calculator.
- **ICT:** Information and Communication Technology, an alternative term used in some of the references for the software, devices and online materials that can be used in teaching and learning.
- **Technology:** term used in this chapter to encompass the range of software, devices and online materials that can be used in teaching and learning.
- **TPACK:** Technological Pedagogical Content Knowledge, a model that can be used to represent the interaction of technology in Shulman's PCK model.
- **Virtual learning:** the use of online materials to facilitate student learning.

Why use technology in the teaching and learning of mathematics?

The use of technology in the teaching and learning of mathematics is important for a number of reasons. Research evidence (Slavin, Lake & Groff, 2009; Rakes et al., 2010; Cheung & Slavin, 2013) concludes that there are modest gains in student achievement when technology is used. Students expect teachers to use technology. Ken Clements, in the introduction to the *Third International Handbook of Mathematics Education*, noted 'the world of mathematics education is changing very rapidly, and [that] technology is a major factor influencing the directions of change' (Clements et al., 2013, p. viii). As such, it is critical for you as pre-service teachers to be aware of the impact of technology. Technology also enables you as a teacher, and your students, to engage with mathematics in different ways, such as modelling, simulations, dynamic geometry and large statistical data sets, which are all difficult to do without the use of technology. Overlaying a function plot onto a picture of a bridge, for example, can be accomplished using Google maps to find areas.

What is also clear from the research (Hoyles, 2010) is that simply having the technology at your disposal is not enough. Rather, it is the *way* it is used in the classroom that is crucial to its successful implementation and many teachers struggle with this (Zbiek & Hollebrands, 2008).

There have been numerous studies on the use of technology in mathematics education. For example, a meta-analysis by Cheung and Slavin (2013) looked at 74 studies carried out in K–12 schools and found small but significant increases in student achievement when technology-based programs or applications were used to support the learning of mathematics in K–12 classrooms. Interestingly, they found the effect size was greater for primary students than for secondary students. An unexpected finding was that with newer studies, the effect size decreased. Other studies indicate a better attitude towards mathematics (Ellington, 2003) as well as improvement in students' problem-solving skills in mathematics (Ellington, 2003; Goos, 2010). A number of studies have looked at the use of calculators (Ellington, 2003; Burrill et al., 2002) and again these have indicated small but positive gains in student outcomes, improved problem-solving skills and a better understanding of mathematical concepts.

What the policy and curriculum documents require

Technology use has been embedded in mathematics curriculum documents (ACARA, 2015a) as well as the Professional Standards for Teachers (AITSL, 2013; NCTM, 2014) The Australian Curriculum Mathematics (ACARA, 2015a), for example, has the following statement in the Rationale referring to the use of digital technologies: 'Mathematical ideas have evolved across all cultures over thousands of years, and are constantly developing. Digital technologies are facilitating this expansion of ideas and providing access to new tools for continuing mathematical exploration and invention' (ACARA, 2015a).

Historically, we have been using a range of devices for mathematical exploration, from the Sumerians using clay tablets to keep track of business transactions, the abacus logarithms, and the slide rule through to modern computers and calculators. Mathematics has long been at the cutting edge of using technology.

Digital technologies are also embedded in a number of the content strands of the Australian Curriculum Mathematics (ACARA, 2015b). For example, 'Carry out the four operations with rational numbers and integers, using efficient mental and written strategies and appropriate digital technologies (ACMNA183)' and 'Solve problems involving profit and loss, with and without digital technologies (ACMNA189)'. These are typical of the descriptors used and, as indicated by Goos (2010), at times the use of ICT seems to be somewhat tacked on and not always incorporated in a meaningful way. Also, the technology is used 'to facilitate the traditional content and skills rather than affect the knowledge and possible learning that can occur where the use of technology becomes central' (Atweh & Goos, 2011, p. 226). These statements do not identify any specific technologies and this also highlights the point that the use of technology in mathematics is not a source of study, but rather a tool for solving mathematical problems and teaching and learning mathematics.

The Australian Professional Standards for Teachers identifies three standards (Standards 2.6, 3.4 and 4.5) that explicitly address the importance of teachers using and integrating ICT effectively into their teaching (AITSL, 2013).

The following preamble from the Australian Association of Mathematics Teachers position statement on the use of digital technologies acknowledges that technology changes rapidly and that while it can be useful to support learning, the use of technology does raise issues for curriculum teaching and assessment in mathematics:

> In this document "digital" refers to technologies such as graphics calculators, computers, iPads, mobile devices, tablet computers, 3D printing, internet and future developments in computer software and hardware. Acknowledging that there are rapid advances in technology, it is intended that this policy, dated 2014, will be revised within five years. Digital learning supports and can provide feedback on personalised learning of students in school mathematics at all levels. The responsible use of relevant technologies by students is a significant contribution to enhancing the skills of all members of Australian society, with information and communication technologies as human-centred means of enhancing our personal and working lives. Access to technologies in school mathematics raises important issues for curriculum design, teaching, and assessment of learning, and for the capacity of schools to provide and support the use of such technologies. Social networking tools and online learning in virtual environments are important parts of teaching and learning in mathematics. Classrooms are beyond school walls and this impacts on pedagogies and student learning styles. The web platforms offer opportunities for collaboration, and the proliferation of resources on the internet requires students to make critical judgements on the accuracy of information. (AAMT, 2014)

Similarly, in the National Council of Teachers of Mathematics (NCTM) position statement there is a strong emphasis on the use of technology to support students' understanding of mathematics:

> It is essential that teachers and students have regular access to technologies that support and advance mathematical sense making, reasoning, problem solving, and communication. Effective teachers optimize the potential of technology to develop students' understanding, stimulate their interest, and increase their proficiency in mathematics. When teachers use technology strategically, they can provide greater access to mathematics for all students. (NCTM, 2011)

Technological Pedagogical Content Knowledge (TPACK)

Pedagogical Content Knowledge PCK (Shulman, 1986) was the model developed by Shulman which connected the Content Knowledge (i.e. knowledge of the material being taught) with the Pedagogical Knowledge (i.e. the knowledge about how to teach). The intersection is PCK, which would be how to teach mathematics. There have been a number of models developed from Shulman's initial work. Ball and colleagues (Ball, Hill & Bass, 2005) refined it to look at the Mathematical Knowledge for Teaching (MKT). While the Technological Pedagogical and Content Knowledge (**TPACK**) framework (Koehler & Mishra, 2009) is an extension to include the use of technology. Mishra (Thompson & Mishra, 2008) also used the description of Total PACKage to describe TPACK to indicate that the technology must be integrated into the teaching practice. TPACK adds to the Shulman PCK model by looking at what technology skills teachers need to effectively embed technology into their teaching in a similar vein to PCK. It identifies an additional domain of Technological Knowledge (TK) and three intersections (see Figure 4.1):

1　The intersection of TK with Pedagogical Knowledge (PK) to identify Technological Pedagogical Knowledge (TPK) – how to use technology to teach in general terms.
2　The intersection of TK with CK to describe Technological Content Knowledge (TCK) – using technology to solve mathematical problems.
3　The intersection of all three domains to describe Technological Pedagogical Content Knowledge (TPACK) for the discipline of mathematics is how to teach mathematics with technology.

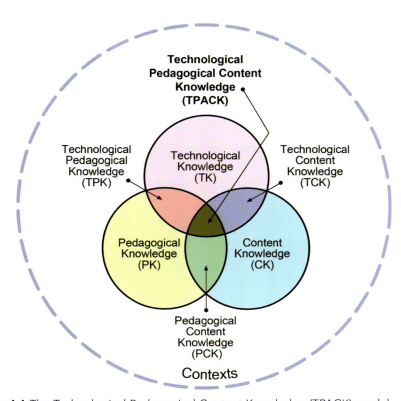

Figure 4.1 The Technological Pedagogical Content Knowledge (TPACK) model

A recent Australian project looked at the readiness of Australian pre-service teachers to use technology in their teaching and found that generally there was a range of confidence levels in how, where and when to use technology (Finger et al., 2015). This was consistent with the data for teaching mathematics (see Figure 4.2). The survey data also showed that generally pre-service teachers believed that technology was useful to support student learning and again this was supported by the data for its perceived usefulness in teaching mathematics (Figure 4.3).

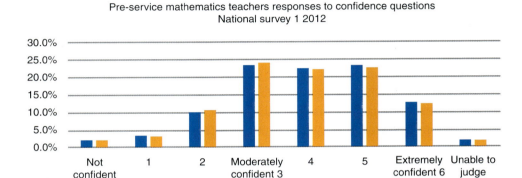

Figure 4.2 Graph showing the student responses from the Teaching Teachers for the Future (TTF) project National survey 1, with regard to confidence in designing and implementing tasks that use technology to teach mathematics

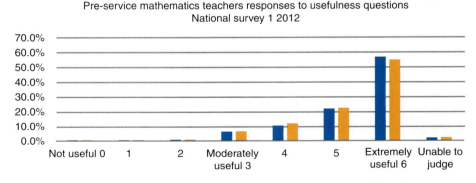

Figure 4.3 Graph showing the student responses from the Teaching Teachers for the Future (TTF) project National survey 1, with regard to perceived usefulness in designing and implementing tasks that use technology to teach mathematics

ACTIVITY 4.1

Research one of the earlier devices that have been used in mathematics. Would a historical perspective of technology support your students' understanding of mathematics?

ACTIVITY 4.2

Compare and contrast the AAMT and the NCTM positions. Do they have a different focus? How does this compare with the implementation of technology in the Australian Curriculum mathematics?

ACTIVITY 4.3

Do a self-assessment of your current skills using the TPACK model. Do you have a strong TK? Perhaps you use a range of technologies in your daily life, including in your study.

REFLECTIVE QUESTION

Take some time to think about your strengths and weaknesses as they relate to the TPACK model, and what you need to do to develop further.

Mathematics software

There is a wide range of computer software available on different operating systems (Windows, Mac OS, Android, etc.) which is useful in mathematics education. There are also different types of online and offline devices available (computer, laptop, tablet, phone etc.). The software falls into two groups: software directly related to the teaching of mathematics and general purpose software that is used across a wide range of areas (for example, communications tools and web browsers). This section will focus on the software directly related to mathematics. There are programs that have been designed as tools to solve problems using mathematics, including Excel, Maple, Mathematica, MATLAB and SPSS, which can also be used to support mathematics teaching in secondary schools. There are also programs that have been designed as tools to support the teaching of mathematics, including GeoGebra, TinkerPlots, Fathom, Nspire, Autograph and Geometers Sketchpad. The diversity is too great to list all of the available software in this chapter and it is also constantly changing. Therefore, the focus here will be on types of software available and while some programs will be named, this is for illustrative purposes only and you will need to keep yourself abreast of the current offerings. One of the questions that you will need to ask is the amount of time required to learn to use the tool versus learning with the tool (Pierce et al., 2011). A tool such as a spreadsheet is multipurpose and used for a variety of types of mathematics problems but may take some time for a student to learn to use effectively. In contrast, a single purpose app for a tablet device may only do one thing, but may take very little time to learn how to use. Students may also confuse learning the software with learning mathematics. This investment of time to learn to use the program effectively is one that must be considered when choosing the software you will use.

Spreadsheets

One of the most commonly used tools used in mathematics is the electronic spreadsheet. Spreadsheet programs have been commonly used since the original VisiCalc was first released in 1979, although the functionality and power of the spreadsheets has increased dramatically, the idea of using cells to hold text, numbers or a formula has remained the same. The computational power of the spreadsheet and its ability to recalculate if one or more cells are changed has led to their being very popular in a range of areas, including the finance world, and as such they are very popular when teaching financial mathematics. They are also able to perform most if not all of the statistical calculations required for secondary school mathematics (depending on the package used), and the ability to do 'what if' type calculations means that they are also very useful for modelling. There is some evidence to suggest that the use of spreadsheets assists in the transition from arithmetic to algebra (Filloy, Rojano & Rubio, 2001) because of the use of cell references.

As a teaching tool, a spreadsheet can be used to set up compound interest problems, for example, that show the rate of growth in a simple yet effective manner. Another example is using the spreadsheet to look at the classic birthday problem (Neal et al., 2014).

A	B	C	D	E
Number of people	Number of available days	Col B value/ 365	Cumulative product of entries in C =p(no matches)	Chance of birthday matches (=1 . Col D value)
1	365	1.0000	1.0000	0%
2	364	0.9973	0.9973	0.27%
3	363	0.9945	0.9918	0.82%
4	362	0.9918	0.9836	1.64%
5	361	0.9890	0.9729	2.71%
6	360	0.9863	0.9595	4.05%
7	359	0.9836	0.9438	5.62%
8	358	0.9808	0.9257	7.43%
9	357	0.9781	0.9054	9.46%
10	356	0.9753	0.8831	11.69%
11	355	0.9726	0.8589	14.11%
12	354	0.9699	0.8330	16.70%
13	353	0.9671	0.8056	19.44%
14	352	0.9644	0.7769	22.31%
15	351	0.9616	0.7471	25.29%
16	350	0.9589	0.7164	28.36%
17	349	0.9562	0.6850	31.50%
18	348	0.9534	0.6531	34.69%
19	347	0.9507	0.6209	37.91%
20	346	0.9479	0.5886	41.14%
21	345	0.9452	0.5563	44.37%
22	344	0.9425	0.5243	47.57%
23	343	0.9397	0.4927	50.73%
24	342	0.9370	0.4617	53.83%
25	341	0.9342	0.4313	56.87%
26	340	0.9315	0.4018	59.82%
27	339	0.9288	0.3731	62.69%
28	338	0.9260	0.3455	65.45%
29	337	0.9233	0.3190	68.10%
30	336	0.9205	0.2937	70.63%

A	B	C	D	E
34	332	0.9096	0.20.47	79.53%
35	331	0.9068	0.1856	81.44%
36	330	0.9041	0.16.78	83.22%
37	329	0.9014	1.1513	84.87%
38	328	0.8986	0.1359	86.41%
39	327	0.8959	0.1218	87.82%
40	326	0.8932	0.1088	89.12%
41	325	0.8904	0.0968	90.32%
48		0.8712	0.0394	96.06%
49	317	0.8685	0.0342	96.58%
50	316	0.8658	0.0296	97.04%
51	315	0.8630	0.0256	97.44%
53			0.0220	97.80%
57	309	0.8466	0.0099	99.91%
58	308	0.8438	0.0083	99.17%
59	307	0.8411	0.0070	99.30%
60	306	0.8384	0.0059	99.41%
61	305	0.8356	0.0049	99.51%
69	297	0.8137	0.0010	99.90%
70	296	0.8110	0.0008	99.92%
71	295	0.8082	0.0007	99.93%
79	287	0.7863	0.0001	99.99%
80	286	0.7836	0.0001	99.99%
81	285	0.7808	0.0001	99.99%
88	278	0.7616	0.0000	100.00%
89	277	0.7589	0.0000	100.00%
90	276	0.7562	0.0000	100.00%
91	275	0.7534	0.0000	100.00%
92	274	0.7507	0.0000	100.00%

Figure 4.4 The probability that there will be at least one shared birthday in a group for different sized groups of people

Source: Neal et al. 2014.

Dynamic geometry software

The ability to use dynamic geometry software (**DGS**) opens up a range of possibilities for the study of geometry, allowing for simple constructions, looking at the properties of shapes and developing conjectures using the features of the DGS software, in particular the ability to measure and drag objects. The use of this software has led to improvements in understanding (Jiang, White & Rosenwasser, 2011; Guven, Baki & Cekmez, 2012; Chan & Leung, 2014.), increased engagement and increased problem-solving skills (Sinclair, 2003). There is a range of software available that is relatively easy to use and combines well with devices that have a touch interface, such as an interactive whiteboard (IWB) or tablet computer. An example of the power of DGS is the simple construction of a triangle, followed by the measurement of the internal angles and a simple sum to total the value of the angles. Dragging any of the points shows how the value of the individual angles may change but the total remains at 180. Battista (2007) hypothesised that students notice this invariance, which he called *transformational-saliency hypothesis*. Students cannot help but notice the invariance. A more complex example would be the demonstration of the Pythagorean equality. Again, a simple construction of a right-angled triangle followed by the measurement of sides and substitution into the equation allows a demonstration of the relationship between the sides, which in turn allows students to form conjectures (Sinclair & Robutti, 2013). These examples serve as a clear demonstration that a relationship does exist, which may lead to students developing conjectures. The dynamic nature of the software and its ability to measure supports students to transition from exploration, i.e. conjecture, to the construction of geometric proof (Sinclair & Robutti, 2013). Exploring real-world examples, for example examining how a scissor lift works, is also made easier because of some of the features of the DGS (Pierce & Stacey, 2010).

Some implementations of the **dynamic geometry software** allow you to export the constructions to the internet so you can create your own interactive learning objects. One example of this is GeoGebra tube (tube.geogebra.org). This website has a range of materials that have been uploaded by others and allows you to upload your materials also.

Statistics software

The advent of specialised statistics software increases the potential for student understanding of this branch of mathematics. There are a number of specialised statistics packages available, including SPSS (www.spss.com), SAS (www.sas.com) and Minitab (www.minitab.com). However, these are rarely used in schools. As previously discussed, spreadsheets also provide the facility to do statistical analyses, as do the graphic and **CAS** calculators that will be discussed in the next section.

Statistics software has meant that students are now able to work with much larger and more realistic data sets and to experiment with the data and model various scenarios. This means that students' understanding of statistics can be further developed. The issue with many of these specialised programs is that they have many more features than are required in schools. This can be confusing for students and you are paying for the features that you would not use.

There are also statistics packages such as TinkerPlots (www.tinkerplots.com) and Fathom (fathom.concord.org) which are aimed specifically at the education market as teaching tools. Biehler et al. (2013) identified the following four requirements for statistics software to make it more useful for teachers:

1 Students can practise graphical and numerical data analysis by developing an exploratory working style.
2 Students can construct models for random experiments and use computer simulation to study them.

3   Students can participate in 'research in statistics': that is to say they participate in constructing, analysing and comparing statistical methods.
4   Students can use, modify and create 'embedded' microworlds in the software for exploring statistical concepts.

Software such as Autograph, TinkerPlots and Fathom implement these through features such as drag-and-drop graphing, dragging points, linked multiple representations and simulations. These dynamic features enable students to explore data, develop connections between chance and data, explore different representations of data (Figure 4.5) and support the transition to formal hypothesis testing (Biehler et al., 2013). In Figure 4.5, which is a screen capture using the TinkerPlots software, the data have been graphically represented in two formats and the mean and median of the data are indicated on the horizontal axis. This multiple representation allows students the opportunity to see the relationship between the box plot and the actual data as well as the two measures of central tendency.

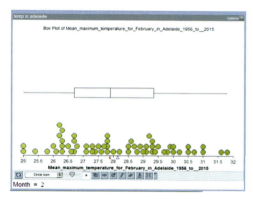

Figure 4.5 Screenshot of TinkerPlots graph: two representations of mean maximum temperatures for February

Access to statistics software and other technologies that allow students to do statistical calculations (such as spreadsheets) means that larger data sets can be used than if the students were doing the calculations manually. It also allows for individual data sets for assessment purposes. Students can create their own data sets in a range of ways, including using dataloggers and online surveys using tools such as SurveyMonkey. Secondary data sources are also readily available from sites such as the Bureau of Meteorology (www.bom.gov.au) or the Australian Bureau of Statistics CensusAtSchool (www.abs.gov.au/censusatschool). While the CensusAtSchool project is no longer collecting data, the data sets that are available are excellent.

Computer algebra systems (CAS) software

CAS software is becoming increasingly available to schools. Computer algebra systems used to be only available through programs such as Mathematica, Maple, MATLAB etc., which were not commonly used in secondary schools. However the CAS is now not only available in software designed for schools (e.g. TI-Nspire), but also on CAS calculators and hand-held devices that are commonly used by schools. CAS software performs symbolic manipulation on algebraic objects, including the solution of algebraic equations, factoring, expanding and finding derivatives. The power of CAS is that it performs these operations accurately and quickly. It means that students and teachers can use CAS as a powerful functional tool to solve mathematics problems with confidence, and this also offers significant pedagogical opportunities because it allows teachers to set tasks that require high levels of mathematical thinking (Pierce & Bardini, 2015). CAS will be further discussed in the CAS calculator section of this chapter.

Multipurpose software

Increasingly, programs are available that offer connections with one or more of the software groups previously identified. These programs offer interconnections between spreadsheet, DGS, CAS, statistics and graphing packages, which increases the opportunity for teachers to highlight the connections between mathematical concepts. The connection between an equation, a spreadsheet of values and its graphical form is just one example of the power of these connections. These combinations support students to have multiple representations of a mathematics problem, which is generally accepted as being important. However, implementation does need careful planning in order to reduce cognitive load, maintain motivation and clear learning outcomes (Pierce et al., 2011).

ACTIVITY 4.4

Access one of the Dynamic Geometry Software packages (there are some that have evaluation downloads and some that are free for personal use). Use the construction tools to create an object(s) to support your teaching of the Australian Curriculum Mathematics content descriptor 'Use similarity to investigate the constancy of the sine, cosine and tangent ratios for a given angle in right-angled triangles (ACMMG223)' (ACARA, 2015c).

ACTIVITY 4.5

Using either Fathom (evaluation copy and support materials from fathom. concord.org) or TinkerPlots (preview copy support materials from www .tinkerplots.com/get), create an activity that will show students how outliers impact on boxplots.

ACTIVITY 4.6

Using a spreadsheet, recreate the birthday problem and then explain how this would support or otherwise a student's understanding of probability.

REFLECTIVE QUESTION

What features would your ideal mathematics software have?

Devices used for teaching, learning and doing mathematics

Mathematics teachers use a range of physical materials to support their teaching. Instruments such as rulers, protractors and compasses, manipulatives such as blocks, dice and spinners are all used to support students develop an understanding of mathematical concepts. There is also a range of technological devices that can be used to both solve mathematical problems and to support the learning of mathematics.

Calculators: graphics and CAS

Calculators are one of the more prevalent technologies used in secondary school mathematics teaching. Graphics/Graphing Calculators (**GC**) and CAS calculators, graphing calculators with an integrated Computer Algebra System (**CAS**), are the two main types of calculators used in schools. The choice is often determined by which ones are approved for use in the senior years' assessments. In jurisdictions where GCs are the only ones approved and the assessments are written with the expectation that students will have access to a GC from an approved list, schools use GCs and very few use CAS calculators. However, in jurisdictions where CAS calculators are able to be used for assessment, schools will generally adopt them because of the advantages they have over the GC.

Research into the use of both GC and CAS calculators (Ellington, 2003; Cheung & Slavin, 2013) has indicated small but positive gains in student outcomes, improved problem-solving skills and better understanding of mathematical concepts. There are also gains recorded in problem-solving ability and increased engagement with mathematics. One of the perennial issues that is raised against the use of calculators, scientific, GC or CAS, is the concern that students will lose the ability to do mathematics 'by hand' and that their mental arithmetic skills will be reduced. This loss of skill can be used as a reason not to implement technology into teaching despite evidence that suggests that this is not the case and that these skills may improve. Pierce & Bardini (2015) indicated that according to the students they surveyed many teachers were frequently not using CAS in the classroom. The students had the skills to use CAS calculators, as they were using them much more often than the teachers, therefore other factors such as loss of 'by hand' skills and insufficient time must be impacting on teachers' choices.

Calculators can be used both as a functional tool for solving mathematical problems and as a pedagogical tool. Graphics calculators are particularly useful for checking, difficult calculations, solving statistics problems, graphical solutions to algebraic problems and matrix algebra. The calculator removes the restriction of using 'nice' whole numbers so that manual calculations are easy. It also means that larger data sets can be used, more meaningful problems can be set by the teacher, and more interesting investigations can be attempted by students. Another example would be when investigating outliers in a statistics topic, the calculator allows the teacher to manipulate the data set and quickly see the impact on mean, median and mode and standard deviation. Figure 4.6 illustrates a typical output from a GC. The speed of the GC also means that students and teachers can also use a variety of graphical representations of data, see how they change when the data changes and also determine which is the best representation. This functionality means that more time can be spent in class on investigating and discussing results, rather than manual calculations.

As a teaching tool, graphics calculators allow students to manipulate variables and look at the outcome very quickly. An example of this is to graph $y = x^2$ then $y = x^2 + 1$, $y = x^2 + 2y = x^2 - 1$ etc. and to look at how the graph changes (Figure 4.7).

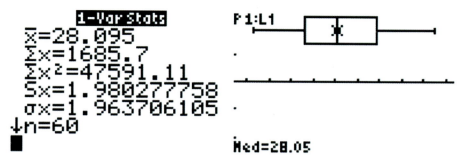

Figure 4.6 GC output of statistics problem

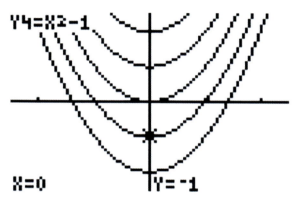

Figure 4.7 Screenshot from GC showing graphs of quadratics

CAS calculators, in addition to the features mentioned, also allow for the manipulation of algebraic functions including solving equations, differentiating functions factoring and expanding.

The use of CAS calculators extends multiple representations and, as indicated in Figure 4.8, allows for images of real objects to be modelled and analysed.

Touch and visual interfaces

The use of visual representations in mathematics is very important (David & Tomaz, 2012) and there are a number of technologies that can support their use in mathematics. Computer keyboards are not set up for mathematics and as such it has been a difficult subject to develop materials for teaching and to communicate with electronically. Mathematical symbols until recently required complex markup languages or keystroke combinations to allow them to be incorporated into electronic communications. Specialised software such

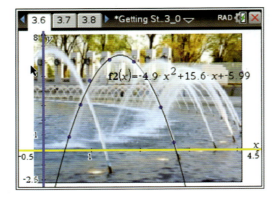

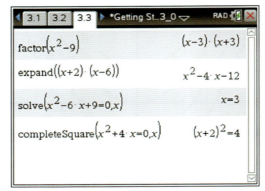

Figure 4.8 Screenshots from CAS calculator showing function plot and sample algebraic solutions

as Equation Editor and MathType have made this much easier in documents, however, emailing a mathematics function can require some ingenuity to compensate for the lack of symbols and layout that is possible. Even writing a simple algebraic fraction can be fraught with difficulty without the use of specialised software. The increasing popularity of tablets and interactive whiteboards has meant that the mathematics can be drawn, an image captured or in some cases converted to an editable object and the image can be shared and worked on in a collaborative environment.

Interactive whiteboards

The interactive whiteboard (IWB) is a multipurpose tool that allows the user to interact on the whiteboard screen as if it were a large touch screen. The ability to interact through the IWB with a range of media, for example video, images and animations, has seen interactive whiteboards being described as a technology hub (Miller & Glover, 2010). This highlights the role of the IWB as a multimodal media tool which, when used appropriately, has the potential to offer rich learning experiences to students (White, Barnes & Lawson, 2012).

The defining feature of an IWB is its touch screen interactivity, which enables the user to physically interact with the program running on the computer. The user can move objects around the screen, create and manipulate objects with software such as GeoGebra (Betcher & Lee, 2009). Another feature that is very useful in the mathematics classroom is the ability to record what happens on the IWB and save this as a movie file. This can then be made available to students to view after the class or, in the case of the flipped classroom (discussed later in the chapter), before the class. This feature is very useful for revision purposes for students in senior years as they are often assessed on material long after they studied it, usually in the form of an end of year exam. This combination of IWB and online availability has shown improved motivation of students in mathematics classrooms (Heemskerk, Kuiper & Meijer, 2014).

Tablets

Tablet or slate devices such as the iPad, and smartphones, are becoming household items in Australia and their use in mathematics classrooms is increasing. They combine the interface of the touch screen with the personal features and portability that calculators offer and the flexibility and the power and interconnectivity of computers in one device. This combination has significant potential as a teaching tool, although, for reasons similar to why CAS calculators have not become the device of choice (i.e. what is allowed in the high-stakes assessment), these devices will be restricted in their use in senior years. However, students often have access to tablet devices either because it is their own or their school's. They do offer some significant advantages over the GC and GAS calculators, screen size being the most obvious one, internet connectivity, multipurpose, inbuilt camera and touch interface being just a few. As a consequence, some of the manufacturers have moved their CAS and GC capabilities to the tablet devices (e.g. TI-Nspire, GeoGebra). There is now software available that allows students to write mathematical equations using a pen, and the software will then not only convert the handwriting, but it will also solve the problem being written. This natural interface on a portable device adds another dimension to what is possible in mathematics classrooms.

ACTIVITY 4.7

Research the current rules in your state with regard to the use of Graphics/CAS calculators in senior years' assessments. Has this impacted on how often and in what ways these calculators are used?

ACTIVITY 4.8

Design a task that could be used to teach 'Solve linear equations using algebraic and graphical techniques. Verify solutions by substitution (ACMNA194)' using either a GC or CAS calculator.

ACTIVITY 4.9

Examine the Year 8 and Year 9 algebra strand in the Australian Curriculum Mathematics. How much of this would become trivial if students had their own CAS calculators?

REFLECTIVE QUESTION

Write a paragraph outlining your position on the use of CAS in Year 8 and Year 9 mathematics.

Mathematics online

The internet has made a wide range of resources available to teachers. Resources such as the NRICH website, SCOOTLE, Maths300, Cambridge HOTmaths and the National Library of Virtual Manipulatives (NLVM) provide activities and teacher resources that can be used as they are or adapted to suit a teaching situation. Some websites allow students to ask questions of experts to get help with their homework and access practice activities for a range of mathematical skills. Another form of resource comes from the WolframAlpha website that allows students to solve simple mathematical problems by typing the question into what it calls a computational knowledge engine. Its stated aim is: 'We aim to collect and curate all objective data; implement every known model, method, and algorithm; and make it possible to compute whatever can be computed about anything' (www.wolframalpha.com/about.html).

This technology forces teachers to examine the mathematics taught, and also opens up options for teachers and students about what is possible. Students can use the website to check their computations and research areas of interest beyond the school curriculum more easily. It does, of course, open up the possibility of students taking shortcuts and not mastering the concepts, similar to the arguments used against the use of calculators, and so requires teachers to look at alternative strategies to ensure that students achieve an understanding of the mathematics required.

Virtual learning environments and blended learning

Virtual learning environments (VLE) is a term that is used to describe the range of technologies that are used to deliver learning via the web. The terms Course Management Systems (CMS) or Learning Management Systems (LMS) are also used to describe these technologies. They usually include features for collaboration, communication, content delivery and assignment processing. These technologies are being increasingly used in schools to support the face-to-face delivery of classes. This combination of face-to-face and online learning is usually called **blended learning**. Blended learning is the 'blend' of physical and virtual environments, with the virtual environment being used to supplement the physical classroom environment. The balance and manner of implementation varies to suit particular circumstances (International Association for K–12 Online Learning, 2008). While there is little research into blended learning in secondary schools (Drysdale et al., 2013), it has been suggested that students appreciate the increased opportunities to learn outside the normal constraints of the classroom, with learning occurring across different mediums and at various times (White & Geer, 2010). Also, Brown (2003) suggested that the benefits of e-learning, such as time efficiency and location convenience, supplement the face-to-face advantages of one-to-one personal understanding and motivation. It takes more time to structure and incorporate the online materials into the blended structure, but some of this time may be recovered in future years (Ahmad, Shafie & Janier, 2008) as teachers take advantage of the blended environment.

There are many forms of blended learning. The flipped classroom is one where the instruction is done online by the students prior to the face-to-face class. Often the instruction is in the form of a video presentation. The advantages given for this approach are that flipping the classroom allows for better use of the face-to-face class time for discussion group work. Some schools that have tried this approach in mathematics teaching (Fulton, 2012) give reasons including that it allows students to move at their own pace, that teachers can customise and update the curriculum, and provide it to students 24/7, that students have access to multiple teachers' expertise and that teachers can make better use of the face-to-face classroom time. They also indicated that teachers could learn from each other's material and so it was good professional learning for them, and that students liked this form of learning.

Essential to the success of blended learning is the availability and use of high-quality online materials. There is a range of materials that are accessible for teachers to use when constructing the virtual environment. Online video tutorials are available from a number of sources, with one of the better known being the Khan Academy website (www.khanacademy.org). There is a range of virtual manipulatives available on websites such as the National Library of Virtual Manipulatives (NLVM), a library of interactive, web-based virtual manipulatives for mathematics teaching and learning (nlvm.usu.edu/en/nav/vlibrary.html) and Scootle (www.scootle.edu.au), which is a national digital learning repository which provides Australian teachers with access to more than 20 000 digital learning items, provided by a wide array of contributors and aligned to core areas of the Australian Curriculum. Publishers have online materials to supplement textbooks, for example the Cambridge HOTmaths materials. These resources can be used in a number of ways: they can be embedded into a virtual environment to supplement the face-to-face teaching, used as one-off activities or as part of a flipped classroom. They offer alternative explanation of concepts and also the ability to interact, with simulation and animations to help conceptual development. The resources generally do not replace the teacher and teachers must carefully look at how they fit within the whole student experience. There are websites that offer a form of programmed instruction where the students do a diagnostic test online and are then presented with learning activities that match their results. Textbooks are often packaged with online materials that match the approach used in the textbook and supplementary materials that are not able to be put into print.

ACTIVITY 4.10

Log in to HOTmaths and spend some time familiarising yourself with the materials. You need to not just browse, but actually work through the activities as you would expect your students to do. The key to successful use is to be aware of the materials and how to fit them within your program.

ACTIVITY 4.11

Create your own video explanation of a concept (using a format similar to those on the Khan Academy website) and use this to evaluate your explanation by trying it out on some students or your peers.

ACTIVITY 4.12

With all of the instructional materials that are available online, what is the role of the mathematics teacher in this virtual school environment?

REFLECTIVE QUESTION

With the development of tools like WolframAlpha, what do you think is the future of school mathematics?

Teacher professional learning

Technology also has a role to play in teacher professional learning. There is a variety of websites that are specifically designed for mathematics teachers. The Teaching Teachers for the Future project website (www.ttf.edu.au) has materials that looks at how to integrate technology into teaching, with video footage of experienced teachers and pre-service teachers trialling the examples on the website. Another example is the Top Drawer materials on the

AAMT website (topdrawer.aamt.edu.au), which has teaching advice and activities for a range of mathematical concepts.

Summary

Understand the reasons why it is important to incorporate technology into the teaching and learning of mathematics

Using technology improves students' understanding of mathematics, increases engagement and improves problem-solving skills. Students and teachers are able to engage with richer and more complex tasks because of the speed and accuracy of technology. While it may not be central to the Australian Curriculum Mathematics, it is embedded in parts and its value is recognised.

Articulate your strengths and areas of development in terms of the TPACK model

All students come to the classroom with a varying level of technological skill. While this needs to be developed, it is the connection with pedagogy and the mathematics content to develop TPACK that is key.

Be able to choose the appropriate technology for the teaching and learning of mathematics and have a range of strategies to implement the technologies

It is essential that pre-service teachers familiarise themselves with a range of software and devices. Doing this will enable teachers to make appropriate choices and to ensure they are aware of a constantly changing field and that they will be required to keep up with current trends.

Find and adapt resources from the internet and know how to use them in class and online

It is important to know where to find quality resources but also how to adapt them and implement them into a teaching situation. The richness of the available materials is matched by their vast volume and so it is essential that teachers become discerning users of these materials.

CHAPTER 5

Inquiry-based learning

Learning outcomes

After studying this chapter, you should be able to:

- identify the benefits and challenges of using student-centred (or learner-centred) teaching approaches in secondary mathematics classrooms
- describe a range of teaching strategies that foster student-centred approaches
- consider how problem solving can be implemented in mathematics classrooms as well as the similarities and differences between problem solving and mathematical modelling
- identify types of inquiry-based learning with appropriate task selection
- plan inquiry-based learning approaches using collaborative or cooperative learning.

Introduction

The purpose of this chapter is to investigate a range of teaching strategies suitable for use in secondary mathematics classrooms that are typically referred to as **student-centred** or **learner-centred teaching approaches**. Such approaches usually involve student inquiry into mathematical ideas and may involve discussion, small-group work, collaborative or cooperative learning, problem solving and mathematical modelling. Student **inquiry-based learning** requires students to make sense of a problem situation, think about the mathematics they might use to solve the problem, make decisions about the strategies they will use, and then carry out the necessary strategies and procedures to find a resolution to the situation. It also requires students to critically analyse, reflect, evaluate and draw conclusions, along with other processes, and is frequently enabled by working with others in collaborative or cooperative learning teams. In this chapter, a range of student-centred approaches is considered, with advice about the best ways to implement such approaches, including examples of suitable mathematics task types. Strategies for implementing cooperative learning will also be discussed. Additionally, we will also explore some of the HOTmaths resources that you could use in your lessons and you will have an opportunity to reflect on their significance.

KEY TERMS

- **Collaborative learning:** '… involves a team of students who learn through working together to share ideas, solve a problem, or accomplish a common goal' (Lahann & Lambdin, 2014, p. 75). In this approach, the students organise how they will work as a group by negotiation.
- **Cooperative learning:** '… involves students working together in small groups to accomplish shared goals' (Gillies, 2007, p. 1). In this approach, the teacher prepares students for the group task by assigning roles and structuring the task.
- **Inquiry-based learning:** requires students to 'observe phenomena, ask questions, look for mathematical ways of how to answer these questions, … interpret and evaluate their solutions, and communicate and discuss their solutions effectively' (Dorier & Maass, 2014, p. 300).
- **Learner-centred teaching:** '… is an approach to mathematics instruction that places heavy emphasis on the students taking responsibility for problem solving and inquiry. The teacher is viewed as a facilitator by posing problems and guiding students as they work with partners toward creating a solution' (Stephan, 2014a, p. 338).
- **Mathematical modelling:** the process of describing a system in the real world using mathematical concepts and language. A model may help to explain a system, to study the effects of different components, and to make predictions about behaviour.
- **Problem solving:** Students develop the ability to make choices, interpret, formulate, model and investigate problem situations, and communicate solutions effectively. They formulate and solve problems when they use mathematics to represent unfamiliar or meaningful situations, design investigations and plan their approaches, apply strategies to seek solutions, and verify that their answers are reasonable (ACARA, 2011).
- **Teacher-centred teaching:** '… is an approach to teaching that places the teacher as the director of learning and is mainly accomplished by lecture, repetitive practice of basic skills, and constructive feedback' (Stephan, 2014b, p. 594).

Student-centred learning in mathematics classrooms

A distinction is frequently made between 'teacher-centred' and 'student-centred' (or 'learner-centred') approaches in secondary mathematics classrooms. The former has been referred to as 'transmission' teaching with the teacher delivering content, demonstrating procedures and directing student learning, usually without negotiation. Students become recipients of knowledge and rely on the teacher to direct their learning, provide practice questions typically requiring standard procedures, check their learning and then deliver the next piece of mathematics content. The focus in this type of classroom is on student performance and answering questions correctly rather than on student learning (Boaler, 2015).

If this style of teaching was viewed as one end of a continuum of practice, the other end would include students taking more ownership of their learning by negotiating meaning, discussing possibilities and sharing understanding with fellow students and the teacher, trying a range of different types of problems and assessing their own understanding. The focus in this type of classroom is on accepting the challenge, being willing to try a task or problem they have not seen before, valuing mistakes as learning opportunities, and not giving up. Students may even pose their own questions and instigate their own inquiries. Contrary to the views of some critics, this alternative does not mean the teacher leaves the learning to students without guidance or advice. The teacher's role is still critical because in order to initiate such practices they need to determine what types of learning situations or tasks might stimulate engagement with mathematical ideas, promote discussion, reveal misunderstandings leading to deeper learning, and then they need to be able to summarise the important mathematical ideas for students. Boaler (2002b) compared the two approaches described here by collecting data over a three-year period from teachers and students in two secondary mathematics departments in England in the 1990s. Her research describes some of the benefits and challenges of using student-centred approaches in secondary mathematics contexts. The year levels referred to here have been renamed to align with the Australian education system.

Comparing the two approaches: a research study

Consider two mathematics departments in two similar secondary schools (based on student performance on standardised tests, socio-economic status and student background characteristics) with competent and dedicated mathematics teachers, but with quite different philosophies of teaching. In the study reported here (Boaler, 2002b), both schools follow the National Curriculum, which includes mathematics content strands and a *using and applying* strand incorporating the processes of application, communication, reasoning, logic and proof. One significant difference between the two approaches adopted in these schools was the level of integration of the *using and applying* strand and its processes into mathematics lessons. Amber Hill school (pseudonyms are used for schools) adopted more teacher-centred teaching approaches (or traditional approaches) and Phoenix Park school adopted more student-centred approaches to teaching (or reform approaches).

At Amber Hill, students were assigned to classes based on ability and teachers believed students would learn mathematics if they were presented with clear explanations of mathematical concepts, shown mathematical procedures on the chalk board, and given large sets of similar questions to practise. Based on the textbook used at the school, some open-ended questions and investigations were occasionally given to the students to meet the requirements of the *using and applying* strand, although teachers complained about being expected to do this. The school was well managed, with qualified, competent, committed and hard-working mathematics teachers, and students were compliant, maintaining a high level of on-task behaviour during mathematics lessons. The mathematics teaching approach used at Amber Hill was typical of secondary mathematics departments in England (Boaler, 2002b; Swan, 2006) and typical of Year 8 mathematics classrooms in Australia (Hollingsworth, Lokan & McCrae, 2003) – an approach described as a deeply cultural phenomenon, pervasive in the Western world (Stigler & Hiebert, 1999).

At Phoenix Park a very different approach to learning mathematics was implemented, with teachers using a project-based, **problem-solving** approach, with each project typically lasting two to three weeks. Beginning with a Year 8 cohort of students in mixed-ability mathematics classes, Boaler tracked them for three years, as they worked on open-ended projects in almost every lesson with little use of textbooks – the *using and applying* strand was embedded in all projects and was evident in most lessons. An example of a project was *Volume 216* – the students were asked to think about what shape would have a volume of 216 – they were expected to build on this idea and pose their own questions and interests, thus encouraging independent thinking and taking ownership of their learning. Sometimes teachers would teach mathematics content before the project began but most of the time, new mathematics content would be taught to individuals and small groups as required. The projects were sufficiently open to allow for differentiation and students had choice about what they might pursue in the project, were given formative feedback as they progressed, with grades only being allocated at the end of each academic year. The project work ceased about four months before the end of Year 10 for final mathematics examination preparation, which was teacher organised and managed.

When the students arrived at Phoenix Park, they were not used to this approach to learning mathematics and not all of them enjoyed the open-ended, less structured approach to learning mathematics, particularly some of the boys. Teachers managed this by offering students different amounts of structure depending on their needs – students were never left to manage on their own – teachers always ensured that all students understood the problem and had some ideas about how to begin. Students needed to learn a new way of working in this environment, including how to explain and justify their thinking, how to extend their ideas and how to take ownership of their learning – teachers taught the students *how to learn* as well as the mathematics content as outlined in the National Curriculum.

Because students often question the purpose of the mathematics they learn at school, and employees have noted that students are not always able to apply their mathematics in real contexts outside school, Boaler (2002b, p. 2) was particularly 'interested to discover whether different teaching approaches would influence the *nature* of the knowledge that students developed and the ways that students approached new and different situations'. She collected data over three years from one cohort of students in each school through classroom observations in both schools as well as interviews with teachers and students as they progressed from Year 8 to Year 10. Data were also collected through a range of assessment tasks, including traditional examinations and some project work, so that Boaler could compare the students' different learning experiences to their achievement data.

For the national mathematics examination system in England in Year 10, the students were entered at one of three levels and each level was allocated a different set of possible grades – higher (grades A*, A, B, C or fail), intermediate (C, D, E or fail) and foundation (D, E, F, G or fail). At Amber Hill, the levels were determined by the ability grouping students were placed in, and for many students this remained the same as when they entered the school and were first assigned to a particular class three years earlier. At Phoenix Park, teachers made decisions about the examination level late in Year 10, providing students with every opportunity to aim for the highest possible grade throughout their learning experiences. Schools also had the option that 20 per cent of the students' final grade could be determined by a project, which they submitted for external moderation. Both schools chose this option, but the Phoenix Park students were able to choose their best project from many, whereas Amber Hill students submitted the one and only project they had completed in a three-week allocated time period during Year 10.

Differences between the two groups of students

While there were no significant differences between the school achievement data of the two cohorts at the beginning of Year 8, by the Year 10 examinations, Phoenix Park students significantly outperformed Amber Hill students, particularly at the foundation level. Boaler speculates that the difference in attitudes between students at the two schools as well as the absence of anxiety about mathematics at Phoenix Park may have also been contributing factors.

There were differences between the two groups of students on engagement and enjoyment of lessons and their views of mathematics. The Amber Hill students reported spending more time *working* during mathematics lessons than the Phoenix Park students, but the reverse was the case regarding time spent *engaged in learning*. More of the Phoenix Park than Amber Hill students enjoyed open-ended learning experiences and they also 'believed mathematics to be an active, inquiry-based discipline' (Boaler, 2002b, p. 77). At Amber Hill, students described mathematics as rule-bound, requiring memorisation, and during lessons they were observed basing their mathematical thinking on what they thought was expected of them, using cues from the teacher or the textbook rather than thinking about the mathematics within a question. If required to apply mathematics to a real-world context, students at Amber Hill would ask for help rather than think about what they knew – they were reliant on the teacher rather than being independent thinkers.

Differences in creativity were evident when Year 9 students from both schools were required to design a flat or apartment occupying a given space for a student, a couple, a family or themselves. Typically, students from both schools included a kitchen, a living room, a bathroom and at least one bedroom. However 33 per cent of the designs at Phoenix Park included unusual rooms (games, studies, hi-fi rooms, children's play rooms, etc.), whereas only three per cent of Amber Hill designs included something different. Boaler (2002b) suggested that the Phoenix Park students included rooms they would *want* to have in their flat whereas the Amber Hill students included rooms they thought they *should* have in a flat – those they thought a teacher would approve of.

Using the data from several different forms of assessment, Boaler (2002b) concluded that the students had developed different kinds of mathematical knowledge. She stated that while 'the Phoenix Park students did not have greater knowledge of facts, rules, and procedures, [they] were more able to make use of the knowledge they did have in different situations' (Boaler, 2002b, p. 104). She described the mathematical knowledge of the Amber Hill students as 'inflexible and inert' – they had difficulty remembering information after a period of time and were particularly challenged when problems required knowledge from two different mathematics topics. By learning one mathematical procedure at a time, they did not learn the important connections between mathematical ideas. They became rule followers, had little agency and were unable to develop identities as users and creators of mathematics. Boaler (2002b) states:

> For if learning mathematics entails more than the construction of cognitive forms, but the development of practices through which identities with the discipline are formed, then repeated and limited practices of procedure repetition will limit the identities of all students who do not go beyond such practices. (p. 133)

The impact of the approaches on male and female students

One important aspect of Boaler's (2002b) research was the impact of the different teaching approaches on male and female students. Based on earlier research noting that female students prefer 'connected knowing characterised by intuition, creativity, and experience' and male students 'value separate knowing, characterised by logic, rigor and rationality' (Boaler, 2002b, p. 138), she sought to explore whether the different teaching approaches supported these findings. The differences between female and male students at Amber Hill were described as a 'quest for understanding' versus 'playing a kind of school mathematics *game*'. When given the opportunity to do the open-ended questions in Amber Hill classrooms, more of the female students valued the chance to discuss ideas, work in groups, think creatively and work at their own pace – Boaler noted that the Amber Hill lessons were frequently fast-paced, with students completing many practice questions in each lesson. The male students also liked doing the occasional open-ended work, not because it was a valued learning experience but because it was a change from their normal routine. One striking finding that was established in subsequent studies, revealed the female students from the top ability class at Amber Hill did not like the fast-paced, procedural approach and many wanted to 'move down' in class even though they were able to do and understand the mathematics, and they knew it would impact on their final grades.

Another important finding revealed male students outperformed female students at Amber Hill in the final examinations in Year 10, whereas at Phoenix Park there were no significant gender differences in performance. The girls at Amber Hill were particularly disaffected, not because they thought they were not capable, but because they believed they were unable to improve their situation because of the pedagogical approach at the school. From her data, Boaler suggests that the teacher-centred teaching approach at Amber Hill privileged male over female students, thus presenting an inequitable learning environment – this was most acute for the female students in the top ability class.

Recommendations from the research

As in all research findings, these are generalisations and Boaler (2002b) reiterates that some students at Amber Hill were able to apply their mathematics to real-world problems and not all students at Phoenix Park were able to apply their knowledge in unfamiliar contexts – there was a range in both schools but the differences overall were still striking and evident from all data sources. She notes:

> The two approaches are not at opposite ends of a spectrum of mathematical *effectiveness*, but the differences between the approaches do serve to illuminate the potential of the different methods of teaching for the development of different forms of knowledge and the cultivation of different identities as learners and users of mathematics. (Boaler, 2002b, p. 136)

Boaler concludes with advice based on the approach at Phoenix Park. First, teachers of mixed-ability classes must provide differentiated work – this can be achieved by differentiating by task or outcome. She indicates both approaches can be successful, but she recommends differentiating by outcome, whereby all students begin with the same task but the teacher provides appropriate scaffolding and adapts the task for students as they progress. This is not a trivial exercise and requires careful planning by teachers with judicious selection of open-ended tasks to meet the needs of students. Not all tasks need to have a real-world focus but all should involve important mathematical concepts.

Second, students need to develop appropriate practices of doing mathematics just as much as they need to learn mathematics content. If one of the outcomes of schooling is for students to be able to think independently, be creative, choose appropriate methods, connect mathematical ideas and use mathematics to solve real-world problems, then they need to have these experiences within the mathematics classroom, and they need to have these experiences on a regular basis. Third, teachers need to believe that all students are able to learn mathematics at a conceptual level and all must be challenged with probing questions and not 'spoon fed' – it can be tempting to 'tell' too soon and do the thinking for the students. Fourth, teachers should focus on student engagement through worthwhile activities or tasks that students find interesting. Boaler's study revealed that it is not the amount of work that is completed which leads to the most improved mathematics learning outcomes but the types of tasks and student-centred practices.

Tensions between using the two approaches

At this point it should be noted that very few secondary mathematics classrooms fit neatly into either one or the other of those described above. Most mathematics teachers do implement some teacher-centred teaching approaches and some student-centred approaches to teaching mathematics. However, the balance of each in any one particular classroom might be quite different based on several factors, including the perceived level of ability or behaviour of the students, type of school, year level, impending large-scale or high-stakes assessments, parent expectations, the knowledge, experience and confidence of the teacher, and their beliefs about the nature of mathematics, as well as about mathematics teaching and learning (Anderson, White & Sullivan, 2005). Boaler's study has provided considerable evidence of the outcomes of different approaches to teaching mathematics and encourages us to reflect on the types of approaches we should implement when we teach secondary students.

Further, when teachers are committed to using both approaches in the classroom, tensions or incompatibilities occur which may be difficult to resolve. Swan (2006) describes three that need to be considered – creativity versus coverage, openness versus convergence and autonomy and challenge. Teachers frequently feel the pressure of 'covering' the curriculum, particularly when high-stakes examinations are looming, whereas providing students with opportunities to explore, explain and reason as well as consider a range of alternative approaches to solving problems takes time. Also, teachers know which traditional procedures are recommended, while using open problems may lead to a range of procedures as students choose their preferred methods. Finally, teachers may want their students to be challenged, whereas when given a choice, students may only choose simpler procedures and strategies that they feel more confident using. These tensions can be resolved, but they require careful planning and deliberate action, as the teacher needs to discuss why he or she is using particular approaches and 'at some point, we owe it to students to share the values and purposes we hold' (Swan, 2006, p. 48). The following section provides advice about implementing some commonly used teaching strategies in student-centred classrooms.

ACTIVITY 5.1

Log in to HOTmaths and select 'HOTmaths Global' from the Course list and 'Middle Secondary' from the Course. From the Topic list choose 'Algebraic expressions', then the 'A summary of working with algebraic expressions' Lesson. From the Resources tab, select the 'Playing with expressions' HOTsheet.

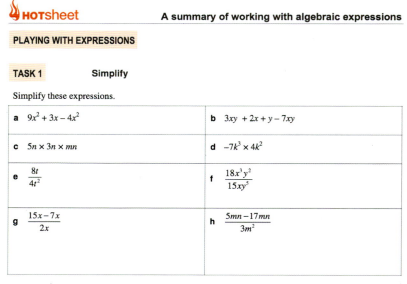

🔥 **HOT**sheet   **A summary of working with algebraic expressions**

PLAYING WITH EXPRESSIONS

TASK 1   **Simplify**

Simplify these expressions.

a $9x^2 + 3x - 4x^2$	**b** $3xy + 2x + y - 7xy$
c $5n \times 3n \times mn$	**d** $-7k^3 \times 4k^2$
e $\dfrac{8t}{4t^2}$	**f** $\dfrac{18x^3y^2}{15xy^5}$
g $\dfrac{15x - 7x}{2x}$	**h** $\dfrac{5mn - 17mn}{3m^2}$

Figure. 5.1 Playing with expressions HOTsheet

For the same content, develop an alternative student task by creating several open-ended questions so that each has a range of possible solutions. List the similarities and differences between the two activities and list the challenges for teachers if they were to use the open-ended task to replace the practice-based worksheet.

REFLECTIVE QUESTIONS

Now consider the following:

- Why is it important to use open-ended tasks in the teaching and learning of mathematics?
- What strategies could teachers use to convince students (and parents) that such tasks are an important part of learning mathematics?
- How does the activity promote problem solving and reasoning?

Teaching strategies that foster student-centred approaches

While the literature is replete with many different types of student-centred teaching approaches, there are five that help to foster student autonomy in secondary mathematics classrooms – discussion, small group work, cooperative or **collaborative learning**, problem solving and modelling, and student inquiry. While these strategies are clearly connected and overlapping, each will be discussed briefly but problem solving and modelling, inquiry-based learning, and **cooperative learning** will be considered in detail in subsequent sections of this chapter.

Discussion

Discussion in classrooms may involve whole class discussion or it may take place in small groups. Whole class discussion is a useful teaching strategy to encourage students to listen to a variety of approaches to solving mathematics problems, to reflect on their own understanding and develop useful meta-cognitive strategies, and to evaluate different processes and procedures. Such discussions need to be carefully orchestrated by the teacher with purposeful selection of student responses so that the solutions offered can be compared and contrasted, thus highlighting either misconceptions or key mathematical ideas for further elaboration. In addition, whole class discussion enables the teacher to evaluate student thinking, affirm important ideas, highlight useful strategies and identify where the next learning focus should be. However, for whole class discussion to be successful, students need time to think about the problem and prepare a possible solution first, teachers need to guard against a small number of students dominating, and the ideas offered by students need to be focused and on topic – this can be time consuming if lots of students wish to share their ideas (Killen, 2013). Teachers can prepare for successful whole class discussion if they allow for small group discussion first. By listening to the conversations, selecting the groups to provide input, and determining the ideal order of presentation of ideas to build up from simple strategies to more complex thinking, successful whole class discussion is more likely to occur. Finally, the teacher needs to summarise the discussion to highlight key points.

Small group discussion can be organised as two or more students work together to share understandings, develop strategies for proceeding on a mathematical task, evaluate their ideas, and consider the best ways of communicating their solution. For small group discussion to be effective, it is critical to select mathematics tasks which will generate ideas, that no one student can answer quickly and hence stifle conversations, and which ideally have either several solutions or several methods of solution. For example, the tasks known as 'always, sometimes, or never true' that encourage the students to evaluate the validity of statements and generalisations provide good discussion starters for small groups (Swan, 2006). The following examples are content specific and should be used when appropriate:

- multiplication makes bigger
- $a^2 > a$
- parallelograms have no axes of symmetry (for further examples see Bills et al., 2004).

These statements can be presented to students to discuss, with teachers encouraging students to determine the conditions when they are true with counter examples demonstrating when they are false. Ultimately, we want students to be able to generalise and justify their thinking.

Small group work

Students working in small groups together shifts the focus from passive learning to more active participation, teaches students to be less reliant on the teacher, encourages students to verbalise their thinking, enables sharing of different strategies and approaches to solving problems, and encourages cooperation and the learning of important classroom norms (Killen, 2013). Such social norms might include:

1 Explain and justify your solutions and methods.
2 Attempt to make sense of others' explanations.
3 Indicate agreement or disagreement.
4 Ask clarifying questions when the need arises. (Yackel & Cobb, 1996)

For small group work to be effective, teachers need to select tasks that require discussion and negotiation, that will take some time to complete and that will be easier to do with more participants sharing the load. The size of the group should reflect the complexity of the task or the amount of work required in solving the problem – some teachers find allocating group roles beneficial (e.g. recorder, reporter, resource manager, etc.). Monitoring group efforts is essential to providing feedback, encouraging further exploration and explanation, extending and challenging when necessary, and identifying misconceptions and resolving disagreements. After listening to group solutions, teachers are then able to plan the sharing session in order for the whole class to benefit from the different approaches and solution strategies. As noted earlier, it is best to carefully select the order of presentation in order to build from simple to more complex solutions. Some schools have whiteboards around the walls in classrooms so that groups can display their strategies for ease of sharing.

A more structured form of small group work involves collaborative or cooperative learning teams – while these terms are frequently used inter-changeably, some authors suggest collaboration requires giving the students considerable autonomy, whereas cooperative learning is more clearly man-aged by the teacher with students trained in the approach with assigned roles. For cooperative learning to be effective, teachers need to determine the struc-ture of the groups (mixed ability is preferable) and how group members will interact, the level of challenge of the task, group roles and mutual account-ability, and possible assessment approaches so that learning outcomes are made clear. Careful selection of the mathematics task or project is required so that all members are able to make an active contribution. Further advice about using this teaching approach is provided later in this chapter.

Problem solving

There are three approaches to teaching problem solving – teaching for problem solving, teaching about problem solving and teaching through problem solv-ing (Stein, Boaler & Silver, 2003). The first two approaches are more teacher-centred – in the first approach, the teacher focuses on the mathematics before students attempt problems (usually connected to the content) and in the sec-ond approach, the teacher teaches the students about the problem-solving process as well as about useful problem-solving strategies (e.g. draw a diagram or table, work backwards, try a simpler example). The third approach is more student-centred since the students are able to learn new mathematics by con-fronting problematic situations. This approach is most difficult to implement, but if challenging tasks are used and the challenge can be sustained, concep-tual understanding is more likely to develop. In a range of research studies where teaching *through* problem solving has been investigated, the most suc-cessful classrooms revealed a set of important teacher actions including:

- scaffolding of students' thinking
- a sustained press for students' explanations
- thoughtful probing of students' strategies and solutions
- helping students accept responsibility for, and gain facility with, learning in a more open way, and
- attending to issues of equity in the classroom (Stein, Boaler & Silver, 2003, p. 253).

These findings into mathematics classrooms where teaching *through* problem solving is a frequently used teaching strategy reveal the important role of the teacher – students are *not* left to figure things out on their own or to 'discover' mathematics without assistance.

The Australian Curriculum: Mathematics F–10 (2011b) includes problem solving as one of the four proficiencies and recommends that students 'make choices, interpret, formulate, model and investigate problem situations'. Clearly, students need opportunities to solve problems in mathematics classrooms if this proficiency is to be developed. Whether they are asked to apply mathematics they have already learnt to problem situations or whether they are asked to struggle with unfamiliar problem contexts in order to develop new mathematical ideas, it is important that students are given as many opportunities as possible to develop problem-solving skills and competencies.

In addition to problem solving, the curriculum recommends that students have opportunities to model real-world situations. **Mathematical modelling** involves solving problems set in real-world contexts whereby the real world problem is simplified in order to build a real model of the situation – this process is often referred to as 'mathematisation'. The mathematical modelling process is usually represented as a cycle of stages and like problem solving, it requires a range of competencies for students to be successful. Further information about problem solving and modelling is presented later in this chapter.

Another term, which may be used to refer to implementing problem solving in the classroom, is **problem-based learning** (PBL). Problem-based learning is an approach to curriculum design rather than a teaching strategy that is designed around a comprehensive set of real, complex problems that enable students to learn the required knowledge, skills and understandings outlined in curriculum outcomes and objectives – the problems become the curriculum. This approach requires the careful selection and sequencing of suitable problems that allow students to engage in sustained inquiry and thinking using collaboration (Killen, 2013).

Inquiry-based teaching

The use of inquiry-based teaching approaches incorporates many of the teaching approaches already discussed in this section. By using small-scale student research projects, teachers can blend a combination of discussion (whole class and small group), collaborative learning, and problem solving and modelling into mathematics learning experiences for students. Inquiry-based learning projects may be structured or guided so that students learn mathematics content as well as inquiry skills, but they are typically much more open-ended, with students encouraged to pose their own inquiry questions.

Initially attributed to John Dewey (1859–1952), inquiry-based learning was considered to tap into children's natural curiosity, fertile imagination and willingness to play and try to see how things work (Dewey, 1910). By using projects which students find interesting, inquiry-based learning engages students in experimental or scientific inquiry that requires creating questions, giving priority to evidence, formulating explanations from evidence, connecting explanations to what is known and communicating and justifying conclusions. While initially applied to scientific inquiry, inquiry-based learning has more recently been used in mathematics education as it connects to problem solving and reasoning. However, the current research into inquiry-based learning in secondary mathematics education is sparse and inconclusive (Bruder & Prescott, 2013).

One form of inquiry-based learning providing links to real-world mathematics applications is **project-based learning**, which involves cross-disciplinary, multifaceted, open-ended tasks, usually set in a real-world context, with results presented via oral or written presentation (Lahann & Lambdin, 2014). Such tasks may take several lessons as students need to define the task, plan the project, collect data, analyse and draw conclusions and determine how best to present the results. The next three sections of this chapter elaborate on the problem solving and modelling, inquiry-based and collaborative learning teaching approaches, and provide examples of tasks suitable for use in secondary mathematics classrooms.

Problem solving and modelling in mathematics classrooms

While problem solving and modelling require similar skills and dispositions and have characteristics in common, there are important differences, which impact on the types of tasks teachers choose and the way teachers engage students in such learning experiences. This section outlines the wide range of skills required for problem-solving success, discusses the similarities and differences between problem solving and modelling, and presents the characteristics of modelling tasks.

Problem solving

Real problem solving in mathematics requires doing a question which is typically unfamiliar and which the problem solver cannot answer immediately from known facts or procedures. In addition, problem solvers are usually motivated to find a solution. If problem solving is to reflect a 'real world' approach, then problem solvers will usually know why they need to solve the problem, they may not have all of the skills required to solve the problem, and the problem may be ill-defined (Killen, 2013). Real problems do not come labelled with the appropriate mathematics topics to be used and they may also contain insufficient information or too much information.

Problem solving requires time and effort to understand the problem, devise a plan to solve the problem, carry out the plan, and look back to evaluate the solution (Polya, 1957). Problem solving heuristics such as draw a diagram, work backwards, and think of a simpler example, frequently help students to solve unfamiliar problems. But being familiar with the problem-solving process advocated by Polya (1957) and having had practice in using a range of heuristics is not sufficient to become a competent problem solver. According to Stacey (2005), there are many skills required (see Figure 5.2). Students need deep mathematical knowledge and general reasoning ability as well as helpful beliefs and personal attributes for organising and directing their efforts. Coupled with this, students should develop good communication skills and the ability to work in cooperative groups. In addition, students need regular opportunities to experience problem situations in mathematics lessons.

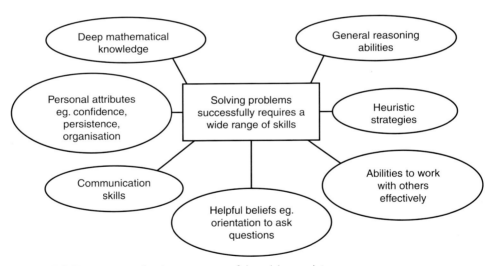

Figure 5.2 Factors contributing to successful problem solving
Source: Stacey 2005, p. 342.

One challenge for teachers is choosing the most appropriate problems for the full range of students in the classroom. Sullivan (2011) recommends we choose challenging tasks that are accessible to all students by offering enabling prompts to those who have difficulty starting, and extending prompts to those who may find a solution more quickly. In a similar manner, others have described the need to use tasks that have a 'low floor' and 'high ceiling' (e.g. Boaler, 2015).

Some authors classify mathematics tasks according to particular characteristics including level of complexity. Sullivan (2011) describes tasks that focus on developing procedural fluency, those that use a model or representation, those that use authentic contexts, and those which are more open-ended. An example of a challenging, open-ended problem with enabling and extending prompts (Sullivan, 2011, p. 47) is:

> Five people went fishing. The mean number of fish caught was 3 and the median number of fish caught was 2. How many fish might each person have caught?

An enabling prompt might be:

> Work on this problem: 'Five people went fishing. The mean number of fish caught is 3. How many fish might each person have caught?'

Some extending prompts include:

> How many different answers are possible?
> What if the mode number of fish caught was 1?
> If only four people went fishing, what difference would that make to the mean?

Fermi problems are open-ended, challenging tasks that are usually based in real-world contexts but are ill-defined and ideal for small groups working together to share knowledge and understanding (Taggart et al., 2007, p. 165). They have been defined as 'the estimation of rough but quantitative answers to unexpected questions about many aspects of the natural world' and hence they encourage conjecture, estimation, critical communication, and evaluation skills. Some examples include:

- How many piano tuners are there in Sydney?
- How many drops of water are there in a nearby dam?
- How many balloons would fill the school hall?

Problem solving and modelling: similarities and differences

Problem solving is usually defined with respect to the problem solver and the process of problem solving involves a search for a means to solve the problem, usually with a focus on correct procedures and correct solutions (Zawojewski, 2007). However, in mathematical modelling, the nature of the tasks posed is now the focus so that appropriate tasks require interpretation of the information and interpretation of the desired outcomes. This is best achieved in cooperative groups of students as they design and identify flaws in proposed models, understand limitations, as well as test and revise the model they choose for the task.

Problem solving and modelling are complex processes, but to support students' problem solving and modelling efforts, approaches have been developed to guide student thinking and promote metacognitive processes. Polya's (1957) problem-solving stages include: understand the problem, devise a plan, carry out the plan, and look back and examine the solution. This is not to suggest problem solving is a linear progression from 'givens to goals' but rather a cyclic process with students often having to backtrack to earlier stages to check information or refine strategies. Swetz and Hartzler (1991) recommend the following four stages in the modelling process, although they should be used in a cyclic manner as students may need to revisit stages as they evaluate their efforts:

- observe and identify the problem
- conjecture how factors are related and interpret them mathematically (mathematising)
- apply mathematical processes and procedures to the model, and
- obtain and interpret results in the context of the problem.

While there are some similarities between problem solving and modelling, the development of a mathematical model from a situation, which seems to be 'seemingly non-mathematical in context', is unique to mathematical modelling (Swetz & Hartzler, 1991, p. 2).

At the same time, there are sufficient common skills and dispositions to support the development of successful problem solving. Zawojewski, Lesh and English (2003) suggest these common skills include teamwork, breaking complex situations into simpler parts, communication and planning, monitoring and evaluating. These are all necessary outcomes of the mathematics curriculum so teachers need to select appropriate modelling tasks to support the development of these skills and dispositions.

Characteristics of modelling tasks

Rich model-eliciting tasks have particular characteristics that involve collaborative problem solving but also require students to mathematise a situation within a real-life context. Finding an appropriate context relevant and meaningful to students is another consideration when choosing tasks. Some teachers have chosen situations within the school or local community to design tasks for students to consider – for example, designing a new parking area to maximise the number of parking spaces, examining the recycling in the school and designing ways to improve and increase recycling opportunities, planning an event for the whole school with timetabling and budgetary requirements. All of these situations require the gathering of information, modelling through mathematising, drawing conclusions and making predictions. While the teacher may provide some of the information and place restrictions on what the situation requires, the best opportunities are created when students work in cooperative groups to ask important questions, determine the type of information they require and then work together to formulate a model.

Inquiry-based learning in mathematics classrooms

It can be argued that if the task is sufficiently open and there is limited guidance from the teacher about strategies and procedures, problem solving and modelling are examples of inquiry-based learning. Typically, the inquiry begins with a *compelling question,* which, ideally, is posed by the students. By considering how they might go about answering the question, and carefully considering their assumptions, students design an *investigation* and plan the data they may need to collect and how it might be analysed to answer the question. As the investigation progresses, students enter a *creative phase* where they are encouraged to evaluate what they are discovering, adapt their strategy to incorporate what they are learning, and communicate their thinking. Many investigations lead to new questions, which need to be accommodated throughout the inquiry. Students need to develop strategies to *discuss* and communicate their findings with others and finally *reflect* on the initial question, their strategies, their findings, and draw conclusions (Killen, 2013).

Types of inquiry-based learning

Inquiry-based learning may vary depending on the level of support given to students. If using inquiry-based learning with students who are not familiar with this form of learning mathematics, it is recommended you begin with a more structured approach. Teachers will need to carefully plan the inquiry to accommodate students' needs, including support for reading, writing, collaboration, communication, self-direction and the mathematics content. As in the Boaler (2002b) study, some students may not like the openness of the approach and will require additional support and a gradual introduction to this way of working mathematically. Chan (2007) suggests three types of inquiry-based learning, including:

- structured inquiry – students are given the problem, as well as the method and materials
- guided inquiry – students are given the problem and materials but not the strategies or methods
- open inquiry – students find their own problems as well as methods and materials (this approach appears to be more common in science classrooms).

Maker (2007) prefers to use ill-structured and open-ended tasks for which the initial conditions and perhaps even the goals are ambiguous in her inquiry-based research with teachers and students. She suggests the teaching of mathematical inquiry requires the need to embrace uncertainty, to support student decision-making, encourage flexible thinking, and a tolerance for noise and disorganisation. Teachers need to be adaptable and able to balance innovation with efficiency, and it is also helpful to be able to balance mathematical knowledge with contextual knowledge. This approach may be the antithesis of what teachers believe to be a well-organised, orderly mathematics-learning environment.

Many of the types of inquiry questions used in mathematics classrooms require the use of statistics because of the natural link to contextual problems. For example:

- Are athletes getting faster over time?
- Is there a typical Year 8 student?
- Do left-handed students have faster reflexes than right-handed students?
- Does talking on your mobile phone affect concentration on other activities?
- Can a person's height be predicted from body part measurements?
- How much water does fixing dripping taps save?
- How much food is 'thrown away' in the canteen in a year?

An inquiry-based task that requires the application of other mathematics topics and also allows for important connections to be made between topics is the *cereal box problem* (Masingila, Lester & Raymond, 2002). The problem is presented in two parts as follows:

Part 1: A store manager told a sales clerk that 45 cereal boxes had to be stacked in a display window and that all the boxes had to be used. The manager also told the clerk that all the boxes had to be set up in a triangle as shown in Figure 5.3. The sales clerk wondered how many boxes would have to be placed on the bottom row to build a triangle that would use all the boxes. *Part* 2: What if the clerk had to use 200 boxes in the display? How many boxes would have to be placed on the bottom row to build the triangle? (Manouchehri, 2007, p. 291)

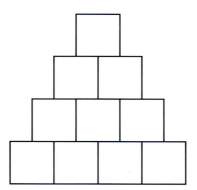

Figure 5.3 Cereal box display

Manouchehri (2007) reports on the use of this problem in a Year 9 class with the expectation that students would use a variety of problem-solving heuristics (use a table, look for a pattern) to solve the problem followed by generalisation of their results. She had planned to use the results to discuss Gauss's method for finding the sums of consecutive positive integers and also to introduce triangular numbers, but what followed led to a complete change of plans. The students interpreted the problem in unexpected ways, leading to quite different solutions and the opportunity for the problem to be extended into new investigations. The task led to combining patterns and functions, multiple representations, geometry, graphing and rates of change, and algebraic reasoning and proof. If Manouchehri had stayed with her original plan, the temptation would have been to dismiss students' different interpretations, and limit the possibility of new inquiries – she needed to be flexible and responsive and know when to 'tell' or let the group conversations continue. This approach can be threatening for teachers, particularly if students pose questions or develop mathematical ideas with which they are unfamiliar. While Manouchehri allowed the inquiries to continue for two weeks, she argues that students learnt far more than she anticipated and covered many more mathematical ideas from the curriculum in a deep and meaningful way – they were also highly engaged.

Choosing tasks for inquiry-based learning

Task choice is critical for successful inquiry-based learning, particularly in mathematics. Suitable tasks must provide challenge and hence have a high cognitive demand. Table 5.1 summarises the characteristics of high cognitive demand tasks and compares them with the characteristics of low cognitive demand tasks – this summary is adapted from Henningsen and Stein (1997).

Table 5.1 Classifying mathematical tasks by level of cognitive demand

Low cognitive demand tasks and characteristics

Memorisation tasks:

- Reproducing or memorising facts, rules, formulas or definitions
- Cannot be solved using procedures
- Non-ambiguous, involves exact reproduction of previously seen material
- No connection with the concepts or meaning that underlie procedure being used.

Procedures without connections tasks:

- Entirely algorithmic
- Use of procedure is called for or evident from prior instruction
- Limited cognitive demand, little ambiguity about what needs to be done and how
- No connection to the concepts or meaning that underlie procedure being used
- Focused on producing correct answers
- Requires no explanations or explanations focus on describing a procedure that was used.

High cognitive demand tasks and characteristics

Procedures with connections tasks:

- Focuses attention on the use of procedures for the purpose of developing deeper levels of understanding of mathematical concepts and ideas
- Suggests pathways to follow that have close connections to underlying conceptual ideas instead of narrow algorithms
- Uses multiple representations
- Requires some degree of cognitive effort, procedures cannot be followed mindlessly
- Students need to engage with conceptual ideas and deep content understandings.

Doing mathematics tasks:

- Requires complex and non-algorithmic thinking, conjecturing, justifying and reasoning
- Requires students to explore and understand the nature of mathematical concepts, processes or relationships
- Demands self-monitoring or self-regulation of one's own cognitive process
- Requires students to access and decide on appropriate use of relevant knowledge
- Requires analysis and interpretation of the task
- Requires considerable cognitive effort
- Solution process is unpredictable or ambiguous.

Adapted from Henningsen & Stein, 1997.

ACTIVITY 5.2

Choose one of the inquiry questions or create one of your own and list the challenges students may have in planning the inquiry. Choose a year level suitable for the task and use the curriculum to determine the key mathematical understandings required to do the inquiry. What support might be required to encourage student collaboration to do such a task? List other questions students might pose that are related to this inquiry.

REFLECTIVE QUESTIONS

Consider the following:

- Why is it important for students to experience using their mathematics in such open-ended inquiry-based learning approaches?
- How could students present their findings to such an inquiry?
- How would you provide feedback to students and how would you grade or assess the outcomes of students' inquiry-based learning projects?

Collaborative and cooperative learning

Since inquiry-based learning tasks usually require the gathering of information, and the need to pose questions and consider different perspectives and points of view, they are best attempted by small groups of students, but this process takes time. However, given the dynamic nature of the inquiry process and the different knowledge and understandings students will bring to the situation, working together can save time as well. Organising groups, allocating roles and responsibilities, and managing group efforts need to be explicitly addressed before implementing problem solving or modelling with students. For groups to successfully cooperate they need to be structured so that:

- all group members work together to complete the task
- there are face-to-face discussions between students
- individual accountability is required
- interpersonal and small-group skills are employed, and
- group members reflect on their work to monitor their progress (Gillies, 2007).

Ideally, groups should include three to four students with mixed knowledge, skills and understandings. If the students are not used to working in cooperative groups, begin with pairs of students working together on shorter tasks and spend time identifying key 'rules' or social norms (Mercer, Wegerif & Dawes, 1999) for effective group work which includes members taking responsibility to share information, reach agreement, provide reasons, challenge ideas, discuss alternatives, and include everyone. The teacher's role is critical while students are working in groups – to probe and clarify issues which arise, acknowledge and validate group efforts, confront discrepancies and clarify opinions and offer tentative suggestions, but only when absolutely necessary (Gillies, 2007). It is best not to provide answers, since this encourages 'asking the teacher' and discourages group autonomy in decision-making.

Some authors distinguish between collaborative and cooperative learning that occurs when group groups of students work together. The similarities between these include:

- stressing the importance of active learning
- facilitating learning by the teacher
- teaching and learning shared by both students and teacher
- enhancing higher-order cognitive skills
- emphasising student responsibility for their own learning
- involving situations where students must articulate ideas to the group
- helping students develop social and teambuilding skills, and
- utilising student diversity.

The differences between collaborative and cooperative learning are summarised in Table 5.2.

Table 5.2 The differences between cooperative learning and collaborative learning

Cooperative learning	Collaborative learning
Students receive training in social skills in small groups.	Students already have the necessary social skills which they use to reach goals.
Activities are structured with each student taking a particular role.	Students organise and negotiate their efforts themselves.
The teacher observes, listens and intervenes when necessary.	When questions arise, the teacher guides, usually by posing further questions.
Students submit work at end of class for evaluation.	Students retain drafts to complete further work.
Students assess individual and group performance.	Students assess individual and group work.

Adapted from Matthews et al., 1995.

Collaborative and cooperative learning best occurs when teachers choose 'group worthy' tasks. If the task only requires students to exchange information and explanations or request assistance, this will lead to low levels of cooperation, lower levels of discussion and debate, and potentially lower levels of thinking. Group worthy tasks are more open-ended, multi-dimensional, have high cognitive demand, require the efforts of several students working together, and have clear outcomes. See Table 5.3 for the features of group-worthy tasks.

Table 5.3 The features of group-worthy tasks

Task features	Characteristics
Open	• Genuine dilemmas and authentic problems with real-life uncertainties and ambiguities. • Many possible solutions. • Accessible problem-solving context.
Multi-dimensional	• Many ways to approach the problem. • Many ways to model the problem (picture, table, graph, etc.). • Requires a variety of skills and mathematical competencies to solve the problem. • Many ways to demonstrate mathematical competence or 'smartness'.
High cognitive demand	• Meets criteria for high cognitive demand tasks. • Highlights a big mathematical idea. • Opportunities for justification and reasoning.
Positive interdependence	• Requires students to work together to complete the tasks. • Group and individual accountability. • Multiple abilities and mathematical competencies needed.
Clear evaluation criteria	• Clearly stated outcomes or product. • Clear deliverables. • Could use a rubric or performance assessment.

Source: Lotan, 2002.

In summary, inquiry-based learning encourages students to develop an important set of 'mathematical habits of mind' (Horn, 2012). In order to develop such habits of mind, students need to be given challenging mathematics tasks that require them to engage in exploring important mathematical ideas, orienting and organising their thinking, representing, justifying, generalising, checking for reasonableness and using mathematical language.

ACTIVITY 5.3

Now that you have read the chapter, consider the following discussion questions:

- What are the benefits of using student-centred teaching approaches rather than teacher-centred teaching approaches?
- If you wanted to introduce more student-centred teaching approaches to a mathematics class that was not familiar with these approaches, how should you begin?
- How well do typical secondary mathematics textbooks prepare students for inquiry-based learning?
- Problem solving has been included in mathematics curriculum documents for at least 25 years. Why is there still little evidence of genuine problem solving in secondary mathematics classrooms?
- What are the assessment implications of using more open-ended, inquiry-based learning approaches in secondary mathematics classrooms?

Summary

The benefits and challenges of using student-centred teaching approaches in secondary mathematics classrooms

The chapter began with an overview of an important and highly regarded research project conducted in the 1990s by Professor Jo Boaler in England. She compared the teaching practices in two schools – one where the teachers used more teacher-centred teaching approaches (Amber Hill) and another where the teachers used more student-centred teaching approaches (Phoenix Park). While the schools had similar characteristics and students had comparable mathematics achievement at the beginning of the research, after three years there were significant differences in learning outcomes, beliefs about the usefulness of mathematics and willingness to engage in unfamiliar mathematics tasks.

Importantly, both schools had dedicated and hard working teachers who were committed to supporting their students to achieve their best in the Year 10 final examination. While the Phoenix Park teachers had a quite different philosophy of teaching from those at Amber Hill, their students did not readily respond to the open-ended inquiry-based approach to learning mathematics and so they needed to develop strategies to develop their confidence and willingness to participate in this new way of learning mathematics. However, having adapted and having learned how to ask questions and justify their thinking, the Phoenix Park students readily engaged with unfamiliar mathematics problems and were able to use their knowledge flexibly.

Teaching strategies that foster student-centred approaches

There are several student-centred teaching approaches which support inquiry-based learning. Discussion as a whole class after students have been presented with a challenging task, or discussion in small groups (two or more students) can facilitate sharing of a range of ideas, learning to listen to others, developing a critical stance, and presenting well-argued strategies and solutions. Problem solving and modelling teaching approaches allow students to engage in posing questions, as well as planning, monitoring and assessing results, and are ideally conducted in small groups. Inquiry-based teaching approaches involve open-ended challenges for students, and can range in the level of support provided by the teacher, but authentic inquiry is ill-structured and can be ambiguous. In a more student-centred classroom, students have more autonomy and are provided with more opportunities to make choices. However, teachers still have a key role to play in organising the learning and choosing appropriate mathematical tasks.

How problem solving can be implemented in mathematics classrooms as well as the similarities and differences between problem solving and mathematical modelling

Problem solving is one of the four proficiencies in the national curriculum but there are different ways that problem solving can be implemented in classrooms. Teaching *for* problem solving is a more teacher-centred teaching approach, with students introduced to problem solving after learning the mathematics – such problems are typically presented at the end of the chapter in a textbook so that students are aware what mathematics they need to use to solve the problem. Teaching *about* problem solving is also more teacher-centred as teachers guide or instruct students in a set of steps in the problem-solving process as well as a set of suitable problem-solving strategies or heuristics. The most challenging approach for teachers to use in secondary mathematics classrooms is teaching *through* problem solving, but this approach is much more student-centred, with students learning the mathematics when it arises in the problem – this approach best represents an authentic inquiry-based learning approach, particularly if the problem has a high level of challenge.

Mathematical modelling provides students with the opportunity to mathematise a situation which might appear to be quite unmathematical. Such situations are usually set in real-world contexts and arise naturally – they encourage students to be curious and pose their own questions about the environment. The overall processes of problem solving and modelling are somewhat similar in that students need to define the problem and plan a solution, and they are best attempted through collaboration.

Types of inquiry-based learning with appropriate task selection

Task selection is quite challenging for teachers, as they need to select questions or problems that require considerable thought and effort, and this may vary considerably depending on the group of students they are teaching. Such problems are usually open-ended and may be contextualised to enable students to see links to a real-life context. It is best that tasks have a high cognitive demand and as such are best attempted in collaborative or cooperative teams. If students are struggling, they may require enabling prompts, and if they finish quickly, it is important that teachers have prepared extending prompts. Model-eliciting tasks are set in real contexts and allow students to mathematise non-mathematical situations.

Planning inquiry-based learning approaches using collaborative or cooperative learning

Because of the high cognitive demand of most tasks which involve inquiry-based learning, they are best attempted in small groups, but putting a group of students together to solve a problem does not mean they will automatically know how best to work as a team. They need to learn social norms about listening to each other, providing constructive feedback, negotiating strategies and approaches and accepting individual accountability. While collaborative groups have more autonomy to decide on how they will work as a group and are allowed to negotiate their efforts with little teacher intervention, cooperative groups are taught how to work as a team, they are given roles and are provided with more feedback during task completion. The level of support for group efforts may depend on their experience of working in groups, as well as on the level of challenge of the task. Group-worthy tasks have particular characteristics which support the involvement of all team members.

CHAPTER 6

Gender, culture and diversity in the mathematics classroom

Learning outcomes

After studying this chapter, you should be able to:

- offer explanations to account for gender differences in motivation and achievement in mathematics
- identify instructional approaches to be used in making a mathematics classroom gender equitable
- outline instructional guidelines for working with students from different cultural backgrounds
- justify the need for diverse learners to receive instruction according to their particular learning needs
- delineate instructional approaches to be used with diverse learners.

Introduction

The Melbourne Declaration on Educational Goals for Young Australians (MCEETYA, 2008a) provides the policy framework for the Australian Curriculum. This framework includes two goals:

Goal 1: Australian schooling promotes equity and excellence.

Goal 2: All young Australians become successful learners, confident and creative individuals and active and informed citizens.

In addressing these goals, the Australian Curriculum has been developed to be appropriate and accessible for all students (ACARA, 2015e). The propositions that have shaped the development of the Australian Curriculum are:

- that each student can learn and that the needs of every student are important
- that each student is entitled to knowledge, understanding and skills that provide a foundation for successful and lifelong learning and participation in the Australian community
- that high expectations should be set for each student as teachers account for the current level of learning of individual students and the different rates at which students develop
- that the needs and interests of students will vary, and that schools and teachers will plan from the curriculum in ways that respond to those needs and interests.

With specific reference to student diversity, ACARA demonstrates a clear commitment to the development of a high-quality curriculum for all Australian students that promotes excellence and equity in education. The Teacher Education Ministerial Advisory Group (TEMAG) (2014) recommends that teachers be suitably equipped with the pedagogical knowledge that will allow them to effectively address the learning and development needs of all students in their class. Moreover, the TEMAG (2014, p. 16) cited that 'a growing body of research acknowledges that teachers need a broad range of skills and strategies to maximise the learning of diverse student populations'. As a corollary to this research, the TEMAG has suggested that an 'ability to work effectively with special needs students, and in particular students with disability and learning difficulties, needs to be considered a core requirement of all teachers rather than a specialisation' (TEMAG, 2014, p. 17). In Australian schools, the term *student diversity* can include those students who are culturally and linguistically diverse, those who have specific learning needs, difficulties or disabilities. This chapter is underpinned by the notion that all students require opportunities to advance their mathematical knowledge through teaching approaches that are attentive and responsive to their learning needs. Such approaches strive to provide educational equity, which has been described by scholars as much more than providing students with an equal opportunity to learn mathematics. Rather, educational equity 'attempts to attain equal outcomes for all students by being sensitive to individual differences' (Van de Walle, Karp & Bay-Williams, 2014, p. 100). In this chapter, the following topics for student diversity in the secondary mathematics classroom will be explored: gender, culture and special needs learners.

KEY TERMS

- *Corpus callosum:* the bundle of nerves that connects the cerebral hemispheres of the brain.
- **Individualised Educational Plan (IEP):** a written education plan designed to meet a child's specific learning needs.
- **Least restrictive environment:** the requirement that, where appropriate, students with disabilities be educated in settings with students without disabilities.
- **Stereotype threat:** this phenomenon occurs when people believe they will be evaluated on societal stereotypes about their particular group.

Gender in the secondary mathematics classroom

Over the last three decades, research has contributed towards a significant shift in male and female students' secondary mathematics achievement and motivation (Bond et al., 2014; Forgasz & Leder, 2001; Huetinck & Munshin, 2008). Before the 1990s a widely held societal belief was that boys were better at mathematics than girls, and up until this time in Australian schools the proportion of boys was higher than girls in advanced levels of mathematics in the senior years (Siemon et al., 2011). In the early 1980s researchers and teachers began to discern decreasing gender differences in mathematics and science, where previously female students consistently had performed more poorly than male students (Huetinck & Munshin, 2008). Such discernment prompted researchers and practitioners to address the question 'What is wrong with the girls?' In the late 1980s researchers began to realise that the ways in which mathematics classes were taught was affecting the performance of girls, and that curriculum content and delivery could be modified to accommodate girls (Siemon et al., 2011). Moreover, Vale and Bartholomew (2008) have argued that male and female students may be shaped by the contexts they are in. Specifically these authors contend that for teachers 'paying attention to the relationships within the classroom, the different identities, and hence the different needs of students in the mathematics classroom, are central to equity' (Vale & Bartholomew, 2008, p. 273). As a result of heightened attention of issues affecting gender performance, various approaches have become standard practice in creating greater equality of performance. Siemon et al. (2011) noted that these approaches include an emphasis on:

> collaborative learning, discussion in small and whole-class groups, using applications of mathematics to social contexts, making contexts clear (for example, descriptions of sporting contexts made overt, rather than assuming learners have played particular sports), and including contexts that would appeal to girls and boys. (pp. 156–57)

More recently, the notion that mathematics is a 'male domain' has dissipated significantly and the mathematical potential of girls is no longer regarded as inferior to boys (Forgasz & Leder, 2001). Perhaps as a corollary to this dissipation, attention has been focused on the apparent underperformance of adolescent boys (Booker et al., 2014) where scholars posit that the most successful boys continue to achieve scholastic success, but increasingly other boys are not (Burton, 2001; Siemon et al., 2011). Following an inquiry into the education of boys, the Australian House of Representatives Standing Committee on Education and Training concluded in their final report:

> While it is dangerous to generalise, boys and girls do tend to prefer different learning styles. Boys tend to respond better to structured activity, clearly defined objectives and instructions, short-term challenging tasks and visual, logical and analytical approaches to learning. They tend not to respond as well as girls to verbal, linguistic approaches. Good teachers respond to the different learning styles of their students and utilise students' preferred learning styles while also aiming to develop the full range of capacities in each student. (House of Representatives Standing Committee on Education and Training, 2002, p. xviii)

Despite these conclusions, more male than female students continue to study the most demanding mathematics courses offered, and male students still dominate in science, technology, engineering and mathematics (STEM) related careers (Siemon et al., 2011; Thomson, 2014). As such, there remains a continuing need to encourage girls to participate in mathematics at all levels, and for educators to engage girls with the science and technology courses that lead to more high status and influential careers (Burton, 2001; Thomson, 2014).

The gender gap

For the past 20 years, results from large-scale international testing, such as the Trends in International Mathematics and Science Study (TIMSS), have revealed consistently no gender differences in mathematics achievement in Australia, with the exception of TIMSS 2007 (Thomson, 2014). In a similar vein, Thomson noted that there were no gender differences apparent during the measurement of mathematical literacy in the 2003 Programme for International Student Assessment (PISA). More recently, however, this author has emphasised that PISA 2012

> found that, while average scores in mathematics had declined in Australia, males in Australia were significantly outperforming females, and females had significantly higher average levels of anxiety about and significantly lower levels of confidence in mathematics. (Thomson, 2014, p. 59)

Several years earlier Vale and Bartholomew (2008) analysed the 2006 PISA survey, concluding that boys were more often among the highest achievers, and:

> there remains a difference in the ways that male and female students respond to their own mathematical experiences ... boys reported higher levels of enjoyment, interest and self-efficacy in mathematics than girls, and boys more highly valued the use of technology in mathematics. (p. 286)

Vale and Bartholomew (2008) also suggested that these findings led to boys' enrolments in higher level mathematics outnumbering those of girls, largely due to a recognised positive relationship between affective factors and enrolment.

Typically, the gender gap has been attributed to outmoded social stereotypes. However, recent neurological discoveries have been used as potential explanations to assist in mathematics learning as it pertains to gender. Sousa (2008) drew attention to these neurological findings, stating:

> male brains are about 6 to 8 percent larger than female brains. But males are on the average about 6 to 8 percent taller than females, which could also explain the similar differences in brain sizes. And brain imaging studies show that males seem to have an advantage in visual-spatial ability (the ability to rotate objects in their heads) while females are more adept at language processing. (p. 65)

Additionally, Sousa (2008) pointed out that the **corpus callosum** in women is proportionally larger than in men, resulting in more efficient communication between brain hemispheres. However, the male brain appears to communicate more efficiently within a hemisphere. Despite these neurological differences, no genetic advantage for mathematical processing or learning has yet been determined (Sousa, 2008). In a meta-analysis of over 100 academic studies and papers, Spelke (2005) found that most suggested that men and women have an equal aptitude for mathematics and science. As such, it is important for teachers to know about gender differences – especially neurological development – as a factor related to students' perceptions, participation and achievement in mathematics (Booker et al., 2014; Sousa, 2008).

Looking at gender issues abroad: the United States

Consistent with findings drawn from an Australian context, a general international trend has been that gender differences in mathematics achievement are declining (Hyde et al., 2008). In the United States, the gap between boys' and girls' high school mathematics course enrolment has narrowed, as well as the difference on standardised tests (Ellison & Swanson, 2010). Huetinck and Munshin (2008) commented on the summary of more than 600 programs funded by the National Science Foundation (NSF) and the American Association for the Advancement of Science (AAAS) from 1966 to 1982. This summary revealed real gains in helping female students to excel in science and mathematics, where common elements of the more effective programs included: academic emphasis, multiple strategies and systems approaches. Huetinck and Munshin noted that in particular, 'achievement in mathematics was nearly equal for males and females until the fifth grade, when females began falling behind' (2008, p. 340). Additionally, the gender differences became more pronounced in high school (Huetinck, 1990). A large-scale study analysing standardised test scores from more than 7.2 million students in the United States in grades 2–11 revealed that there were no differences in mathematics scores between girls and boys (Hyde et al., 2008).

Despite the acknowledged decline of gender differences in mathematics achievement, there are still significant differences at the advanced course levels (American Association of University Women, 1992; Sousa, 2008), university entrance examinations (Sousa, 2008; Wai et al., 2010), large-scale international testing (Guiso et al., 2008) and mathematics competitions (Ellison & Swanson, 2010). Following high school, more male than female students enter fields of study that emphasise STEM areas (Ceci & Williams, 2010). To this end, Tortolani (2007) cited the president of the Society of Women Engineers as asking rhetorically 'Why, while girls comprise 55 per cent of undergraduate students, do they account for only 20 per cent of engineering majors, and boys remain four times more likely to enrol in undergraduate engineering programs?' Looking at high-achieving students, Ellison and Swanson stated that

> there is a 2.1 to 1 male–female ratio among students scoring 800 on the math SAT, and a ratio of at least 1.6 to 1 among students scoring in the 99th percentile on the PISA test in 36 of the 40 countries. (2010, p. 109)

An analysis of the American Mathematics Competition (AMC) data provides three key findings concerning the magnitude of the gender gap at very high performance levels (Ellison & Swanson, 2010). First, the gender gap appears to widen substantially at percentiles beyond the 99th, and at the very high end of analysed data, the male–female ratio exceeds 10 to 1. Second, and although some gender variation was found across all participating schools, there was enough variation

> from school to school to suggest that the number of girls reaching high performance levels would increase substantially if all school environments could somehow be made to resemble those where girls are currently doing relatively well. (Ellison & Swanson, 2010, p. 110)

Third, an examination of extreme high-achieving students chosen to represent the United States in international competitions revealed that the highest-scoring boys and the highest-scoring girls appeared to be drawn from very different pools. While the boys came from a variety of backgrounds, the top-scoring girls were almost exclusively drawn from a remarkably small set of super-elite schools. According to Ellison and Swanson (2010), this finding suggests that almost all American girls with extreme mathematical ability are not developing their mathematical talents to the degree necessary to reach the extreme top percentiles of these contests. To assist with this development – particularly in the formative years – teachers are reminded to challenge pervasive gender stereotypes, create gender-friendly learning environments, and stimulate all students' interest in pursuing university studies and careers in mathematics-related fields (Van de Walle, Karp & Bay-Williams, 2014).

Gender differences explained: achievement and motivation

Current literature suggests several possible explanations to account for gender differences in motivation and achievement in mathematics. These explanations are socially and culturally constructed, and include pervasive belief systems, teacher behaviour and attitudes, and student attitudes. In examining the social and cultural causes of gender differences, teachers can create gender-equitable mathematics instruction for boys and girls (Van de Walle, Karp & Bay-Williams, 2014).

Pervasive belief systems

A pervasive, stereotypical view held by parents and society generally is that mathematics is a male activity which is incongruous with femininity

(Else-Quest, Hyde & Linn, 2010; Hall, 2012). According to Hall (2012, p. 70), these 'deeply entrenched societal views and stereotypes will be difficult to change and may take many years to slowly evolve; however, change can begin now from within the classroom'. Within the classroom, Nosek, Banaji and Greenwald (2002) contended that stereotypical views (i.e. boys are better in mathematics) shape girls' self-perceptions and motivations. As a consequence of these views being upheld, Van de Walle, Karp and Bay-Williams (2014) conjectured that girls' emerging interest in mathematics would decrease. Research from Stevens et al. (2007) indicated that female students report subject interest as a very influential factor when deciding to pursue higher level mathematics courses. In a Victorian context, Year 12 girls enrolled in either Further Mathematics or Mathematical Methods courses were significantly less likely than boys to perceive mathematics as relevant and useful for the future (Helme & Teese, 2011). Furthermore, these commentators found:

> Female Further Mathematics students were significantly less likely than their male counterparts to agree that their teacher understands how they learn, and significantly more likely to report that the pace of learning is too fast. Female Mathematical Methods students were significantly less likely than their male counterparts to agree that they enjoy the subject. (Helme & Teese, 2011, n.p.)

At a later milestone in life, Sousa (2008, pp. 65–66) underscored how differences in career choices are made not due to 'differing abilities in mathematics but to cultural factors, such as subtle but pervasive gender expectations that emerge in high school'. Interestingly, female students often express that they are less proficient than their male classmates at mathematics, even when they perform at similar levels (Correll, 2001). Perhaps an examination of **stereotype threat** can illuminate the basis for such a belief system. For instance, one research project was conducted to determine whether merely telling female students that a mathematics test often shows gender differences was enough to hurt their performance (Spencer, Steele & Quinn, 1999). After giving a mathematics test to male and female students, the researcher told half of the female students that the test would reveal gender differences and the remaining half that the test would find none. There were two key findings to this research: first, those female students who expected gender differences on the test performed significantly worse than the male students. Second, the female students who were told the test would reveal no gender disparity performed equally to the male students. Moreover, the experiment was conducted with high performing female mathematics students (Spencer, Steele & Quinn, 1999).

Teacher behaviour and attitudes

Although teachers may not intentionally set out to stereotype students by gender, the gender-based biases of wider society can affect teacher–student interactions (Van de Walle, Karp & Bay-Williams, 2014). In describing observations of teachers' gender-specific interactions in the classroom, Campbell (1995) noted that boys receive both more attention and different kinds of attention than girls. Furthermore, boys tend to be more involved in discipline-related attention (Campbell, 1995). Another study concluded that female students in mathematics classes go unobserved, and are known as 'quiet achievers' (Clarke et al., 2001). Research from the United Kingdom revealed that most mathematics teachers held different beliefs about students, based on students' gender (Soro, 2002). Although some teachers did not hold gendered beliefs, girls were perceived to use inferior cognitive skills and succeed because of their diligence, while boys were seen to be talented in mathematics but lacking in effort (Soro, 2002). In California, Marshall (1984) hypothesised that girls outperformed boys in elementary school only if there were algorithms to follow in problem solving. After enrolling in higher-level courses, girls were less able to use novel approaches – perhaps due to internalising rules so well their creativity suffered. These gender-based perceptions are incongruent with findings suggesting that female students achieve as good or better grades in mathematics than male students (Gallagher & Kaufmann, 2005; Riegle-Crumb, 2006). However, female teachers with mathematics anxiety can negatively influence female students' mathematics achievement – even over the course of one year (Beilock et al., 2010).

Student attitudes

Although the gender gap concerning mathematics achievement is narrowing, it is still considerably wide in terms of students' attitudes towards mathematics (Hall, 2012; Hannula, 2009). Studies conducted over time and across levels of education have generally found that boys hold a more positive attitude towards mathematics (Fennema & Sherman, 1977; Hall, 2012; Saranen, 1992). In particular, scholars from Finland have discerned that gender differences are pronounced with regard to how difficult mathematics is perceived (Kangasniemi, 1989) and how these perceptions affect students' self-confidence (Hannula & Malmivuori, 1997; Hannula et al., 2005). Hall (2012) examined data presented on students' mathematics achievement, attitudes and participation from the United States, Australia, New Zealand and Canada. Over time and across all educational levels (from primary school to university), Hall (2012) challenged the notion that gender issues in mathematics have been 'solved' in those countries. To support this challenge, gender gaps in achievement, positive attitude and greater self-confidence favoured boys at primary and secondary school levels. Furthermore, Hall noted that scholarly work has shown that student attitudes, achievement and participation are highly inter-related factors: 'Students with positive attitudes toward mathematics have greater achievement, and both of these factors are related to participation in mathematics' (2012, p. 70).

Promoting gender equity in secondary mathematics

Researchers and practitioners offer suggestions for secondary teachers to promote gender equity within mathematics classrooms. Specifically, these suggestions require teachers to develop a heightened awareness of treating male and female students equitably (Van de Walle, Karp & Bay-Williams, 2014) so that instructional practices to involve all students can be implemented (Hall, 2012; Sousa, 2008).

Awareness of gender equity

Van de Walle, Karp and Bay-Williams (2014, p. 114) suggest that when interacting with students, teachers should be sensitive to the following:

- number and type of questions asked
- ability of students to act out or model mathematical situations or concepts with movements and gestures
- amount of attention given to disturbances
- kinds and topics of projects and activities assigned
- praise given in response to students' participation
- makeup and use of groups
- context of problems
- discussions of STEM careers to increase students' interest in these fields.

Additionally, Koontz (1997) recommended that teachers promote gender equity by modelling an attitude of acceptance for both sexes' participation in lessons (e.g. acknowledging all responses, giving supportive comments to incorrect answers) and by using gender-fair language (e.g. men and women scientists/mathematicians/doctors). Sensitivity to these issues can heighten teachers' awareness of their own gender-specific actions. Although Van de Walle, Karp and Bay-Williams (2014) acknowledge that detecting this awareness may be difficult initially, analysing video-recorded instructional lessons may be an instructive exercise. During an analysis, these authors recommend looking at the number of questions asked of male and female students, noting which students ask questions and what kinds of questions are being asked, and examining the types of feedback given.

Instructional practices to involve all students

To minimise gender differences in learning, a variety of teaching approaches and strategies should be used. For instance, Sousa (2008) proposed that educators take into account the learning styles of students, address multiple intelligences, consider teaching styles available, and examine how students think about mathematics in an attempt to make mathematics education equitable for all learners. Van de Walle, Karp and Bay-Williams (2014) recommend teachers find ways to involve all students in their classes, and not just those who appear to be eager. These commentators acknowledge that boys and girls alike may tend to avoid class involvement, lack motivation or seem reluctant in requesting teacher assistance. In a similar vein, Mau and Leitze (2001) offered that when teachers are in a 'show-and-tell mode', there are significantly more opportunities to reinforce boys' more overt behaviours and girls' more passive behaviours. Instead of reinforcing these behaviours, teachers must expect all students to speak, listen and share their thinking with others (Mau & Leitze, 2001). Hall (2012) suggested that making mathematics accessible to both genders can be achieved by avoiding traditional teaching approaches such as memorisation and rule following. Alternatively, planning lessons that focus more on conceptual understanding and connecting mathematics to students' lives may help both male and female students relate better and understand the subject (Becker, 1995; Belenky et al., 1986; Morrow & Morrow, 1995). According to Hall (2012, p. 70) 'these approaches have been shown to have positive outcomes with respect to all students' achievement, attitudes, and participation in mathematics, but particularly for girls and women'. Additionally, Hall noted that various researchers (Boaler, 1997b; Burton, 1995) advocate for creating supportive, inquiry-focused mathematics classroom environments in which female voices are heard and many approaches to solving problems are valued. Various instructional strategies teachers can employ to assist girls and boys in learning mathematics can be found in Table 6.1.

Table 6.1 Strategies for assisting boys and girls to learn mathematics

Girls	Boys
Include group work.	Structure activities.
Include opportunities for discussion in both small groups and whole-class settings.	Assist learners to break tasks into achievable steps, particularly extended tasks such as investigations.
Include contexts of interest to girls.	
Provide assessment opportunities that allow for a variety of responses (e.g. open-ended investigations). Avoid multiple choice questions as the only assessment approach.	Structure assignments with long deadlines (e.g. over several weeks) to include regular class work and milestones that are achievable (this also assists a teacher to be aware of ownership of the work).
	If homework is set, ensure it is checked regularly.
Ensure contexts are understood (e.g. if football is used as a context, ensure all students are familiar with the context).	Employ clearly defined objectives and instructions.
	Ensure all students in the class are clear about what they are required to do.
	Put tasks on the whiteboard so they are retrievable. If a student has not heard the instructions, or joined the class after the task was set, the student will be able to catch up if the task has been recorded clearly.
	Use the 'tour to be sure' strategy: when a task has been set, wait for a few minutes for students to begin and then move around the students to check for understanding of the task.
Use contextual tasks. Include tasks that show the value of mathematics to the solution of social problems. These can include looking at environmental and social justice issues.	Set short-term challenging tasks.
	Include tasks with a focus on mathematical thinking for all learners in the class. Open-ended tasks allow challenge for all learners.
Use verbal and linguistic approaches.	Use visual, logical and analytical approaches.
Communication is an important part of mathematics and needs to be taught explicitly and emphasised for all learners.	A great deal of mathematics can be accessed using visual representations.
Girls may particularly enjoy opportunities to communicate mathematics in linguistic forms.	Allow students opportunities to explain their thinking and to record it in a variety of ways.
Engage learners with problems set in real contexts.	

Source: Siemon et al., 2011, p. 157.

ACTIVITY 6.1

Log in to HOTmaths and select the 'HOTmaths Global' Course list and then choose the 'Early Secondary' Course. From the Topic list choose 'Length, perimeter & area', then the 'Areas of rectangles' Lesson. This lesson compares perimeter with area.

Looking at the Resources, devise a lesson plan where instructional strategies for boys' and girls' mathematical learning are considered. What will your strategies be to maximise learning for both genders? Assume that you are teaching a co-educational Year 7 class.

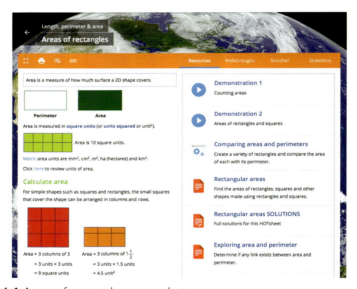

Figure 6.1 Areas of rectangles screenshot

ACTIVITY 6.2

Log in to HOTmaths and select the 'HOTmaths Global' Course list and then choose the 'Early Secondary' Course. Then from the Topic list choose 'Collecting & displaying data', then the 'Collecting & describing data' Lesson.

Looking at the Resources, devise a lesson plan where instructional strategies for boys' and girls' mathematical learning are considered. What will your strategies be to maximise learning for both genders? Assume that you are teaching a co-educational Year 11 Essential Mathematics class.

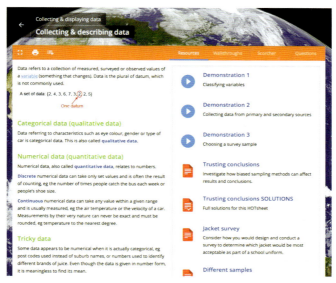

Figure 6.2 Collecting & describing data screenshot

REFLECTIVE QUESTION

Reflect on your secondary school experience, and jot down some recollections as they pertain to how teachers treated both genders in the mathematics classroom. After listing some thoughts, analyse the extent to which boys and girls were treated in terms of: praise and encouragement, selection to offer answers (verbally + demonstrations), amount and quality of attention given, instructional strategies and feedback. In your experience, were boys and girls treated equitably?

Culture in the secondary mathematics classroom

Australia is considered one of the most culturally diverse nations in the world. Churchill et al. (2013) assert that migration is a powerful force in Australia's history, where 'waves of migration from Europe, Asia and Africa have brought many benefits to everyday life in both urban and regional communities in Australia' (p. 147). The 2011 Census of Population and Housing of the Australian Bureau of Statistics (ABS) found that approximately 21 per cent of the population speak a language other than English at home. This statistic represents an approximate decrease of 2.3 per cent from the 2006 Census. The five most commonly spoken languages other than English were Mandarin, Italian, Arabic, Cantonese and Greek, with speakers comprising approximately seven per cent of the total population. Furthermore, approximately 20 per cent of households speak two or more languages at home. The educational implications for this linguistic diversity are considerable, and Marsh (2004, p. 238) states that 'teachers working in Australian schools require a broad understanding of our cultural diversity – diversity due to race, ethnicity, socio-economic status, geographic region, religion and gender'. And while the Australian student population is linguistically and culturally diverse, 'it is significant that the Australian teaching profession is overwhelmingly Anglo-Australian and of middle-class background' (Churchill et al., 2013, p. 147). In light of these data, it must be acknowledged that there are significant regional differences within Australia regarding language and cultural diversity. Within this section, attention is given to teaching mathematics with cultural awareness, and teaching mathematics to English Language Learners (ELLs).

Teaching mathematics with cultural awareness

While mathematics is commonly heralded as a 'universal language', the accuracy of this statement is questionable. For instance, conceptual knowledge (e.g. what multiplication is) is universal, while procedural knowledge (e.g. how to multiply) and symbols are culturally determined, and are non-universal (Van de Walle, Karp & Bay-Williams, 2014). Following research into culturally responsive instructional models within New Zealand mathematics classrooms, Averill et al. (2009) deemed the following conditions as necessary for teachers of culturally diverse students:

> [a] deep mathematical understanding; effective and open relationships, cultural knowledge, opportunities for flexibility of approach and for implementing change, many accessible and non-threatening mathematics learning contexts, involvement of a responsive learning community, and most important, working within a cross-cultural teaching partnership. (p. 180)

Van de Walle, Karp and Bay-Williams (2014) argue that culturally relevant mathematics instruction is for all students, including students from different ethnic groups and socioeconomic status. These authors suggest such instruction requires teachers to focus on 'important mathematics', make content relevant, incorporate students' identities and ensure there is shared power between teacher and student. Teachers should also be careful to select learning materials (e.g. textbooks) that do not exhibit cultural bias (Marsh, 2004). The strategies included in Table 6.2 provide some guidelines for working with students from different cultural backgrounds.

Table 6.2 Strategies for teaching students from different cultural backgrounds

1 Respect the ethnic and racial backgrounds of students – encourage and support.

2 Use role plays to provide student empathy for different culture.

3 Ensure students from different cultures have opportunities to work with others in small group activities.

4 Use curriculum resources that highlight multicultural perspectives.

5 Provide opportunities for students to examine, in depth, particular values, beliefs and points of view relating to cultures.

6 Use media examples to highlight undesirable bias and discrimination.

7 Encourage students to be open and willing to evaluate their values.

Source: Marsh, 2004, p. 238.

Teaching mathematics to English language learners (ELLs)

English language learners (ELLs) enter the mathematics classroom at different ages, and at different stages of learning the English language. ACARA (2015) acknowledges that English as an Additional Language or Dialect (EAL/D) students must achieve the aims of the Australian Curriculum: Mathematics 'while simultaneously learning a new language and learning content and skills through that new language'. To compound the matter for ELLs, there are unique features of the 'mathematics language' that make learning difficult; for instance, certain specific terminology may not be translated easily into another language (Van de Walle, Karp & Bay-Williams, 2014). Moreover, Janzen (2008) highlights that worded problems are particularly difficult for ELLs to complete as the sentences are constructed differently from sentences in conversational English. Khisty (1997) points out that ELLs develop conversational skills in English much faster than their proficiency for benefiting from content lessons taught in English. The corollary of this assertion is that teachers should not assume that because a student can converse competently in English, the same student can 'follow complex, multi-faceted instruction in English requiring her to listen, read, write, speak, and employ other communication structures (e.g. interpret body language and illustrations)' (Cangelosi, 2003, p. 79). In fact, it may take up to seven years for an ELL to learn an academic language, such as mathematics (Cummins, 1994).

When teaching ELLs, one approach is content-based instruction (CBI), which uses specific mathematical content on which to base language instruction. Said another way, in CBI 'the language is taught within the context of a specific academic subject' (TESOL, 2008, p. 1). According to DelliCarpini and Alonso (2014), CBI can enhance the acquisition of both language and content. However, these authors concede that at a secondary level 'to date most of the CBI practice that occurs, whether content-driven or language-driven, does so in the ESL classroom exclusively, and mainstream content teachers are often unprepared or underprepared to work with ELLs in their classrooms' (DelliCarpini & Alonso, 2014, p. 158). Other approaches to assist in language acquisition for ELLs recommended in the literature have been included in Table 6.3. A national EAL/D has been developed to support teachers in making the Australian Curriculum: Foundation to Year 10 in mathematics accessible to EAL/D students (ACARA, 2015e).

Table 6.3 Approaches to support the language acquisition of ELLs

1 Allow students more time to complete tasks.
2 Allow students to work in their native language as required.
3 Provide input and directions at current level of comprehension.
4 Provide additional support to develop students' language needs.
5 Teach specific vocabulary (i.e. the language of mathematics) explicitly.
6 Consider carefully cooperative learning groups to support language acquisition.

Cultural considerations for Indigenous learners

Over the past three decades there have been a number of large-scale literacy and numeracy programs developed to enhance the learning opportunities for Indigenous Australian students (Cooke & Howard, 2009). Some of the more recent programs have included Mathematics in Indigenous Contexts (1999–2005), What Works (2005–2008), Turn the Page: Indigenous

Mathematics and Numeracy (2009–2012), and Make it Count (2009–2012). For instance, the Make it Count program was undertaken by the Australian Association of Mathematics Teachers (AAMT) to address factors contributing to the differences in achievement between Indigenous and non-Indigenous students. The program developed evidence-based, responsive pedagogies and resources to improve learning outcomes of Aboriginal and Torres Strait Islander learners across Australia. Additionally, seven key findings were synthesised as a result of university researchers and eminent Indigenous and mathematics educators working with over 1500 Indigenous students from 35 schools across five states. These summary findings reflect the Australian Professional Standards for Teachers (AITSL, 2015) and underscore the importance of improving current practices used in teaching mathematics to Indigenous students. Herein, the AAMT (2013, p. 2) presented the findings as:

Professional knowledge
1   Know Indigenous learners and know how they learn.
2   Know the mathematics content and know the different ways to teach it effectively to Indigenous learners.

Professional practice
3   Plan for and implement Responsive Mathematics Pedagogy for Indigenous learners that is culturally, academically and socially inclusive.
4   Create and maintain learning environments in which Indigenous learners feel safe and supported.
5   Develop and use tools that assess both affective and cognitive learning outcomes specific to Indigenous learners, provide feedback, and report on student learning.

Professional engagement
6   Engage with colleagues – in professional learning communities in ongoing, action oriented, professional learning – who are prepared to push the boundaries and move outside their comfort zone. Strive for collegial innovation in both Indigenous education and mathematics and numeracy education.
7   Engage with Indigenous parents, families and community in two-way dialogue.

Concomitant with these findings, scholars outline that to successfully teach Aboriginal and Torres Strait Islander students a culturally responsive approach is required (Parkin & Hayes, 2006; Perry & Howard, 2008; Warren & Devries, 2010). Such a response moves away from a transmission model of learning and towards one that requires a 'creative and thoughtful use of each teacher's repertoire of professional skills, and a careful consideration of context' (AAMT, 2013, p. 1). To illustrate, Parkin and Hayes (2006) contend that the linguistic demands of the mathematics curriculum (e.g. mathematics textbooks, worded problems) render mathematics difficult for Indigenous students to access. Teachers can facilitate students' accessibility through several key linguistic strategies; broadly speaking, these strategies assist students in 'sorting out the mathematically important from the unimportant, and in recognising what constituted the mathematical task hidden somewhere in this mass of seemingly irrelevant verbiage' (Parkin & Hayes, 2006, p. 27). Specific strategies include: scaffolded literacy, lower book orientation, higher book orientation and text patterning. Treacy and Frid (2008) conducted research that illustrated how Western mathematics (which is taught in Australian schools) is distinctly different from how traditional Indigenous cultures make sense of, organise and act in their environments. For the research, Treacy and Frid (2008) created three mathematical 'counting' tasks for 18 Aboriginal students (Years 1–11) to complete and examined the extent to which students used the strategy of counting to arrive at their answers. The researchers concluded that while most students demonstrated specific counting knowledge and skills in two of the tasks, there was a preference not to use counting for the third task. Acknowledging that Aboriginal languages do not have many counting words – and as a consequence, many Aboriginal people tend not to count in their everyday situations – the findings from this research suggest that students did not view the third task as one that required counting. Instead, students preferred to use an estimation strategy rather than obtain an exact answer, which supports the idea that precision is much more central to Western society than in most Aboriginal contexts (Malcolm et al., 1999).

ACTIVITY 6.3

Log in to HOTmaths and select the 'HOTmaths Global' Course list and then choose the 'Early Secondary' Course. From the Topic list choose 'Further fractions & decimals', then the 'Understanding decimals' Lesson.

Looking at the Resources, list some strategies where the educational needs for an ELL student are considered. What content and strategies will you use to scaffold learning effectively for this student? Assume that you are teaching this student within a co-educational Year 8 class.

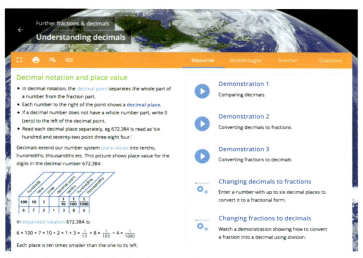

Figure 6.3 Understanding decimals screenshot

ACTIVITY 6.4

Log in to HOTmaths and select the 'HOTmaths Global' Course list and then choose the 'Middle Secondary' Course. From the Topic list choose 'Polynomials', then the 'Introducing polynomials' Lesson.

Looking at the Resources, list some strategies where the educational needs for an ELL student are considered. What content and strategies will you use to scaffold learning effectively for this student? Assume that you are teaching this student within a co-educational Year 11 Mathematical Methods class.

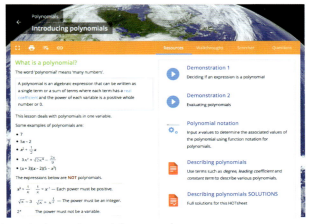

Figure 6.4 Introducing polynomials screenshot

REFLECTIVE QUESTION

Suppose that a student – who is from a culture you know little about, and who speaks a language that you don't understand – is the newest member of your secondary mathematics classroom. After having read this section on culture, what would you do in order to teach this student effectively? Furthermore, what factors will influence the steps you will take in teaching this student?

Special needs learners in the secondary mathematics classroom

Since the mid-1970s the literature base for special needs education has grown considerably. Concomitant to legislation steadily securing the educational rights of students with special needs, research has provided much insight into best instructional practices for these students. Aligning with the belief that all learners – who, with good teaching and the appropriate support – are able to learn mathematics, this section examines two broad groups of special needs learners. These groups are gifted and talented students and students with learning difficulties.

Gifted and talented students

Gifted and talented students generally demonstrate above-average ability, high levels of task commitment, creativeness (Huetinck & Munshin, 2008), and an exceptional level of performance in one or more areas of expression (NAGC, 2007). In addition, students who are mathematically gifted and talented may possess 'an intuitive knowledge of mathematical concepts, whereas others have a passion for the subject even though they may have to work hard to learn it' (Van de Walle, Karp & Bay-Williams, 2014, pp. 115–16). Diezman and Watters (2002, p. 220) have listed characteristics of gifted learners of mathematics to include:

- exceptional reasoning ability
- exceptional memory and concentration span
- preference for abstract and self-directed work
- ability to solve problems in unexpected ways
- ability to identify patterns and relationships
- preference for mathematical activities and puzzles, including enjoyment in posing problems.

Researchers have also identified spatial ability as a characteristic of gifted and talented mathematics students (Perry, Anthony & Diezman, 2004). Students' giftedness can become apparent to parents and teachers through a demonstrated grasp or articulation of mathematical concepts at an age earlier than expected (Van de Walle, Karp & Bay-Williams, 2014). Rotigel and Fello (2005) have noted that these students are often found to easily make connections between topics of study, and frequently are unable to explain how they quickly got an answer. Teachers who have a keen ability to detect giftedness in mathematics note they are students who possess strong number sense or visual/spatial sense (Gavin & Sheffield, 2010). However, gifted mathematics students should not be confused solely with those who are fast with number facts, but those who possess an ability to reason and make sense of mathematical tasks.

Despite the gifts and talents of such learners being recognised by teachers and parents, the students themselves may not embrace comfortably the label of mathematically gifted (Huetinck & Munshin, 2008; Van de Walle, Karp & Bay-Williams, 2014). Popular culture and the media consistently portray talented learners of mathematics as looking strange or acting weird (Sheffield, 1997), socially inept outcasts (Van de Walle, Karp & Bay-Williams, 2014) or as eccentrics with prodigious calculating abilities (Siemon et al., 2011). Such portrayals are unhelpful for gifted adolescent students who 'absorb these powerful negative messages about showing their intelligence in public settings' (Van de Walle, Karp & Bay-Williams, 2014, p. 115). Nevertheless, it is important for teachers to encourage and support mathematically gifted students by showing the class appropriate, successful, 'real world' mathematics role models who appear in popular culture, the media and the real world (Van de Walle, Karp & Bay-Williams, 2014).

Educational strategies for gifted and talented students

According to Huetinck and Munshin (2008), it is generally assumed within education that good teaching can respond to the varying needs of diverse learners – including the gifted and talented. However, educators may tend to feel less inclined to provide for these learners than for students of low ability. A considerably worse outcome is where mathematically talented learners have experienced classrooms where giftedness is not recognised or cultivated appropriately. For instance, such learners may have been given 'more of the same' work for practice and consolidation, given 'free time' or independent time on a technological device, or routinely assigned to assist struggling learners – all of which are considered unsupportive strategies for talented learners (Huetinck & Munshin, 2008; Siemon et al., 2011). Irrespective, teachers should not wait for students to demonstrate their mathematical content; these students require instructional programs designed to assist them reach their full potential (Huetinck & Munshin, 2008). Such programs can include challenging, rich extension tasks (Siemon et al., 2011) and an inquiry-based instructional approach (Van Tassel-Baska & Brown, 2007) which can both promote creative and divergent thinking. To this end, Sheffield (1999) contended that gifted students need to experience the:

> joys and frustrations of thinking deeply about a wide range of original, open-ended, or complex problems that encourage them to respond creatively in ways that are original, fluent, flexible, and elegant. (p. 46)

ACARA recommends that teachers enrich gifted students' learning through the provision of opportunities to work with learning area content in further depth or breadth. This enrichment can take place through teachers:

> emphasising specific aspects of the general capabilities learning continua (for example, the higher order cognitive skills of the Critical and Creative thinking capability); and/or focusing on cross-curriculum priorities. Teachers can also accelerate student learning by drawing on content from later levels in the Australian Curriculum: Mathematics and/or from local state and territory teaching and learning materials. (ACARA, 2015)

Gallagher and Gallagher (1994) suggest four strategies for adapting mathematics content for gifted students: acceleration, enrichment, sophistication and novelty. Within each strategy, these authors contend that planned activities require students to apply learnt information rather than simply acquire it. Additionally, an emphasis on the implementation and extension of ideas must eclipse the acquisition of facts and concepts (Gallagher & Gallagher, 1994).

Four strategies: acceleration, enrichment, sophistication and novelty

Acceleration involves providing students with access to curriculum content different from their year level while demanding more learning independence. It recognises that students may already understand the mathematics content that will be presented; consequently, teachers can either reduce the amount of time these students spend on aspects of the topic or move altogether to more advanced and complex content (Van de Walle, Karp & Bay-Williams, 2014). *Enrichment* activities take further the prescribed, curriculum-based topics of study to extend original mathematical tasks. Alternatively, enrichment activities can involve studying the same topic as the rest of the class while differing on the means and outcomes of the work. Van de Walle, Karp and Bay-Williams (2014, p. 116) suggest that examples of enrichment can include 'group investigations, solving real problems in the community, writing letters to outside audiences, or identifying applications of the mathematics learned'. When educators raise the level of complexity of a topic or have students pursue a task at greater depth, they are using the strategy of *sophistication*. Said another way, students have the opportunity to explore a larger set of ideas within which a mathematics topic exists. For instance, a teacher may ask mathematically gifted students to develop a method for calculating reducible interest on a car loan when the class is studying arithmetic and geometric sequences. *Novelty* requires educators to introduce gifted students to completely different material from the prescribed curriculum, and within their developmental grasp. Alternatively, the novelty approach provides students with various options to demonstrate their mathematical skills and knowledge through demonstrations, inventions, experiments, and oral presentations (Gallagher & Gallagher, 1994).

Students with learning difficulties

The identification and remediation of students whose low ability in mathematics is based on learning difficulties is a complex process (Huetinck & Munshin, 2008). Furthermore, these authors offer that learning difficulties can comprise: intellectual disability, serious emotional disturbances, autism, traumatic brain injury, attention deficit disorder, attention deficit hyperactivity disorders or other specific learning disabilities (Huetinck & Munshin, 2008, p. 343). In turn, students may have very specific difficulties with perceptual or cognitive processing which may affect 'memory; general strategy use; attention; the ability to speak or express ideas in writing; the ability to perceive auditory, visual, or written information; or the ability to integrate abstract ideas' (Van de Walle, Karp & Bay-Williams, 2014, p. 104). Sousa (2008) has identified environmental (e.g. mathematics anxiety) and neurological (e.g. dyscalculia) factors that can impede mathematics learning – as well as mathematics disorders (e.g. procedural, memory), difficulties (arithmetic, number concepts) and deficits (counting skills, visual-spatial). In 1975, a landmark event for special needs education occurred in the United States, where under federal law, the Education for All Handicapped Children Act (EHA) mandated that special education students be placed in the **least restrictive learning environment** for success in learning. This meant that special education students were to remain in educational environments with their non-identified peers unless there was a compelling reason to withdraw them. Since then, Australia has enacted through legislation the *Disability Discrimination Act 1992* (DDA) and more importantly for education, the Disability Standards for Education, Department of Education, Science and Training (DEST, 2006). These standards outline how students with a disability have the right to 'education and training opportunities on the same basis as students without disabilities' (DEST, 2006, p. 42). Over the past three decades the conceptualisation of disability has been described as one of 'deficit', where emphasis was placed on identifying students' learning difficulties in order to 'fix them up' (Siemon et al., 2011). A more recently developed view of disability involves looking at how the curriculum can be adapted to suit the needs of the students. Herein, Diezmann et al. (2004, p. 176) assert:

> the emphasis in the last decade has been less of one seeking to define, diagnose and remediate learning difficulties in individual students to one of identifying ways to make the curriculum more accessible to all learners.

Irrespective of how disability is conceptualised, the strategies used to teach students with disabilities should be based upon the tenets of sound teaching practices that help all students. Some strategies to be used in a secondary mathematics classroom will now be considered.

Teaching students with learning disabilities

At all stages in their learning, students should be able to engage with all topics within the Australian Curriculum: Mathematics. One instructional challenge for the classroom teacher, therefore, is to ensure that all learners – whatever their capability or prior learning – have access to mathematical topics in a meaningful and developmentally appropriate manner. For this access, teachers must make necessary curriculum adjustments by drawing from

> content at different levels along the Foundation to Year 10 sequence. Teachers can draw from content at different levels along the Foundation to Year 10 sequence. Teachers can also use the extended general capabilities learning continua in Literacy, Numeracy and Personal and Social capability to adjust the focus of learning according to individual student need. (ACARA, 2015)

Additionally, teachers can enact considerations for instructional planning, delivery and assessment, including: identification of learning disabilities, using an **Individualised Educational Plan (IEP)** and pedagogical strategies.

Identification of learning disabilities

Initially, identifying students' learning disabilities may be a difficult task. As beginning teachers gain experience in interacting with students of varying ages, maturity levels and mathematical abilities, they become increasingly aware and sensitive to students who may have learning disabilities (Huetinck & Munshin, 2008). While a low assessment result may arouse a teacher's curiosity – and, at the same time, provide valuable information about student weaknesses or possible learning difficulties – there are numerous non-cognitive factors to be considered. Factors including low motivation, prior mathematics instruction, cultural bias of tests and immaturity can be investigated by the teacher before referring a student to a special needs coordinator (Huetinck & Munshin, 2008). Miller and Mercer (1997) posited that students with learning disabilities exhibit problems that contribute to poor mathematics achievement and that are linked to information processing. Following their review of research, these authors suggested various ways in which weaknesses in selected components of information processing may affect mathematics performance. The general problems and accompanying specific weaknesses listed in Table 6.4 could be used by a teacher who suspects a student has learning difficulties in mathematics.

Table 6.4 How components of information processing may affect mathematics performance

Attention deficits	1 Student has difficulty maintaining attention to steps in algorithm of problem solving.
	2 Student has difficulty sustaining attention to critical instruction (e.g. teacher modelling).
Visual–spatial difficulties	1 Student loses place on the worksheet.
	2 Student has difficulty differentiating between numbers, coins, operation symbols, clock hands.
	3 Student has difficulty writing across the paper in a straight line.
	4 Student has difficulty relating to directional aspects of mathematics and number alignment.
	5 Student has difficulty using a number line.
Auditory processing difficulties	1 Student has difficulty doing oral skills.
	2 Student is unable to count on from within a sequence.
Memory problems	1 Student is unable to retain mathematical facts or new information.
	2 Student forgets steps in an algorithm.
	3 Student performs poorly on review lessons or mixed probes.
	4 Student has difficulty telling time.
	5 Student has difficulty solving multi-step word problems.
Motor difficulties/Disabilities	1 Student writes numbers illegibly, slowly and inaccurately.
	2 Student has difficulty writing numbers in small spaces.

Adapted from Miller & Mercer, 1997, p. 50.

Using an individualised education program (IEP)

Once a student's learning disability has been diagnosed accurately, an IEP will be prepared by educators (usually classroom teachers in conjunction with a special needs coordinator), parents and, where appropriate, the student. This document outlines specifically the services the student will receive, including 'positive appropriate behaviour interventions when necessary, assigning certain supplementary learning aids and services, and possibly modifying programs for the youngster' (Huetinck & Munshin, 2008, p. 343). While implementing strategies prescribed within the IEP, teachers are encouraged to actively contribute to the development of the document, seek further advice from relevant professionals (e.g. educational psychologist, special needs coordinator) and work closely with the student's parents (Siemon et al., 2011). Doing so will allow the student's profile of educational strengths to be developed over time.

Pedagogical strategies

Teaching students with learning disabilities is a challenging task, as the classroom teacher requires additional and special skills (Tannock & Martinussen, 2001). The strategies listed in Table 6.5 (Marsh, 2004, p. 236) focus on structured, effective approaches for teaching students with learning disabilities. In offering these general instructional strategies, Marsh acknowledges that learning goals for these students 'tend to be based on basic reading, writing and arithmetic, and on social, vocational and domestic skills' (Marsh, 2004, p. 236). Nonetheless, the strategies provide a solid foundation for any secondary teacher looking to make mathematics accessible to all learners.

Table 6.5 Strategies for teaching intellectually disabled students

1 Carefully develop readiness for each learning task.
2 Present material in small steps.
3 Develop ideas with concrete, manipulative and visually orineted materials.
4 Be prepared for large amounts of practice on the same idea or skill.
5 Relate learnings to familiar experiences and surroundings.
6 Focus on a small number of target behaviours so that students can experience success.
7 Motivate work carefully.
8 Ensure that the material used is appropriate for the physical age of the student and is not demeaning.
9 Every time students complete a task successfully they should be rewarded.

Source: Marsh, 2004, p. 236.

More specific to the discipline of mathematics, Carnine (1997) recommends five key principles of effective instructional design for teachers of students with learning disabilities. These principles include:

1 Teach big ideas

These ideas represent central ideas within a discipline and can make learning subordinate concepts easier and more meaningful. For instance, in the Measurement and Geometry strand students learn various formulas for calculating volume. Using a big idea reduces the number of formulas students must learn through using a single formula (i.e. Volume = Area of Base × Height).

2 Teach conspicuous strategies

Acknowledging that a strategy is 'a series of steps that students follow to achieve some goal' (Carnine, 1997, p. 134), it is an instructional challenge to develop strategies that are 'just right' interventions for students who have difficulty forming them independently. Such strategies are intermediate in generality and facilitate students' use of knowledge to address the *what*, *how* and *when*.

3 Use time efficiently

This principle is centred on teaching students 'all they need to know (in both quantitative and qualitative terms) without 'losing' them by trying to do too much, too quickly' (Carnine, 1997, p. 137). To achieve this, teachers should abandon low-priority learning objectives in favour of focusing on big ideas, ease into the instruction of complex problem-solving strategies, use a strand organisation for lessons and use manipulatives in a time-efficient manner.

4 Clearly and explicitly communicate strategies

Teachers should explain new concepts and strategies in a clear, concise, accurate and comprehensible manner. To achieve this, they must accommodate differences in students' prior knowledge, provide carefully scaffolded examples that will lead to self-directed learning and use corrective feedback consistently.

5 Facilitate retention through practice and review

Keeping in mind that an important goal of any 'mathematics instructional program is to remember and apply increasingly complex concepts and strategies' (Carnine, 1997, p. 138), students need opportunities to carefully practise and review mathematical skills. To facilitate retention, teachers can help students develop automaticity though the provision of concisely delivered demonstrations and the reception of carefully scaffolded practice.

Additionally, Van de Walle, Karp and Bay-Williams (2014) suggest that *explicit instruction*, the *concrete-representational-abstract (CRA) teaching sequence* and *peer-assisted learning* may also be effective in teaching students with learning disabilities. First, explicit instruction involves highly structured, step-by-step, teacher-led explanations of concepts and strategies that focus on 'critical connection building and meaning making that help learners relate new knowledge with concepts they know' (Van de Walle, Karp & Bay-Williams, 2014, p. 104). Using this instructional approach, the teacher uses a tightly scripted sequence that commences with modelling, and then prompts students through the model towards independent practice. Second, the CRA teaching sequence enables students to move away from concrete representations (manipulative materials), through representations (drawings or pictures) and towards abstraction (mathematical symbols) when solving problems. Third, peer-assisted learning allows students to benefit from classmates' modelling and support on an 'as needed' basis. Importantly, tutors and tutees can interchange roles, providing special needs students with a valuable opportunity to explain to another student (Van de Walle, Karp & Bay-Williams, 2014). Additionally, Huetinck and Munshin (2008) note that to succeed, special needs students require considerable encouragement and frequent reminders from the teacher about their progress.

ACTIVITY 6.5

Log in to HOTmaths and select the 'HOTmaths Global' Course list and then choose the 'Upper Primary' Course. From the Topic list choose 'Describing patterns & using symbols', then the 'Use rules to create patterns' Lesson.

Looking at the Resources, devise a lesson plan where the educational needs for a gifted and talented student are considered. What content and strategy will you use to maximise learning for this student? Assume that you are teaching a co-educational Year 7 class.

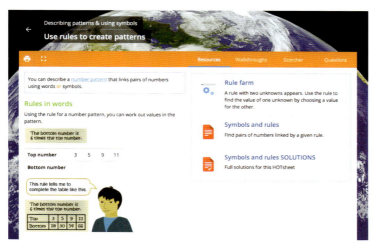

Figure 6.5 Use rules to create patterns screenshot

ACTIVITY 6.6

Log in to HOTmaths and select the 'HOTmaths Global' Course list and then choose the 'Early Secondary' Course. From the Topic list choose 'Introducing algebra', then the 'Number patterns and pronumerals' Lesson.

Looking at the Resources, devise a lesson plan where the educational needs for a student with learning disabilities are considered. What content and strategy will you use to maximise learning for this student? Assume that you are teaching a co-educational Year 12 Essential Mathematics class.

Figure 6.6 Number patterns and pronumerals screenshot

REFLECTIVE QUESTION

Examine your personal philosophy of teaching and learning for statements of how you plan to make mathematics accessible to all children in your class. As a guide, read the Disability Standards for Education (DEST, 2006). From this document, what key points have you included in your personal philosophy?

Summary

Offer explanations to account for gender differences in motivation and achievement in mathematics

Several possible explanations can account for perceived gender differences in motivation and achievement in mathematics. These explanations are socially and culturally constructed and include pervasive belief systems, teacher behaviour and attitudes, and student attitudes. A pervasive, stereotypical belief held by parents and society generally is that mathematics is a male activity that is incongruous with femininity. Research into teacher behaviour indicates that boys receive both more attention and different kinds of attention than girls. Female students in mathematics classes go unobserved, and are known as 'quiet achievers'. Research into teacher attitudes reveals that girls are perceived to use inferior cognitive skills and succeed because of their diligence, while boys are seen to be talented in mathematics but lacking in effort. Despite the narrowing gender gap concerning mathematics achievement, it is still considerably wide in terms of students' attitudes towards mathematics. Studies conducted over time and across levels of education have generally found that boys hold a more positive attitude towards mathematics than girl do.

Identify instructional approaches to be used in making a mathematics classroom gender equitable

To make a classroom gender equitable, a teacher can take into account the learning styles of students, address multiple intelligences, consider teaching styles available and examine how students think about mathematics. Additionally, mathematics can be made accessible to both genders by avoiding traditional teaching approaches such as memorisation and rule following. Instead, planning lessons that focus more on conceptual understanding and connecting mathematics to students' lives may help both male and female students relate better and understand the subject.

Outline instructional guidelines for working with students from different cultural backgrounds

When working with students from different cultural backgrounds, it is important to provide instruction for culturally relevant mathematics. Such instruction requires teachers to focus on 'important mathematics', make content relevant, incorporate students' identities and ensure there is shared power between teacher and student. Additionally, culturally diverse students may be ELLs, which may prompt teachers to allow more time for the completion of set tasks and, when needed, in the students' native language. Teachers should provide input and directions at ELLs' current level of comprehension, and content-specific vocabulary should be taught explicitly.

Justify the need for diverse learners to receive instruction according to their particular learning needs

One instructional challenge for the classroom teacher is to ensure that all students – whatever their capability or prior learning – have access to mathematics learning in a meaningful and developmentally appropriate manner. According to federal legislation, diverse students must receive learning opportunities in the **least restrictive environment**. A common belief held by teachers promoting learner equity is that with good teaching and the right support all students are able to learn mathematics.

Delineate instructional approaches to be used with diverse learners

To make learning accessible for diverse learners, teachers should modify their approaches towards instructional planning, delivery and assessment. For students with learning difficulties, teachers can plan activities focused on big ideas, teach conspicuous strategies, use time efficiently, communicate strategies clearly and explicitly and facilitate retention through practice and review. Teachers of gifted and talented learners can promote creative and divergent thinking through the use of strategies including acceleration, enrichment, sophistication and novelty.

CHAPTER 7

Assessing mathematics learning

Learning outcomes

After studying this chapter, you should be able to:

- define the term 'assessment' and explain the function of assessment in learning
- list and justify assessment types that can be used in mathematics learning
- delineate and rationalise key principles for creating assessment tasks in mathematics learning
- recall appropriate methods of providing feedback to students.

Introduction

You don't fatten a pig by weighing it. (Anonymous)

Nobody ever got taller by being measured. (Professor Wilfred Cockroft)

Not everything that can be counted counts. Not everything that counts can be counted. (Albert Einstein)

These three statements have not been selected randomly at the beginning of a chapter written principally about **assessment** in secondary mathematics. Rather, their prominent inclusion affords educators – from pre-service to veteran, novice to master – some words of wisdom regarding the central issues of *why*, *how* and *what* as they pertain to assessment. In a similar vein, this chapter addresses those issues through an examination of relevant scholarship and an alignment with principles prescribed by the Australian Curriculum and Reporting Authority (ACARA). To begin with, key ideas underpinning the purpose of assessment will be explored. Second, various methods of assessment used currently by secondary mathematics teachers will be outlined and justified. Third, various guidelines for providing effective feedback to students are suggested. Finally, a variety of principles for appropriate assessment practices is offered to guide educators in developing, administering and refining effective tools for measuring student achievement. The theory, research, recommendations and guidelines for best assessment practices offered herein are driven implicitly by the need for mathematical educators to produce and communicate information that is worthwhile and accurate.

KEY TERMS

- **Assessment:** the intentional and unintentional activities undertaken by teachers to gather information about student learning (knowledge, skills, dispositions).
- **Conceptual knowledge:** the knowledge of abstract and general ideas (concepts) and how these ideas can be interconnected.
- **Diagnostic assessment:** those planned processes of getting in-depth information about an individual student's knowledge and mental strategies about concepts.
- **Feedback:** the information gained from formal and informal assessment, provided to students by teachers, about what the student has learnt, and needs to learn, mistaken ideas and directions for improvement.
- **Formative assessment:** those planned processes that regularly monitor students' understanding of instructional activities and during instructional activities.
- **Performance-based task:** a task that is connected to actual problem-solving activities used in instruction.
- **Procedural knowledge:** the knowledge of processes or steps needed to achieve a particular goal.
- **Rich assessment tasks:** those planned tasks ensuring that students can develop concepts in depth, and emphasising a wide range of problem-solving processes in reaching a solution.
- **Summative assessment:** those planned processes which demonstrate an accumulated demonstration of learning by a student over a given period of time.
- **Validity:** the degree to which an assessment item measures what it has been designed to measure.

Why assess?

Teachers and researchers consistently identify assessment as a critical feature of the instructional process (Bobis, Mulligan & Lowrie, 2013; Killen, 2005; Marsh, 2010). Not surprisingly, those involved in mathematics education view the importance of assessment in a similar light (Van de Walle, Karp & Bay-Williams, 2014; Wiliam, 2011). One question driving the need for assessment is: 'What is the purpose of assessment?' or more specifically 'Why assess?' Killen (2005) argued that the purpose of assessment is underpinned by two assumptions. First, there must be *something* (e.g. intelligence, aptitude for a particular job, program quality) that can be measured. The second assumption is 'that this factor can be measured in a way that distinguishes between how much of it is possessed by different individuals, by different groups of learners or by different instructional purposes' (Killen, 2005, pp. 101–02). The purpose of assessment is recognised also by national and international authorities who outline reasons for assessment, and recommend principles for assessing student work. For instance, the ACARA asserts that assessment within the Australian Curriculum (AC) takes place in different levels and for different purposes. These levels and purposes include:

- ongoing **formative assessment** within classrooms for the purposes of monitoring learning and providing feedback, to teachers to inform their teaching, and for students to inform their learning
- **summative assessment** for the purposes of twice-yearly reporting by schools to parents and carers on the progress and achievement of students
- annual testing of students' levels of achievement in aspects of literacy and numeracy in Years 3, 5, 7 and 9, conducted as part of the National Assessment Program – Literacy and Numeracy (NAPLAN)
- periodic sample testing of specific learning areas within the Australian Curriculum as part of the National Assessment Program (NAP) (ACARA, 2015d).

In the United States the National Council of Teachers of Mathematics (1995) posited that teachers assess to monitor student progress, make instructional decisions, evaluate student achievement, and evaluate teaching programs. Churchill et al. (2013) asserted that the purpose of assessment included teachers monitoring instructional effectiveness, providing timely feedback to students and parents, and identifying special needs of students. Additionally, Booker et al. (2014) emphasised how assessment helped teachers identify specific areas of mathematical development in students. In turn, identification of such areas is instrumental in subsequent iterations of instructional planning and in communicating mathematical capabilities to students, parents and the wider community. In their research, Clarke and Clarke (2002) asked many mathematics teachers to respond to the question 'Why do we assess students?' These scholars found that the teachers' responses fell broadly into the following (overlapping) categories:

- To find out what my students know and can do.
- To help me know what to teach next.
- To measure the effectiveness of my teaching.
- To provide feedback to students on their learning.
- To inform parents, employers and interested others of academic progress.
- Because my principal/school/community expects it. (Clarke & Clarke, 2002, p. 1)

Clearly there is much agreement about why teachers engage in the practice of assessment. However, reasons underpinning the selection and implementation of effective assessment options are not always met with the same degree of assonance. For example, Panizzon and Pegg (2007) recorded a self-reported realisation by secondary mathematics educators that their students were rarely given the opportunity to explain what they understood conceptually about a mathematics concept. According to these researchers, one surveyed teacher stated that while his 'students could calculate compound interest, only a small proportion were able to articulate what it was (as a concept) in their own words' (Panizzon & Pegg, 2007, p. 432). In addition, Watt (2005) noted that in a contemporary Australian context where there is a decreased demand for computational skills, 'syllabi now emphasise mathematical process as distinct from product' (p. 22). In a similar vein, Clarke (1987) argued that the reconceptualisation of a 'successful mathematics student' was the impetus for widespread changes in assessment practices within Australian schools. Specifically, this change involves a departure from the view that successful mathematics students can neatly replicate a learned procedure to a routine task in a familiar context. Instead, successful mathematics students can devise problem-solving strategies across various contexts, identify conceptual similarities in different situations, assess the relevance of different procedures to applied contexts, and work productively with others (Clarke cited in Watt, 2005, pp. 22–23).

ACTIVITY 7.1

Access the 'Exploring simple polynomials' HOTsheet by logging into HOTmaths, entering its name in the search field and then selecting the HOTsheets tab.

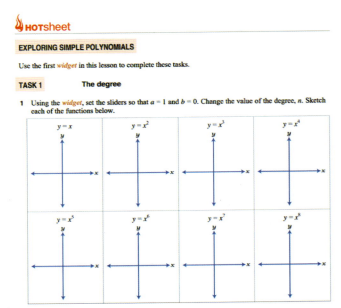

Figure 7.1 Exploring simple polynomials HOTsheet

After completing the test, decide what the classroom teacher was trying to achieve by giving this to his Year 8 general class. Use the following questions and prompts to guide you:

1   What was the AC (Mathematics) strand being assessed?
2   What key sub-strands and topics from the AC (Mathematics) are being assessed?
3   Estimate the number of lessons it would take for the content in this test to be taught adequately.
4   Classify the types of questions found in this test as *elementary, intermediate* or *advanced*.
5   Identify some necessary prior learning Year 8 students would need to master before commencing this topic.
6   What mathematical content and knowledge do you feel students will be ready to learn following the successful completion of this test?

ACTIVITY 7.2

Access the 'Exploring simple polynomials' HOTsheet by logging into HOTmaths, entering its name in the search field and then selecting the HOTsheets tab (see Figure 7.1).

After completing the investigation, decide what the classroom teacher was trying to achieve by giving this to her Year 11 class. Use the following questions and prompts to guide you:

1   Name some of the AC (Mathematics) Upper Secondary units this test could be used in.
2   What key topics from the AC (Mathematics) are being assessed?
3   Estimate the number of lessons it would take for the content in this test to be taught adequately.
4   Classify the types of questions found in this test as *elementary, intermediate* or *advanced*.
5   Identify some necessary prior learning Year 11 students would need to master before commencing this topic.
6   What mathematical content and knowledge do you feel students will be ready to learn following the successful completion of this test?

REFLECTIVE QUESTIONS

In your pre-service education degree or qualification you will be required to develop a *philosophy of teaching and learning*, which encompasses significant statements about the profession you are entering. This philosophical statement is comprised usually of the way(s) you view: knowledge, the purpose of education, the nature of people and approaches towards best instructional practices. As you gain further classroom experience, this statement can undergo editing and refinement. Within this statement, develop a rationale for assessment practices, asking yourself the following questions:

1   What is the purpose of assessment?
2   What is the purpose of assessment in mathematics?
3   How can I ensure that best practices for assessment occur in my secondary mathematics classes?

Ensure that your responses for these statements are kept in an accessible location, so that you can periodically review and revise them.

What is assessment?

Assessment is the term used to describe an individual activity or series of activities undertaken by a teacher in order to gather information about student learning (Marsh, 2010). Earlier in this chapter, some of the reasons teachers assess student learning were outlined. Some of these reasons included: to make informed decisions about children (Sattler, 2008) and to enhance instructional planning (Ohlsen, 2007; Stiggins, 1994). Within the classroom context – which is one characterised by fairly constant formal and informal assessment over time, and across many dimensions of behaviour – teachers have a wide range of classroom assessment methods to select from (Brualdi, 1998). Some of the more common types of assessment used by secondary mathematics teachers are: tests, quizzes, examinations, investigations, extended pieces of work and homework assignments. Irrespective of the type of assessment used, Clarke (1997) argued that it is through selection of assessment that teachers communicate most clearly to students those activities and learning outcomes they value.

Types of assessment

Assessments typically fall into one of three broad categories:

- summative assessments
- formative assessments
- diagnostic assessments.

Each of these categories will be explored below.

Summative assessment

Summative assessment (or assessment *of* learning) is any assessment designed mainly to measure student learning for the purposes of ranking or providing an indication of performance of learning at a fixed point in time (e.g. a score or grade) (Bobis, Mulligan & Lowrie, 2013). According to Van de Walle, Karp and Bay-Williams (2014), summative assessments are 'cumulative evaluations that might generate a single score, such as an end-of-unit test or the standardised test that is used in your state or school districts' (p. 82). These assessments are regarded as those which demonstrate an accumulated demonstration of learning by a student over a given period of time. Van de Walle, Karp and Bay-Williams (2014) note that although summative assessments can be useful for schools, teachers and systems in long-term planning or revising curricula, they are often unhelpful in shaping teaching decisions that require more immediate attention. Such decisions can include the selection of pedagogy regarding particular mathematics topics, or the identification of conceptual misunderstandings that may prevent student growth.

Formative assessment

Formative assessment (or assessment *for* learning) is any assessment for which the main purpose is to enhance learning (Bobis, Mulligan & Lowrie, 2013; Klenowski, 2009; Wiliam, 2011). Formative assessment practices are those methods that assess student learning 'on the go' or 'along the way'. Scholars contend that educators use formative assessments as planned processes to regularly monitor students' understanding of and during instructional activities (Hattie, 2009; Popham, 2008; Wiliam, 2008). Formative assessment practices in secondary mathematics usually take the form of quizzes, chapter tests, student work, standardised exams, as well as questioning or interviewing techniques to inform instruction (Chen et al., 2012).

Diagnostic assessment

As students enter classrooms at varying points along their educational journeys, it is considered an efficient exercise to start a new teaching unit by checking their knowledge and understandings (Marsh, 2010). Conducting a diagnostic assessment can assist teachers in ascertaining knowledge about students' prerequisite skills and dispositions towards the learning area. According to Marsh (2010), diagnostic assessments remind teachers to commence their instruction precisely at the level students have reached. A diagnostic assessment is the means of getting in-depth information about an individual student's knowledge and mental strategies about concepts (Van de Walle, Karp & Bay-Williams, 2014). Moreover, diagnostic assessments can help identify the way in which mathematics is understood and used, locate areas of strength and determine underlying weaknesses causing errors or difficulties (Booker, 2011). Although these assessments (which can be conducted through a teacher–student interview) are often labour-intensive, they are rich methods 'that provide evidence of misunderstandings and explore students' ways of thinking' (Booker, 2011, p. 94). During the assessment a student is given specifically selected problems to complete on a single topic, where the working out or answers supplied can illuminate teachers as to a student's misunderstanding or progress in that topic.

Figure 7.2 displays a diagnostic assessment item developed by Minstrell (2015) that can be used in secondary mathematics and science class-rooms. This item requires the student to answer a question related to the position-time graph given. Following completion of the item, the teacher will apply a particular *facet* (Minstrell, 2001) from the diagnostic key (see Figure 7.3). Each facet corresponds to a *facet cluster* (Minstrell, 2001), which acts as a framework for diagnosing how students understand or misunderstand conceptual material.

Question: 1c

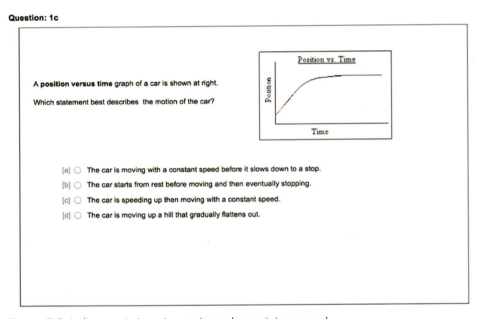

Figure 7.2 A diagnostic item in a unit on determining speed

Key	Facet	Next Question
a	03	2
b	50	2
c	80	2
d	90	2

Figure 7.3 The diagnosis key for the item in Figure 7.2

Facet Cluster - Determining Speed

Facets and facet clusters are a framework for organizing the research on student conceptions so that they are understandable to both discipline experts and teachers. Facet clusters include the explicit learning goals in addition to various sorts of reasoning, conceptual, and procedural difficulties. Each cluster contains the intuitive ideas students have as they move toward scientifically accurate learning targets.

Facets are arranged with the Goal Facets at the top of the page followed by the more problematic facets. Each facet has a two-digit number. The 0X and 1X facets are the learning targets. The facets that begin with the numbers 2X through 9X indicate ideas that have more problematic aspects. In general, the higher facet numbers (e.g., 7X, 8X, 9X) are the more problematic facets. The X0's indicate more general statements of student ideas. Often these are followed by more specific examples, which are coded X1 through X9.

Print Page

Facet Cluster

00 Student correctly describes the motion and interprets the speed of an object for a specific time from the information given. The information may be given in graphs, tables, pictures, or words.

   01 Given speed vs. time data, student correctly identifies an instantaneous speed of an object.

   02 Given position vs. time data, student correctly describes and determines the speed of an object moving uniformly.

   03 Student qualitatively describes the motion of an object from the information given.

40 The student misses a portion of the trip during their discussion of the motion.

   41 The student incorrectly describes or omits the middle portion of an object's motion.

   42 The student incorrectly describes or omits the beginning or end of an object's motion.

50 Student incorrectly describes the initial or final conditions of motion of the object.

   51 Initial speed is incorrectly identified as zero, because at the beginning of trips things are not supposed to be moving.

   52 Final speed is incorrectly identified as zero, because at the end of trips things are not supposed to be moving.

70 When asked for the speed at one instant, the student incorrectly reports another quantity or rate.

   71 Student reports the position, change in position or distance traveled.

   72 Student reports the average speed.

   73 Student reports the change in speed.

   74 Student divides the speed by the final time or change in time.

   75 Student divides the change in speed by the final time or change in time.

   76 Student reports zero, the object cannot be moving at an instant in time.

80 Student confuses position vs. time and speed vs. time graphs or data tables.

   81 Student interprets sloping up (or down) on a position graph to mean the object is speeding up (or slowing down).

   82 Student interprets a flat line segment on a position graph to mean the object is moving at constant speed.

   83 Student interprets sloping up (or down) on a speed graph to mean the object is moving with constant speed away from (or toward) the origin.

   84 Student interprets a flat line segment on a speed graph to mean the object is not moving.

90 Student views a position or speed graph as a map of the actual motion.

   91 Student interprets an upward (or downward) sloping graph to mean the object is going up hill (or downhill).

   92 Student interprets a flat line on the graph to mean the object is moving on a flat surface.

Figure 7.4 The facet cluster used to code student responses to assessment items in the determining speed unit

For instance, if a student writes *(b)* as the correct answer, then the corresponding facet for this response would be 50. In turn, the facet cluster *Student incorrectly describes the initial or final conditions of the motion of the object* describes generally how the student has misunderstood the question. Additionally, the facet numbers 51 and 52 are included to help teachers locate more specifically how the student has applied incorrect reasoning, or experienced conceptual or procedural difficulties. Alternatively, if a student writes *(d)* as the correct answer the facet 90 would be applied, aligning with the facet cluster *Student views a position or speed graph as a map of the actual motion*. Facet numbers 91 and 92 are also available for a more specific diagnosis of conceptual misunderstanding. Minstrell (2001) notes that in relation to facet 50, facet *80* indicates how a student exhibits greater conceptual or procedural difficulties in completing the given problem (See Figure 7.4).

For this diagnostic assessment item, the Australian Curriculum Content Descriptions are:

- Year 7 (ACMNA 180)
- Year 8 (ACMNA188)
- Year 9 (ACMNA 208)
- Year 10 (ACMNA 235)
- General Mathematics (ACMGM 042 & ACMGM 043)
- Mathematics Methods (ACMMM 002 & ACMMM 003).

Within Australia a series of diagnostic assessments (known as 'smart tests') have been developed at the University of Melbourne by Stacey et al. (2013). These smart tests (**s**pecific **m**athematics **a**ssessments that **r**eveal **t**hinking) are designed to provide teachers with an insightful and informative diagnosis of junior secondary students' conceptual understanding of mathematical topics. After signing up at no cost, teachers gain internet access to a repository of online tests for most mathematical topics. Although tests are completed online, students may use paper, pens and a calculator if required. The teacher receives students' results instantly, where interpretive analysis, further teaching and follow-up testing can follow. Parallel tests are also available for teachers who wish to use a 'pre-test/post-test' approach, or for those who prefer to offer alternative tests for particular students. The smart tests can be accessed online at www.smartvic.com/teacher/.

According to Booker (2011), diagnostic assessment can be expressed as a 3-phase cycle. The cycle is comprised of a series of observations, assumptions and probes that are followed until the underlying causes of difficulties are illuminated.

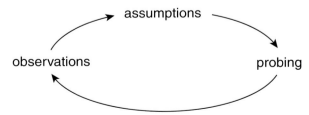

Figure 7.5 3-phase cycle of observations, assumptions, probes
Source: Booker, 2011.

First, teacher observations focus on individual students and their behaviour during teaching and learning activities as students work individually or cooperatively. Such observations take place to try to determine thinking and understanding of specific concepts and processes. Second, ongoing observations lead to assumptions about a student's understanding and the progress of the class as a whole. These assumptions reflect 'a teacher's knowledge of both the underlying mathematical ideas and experience with a range of children in similar situations' (Booker et al., 2014, p. 32). As a result of such assumptions, teachers form conjectures regarding student mathematical thinking (evidenced in explanations, representations or drawings), and understanding (use of processes, display of **conceptual knowledge**). Finally, any assumptions made then need to be confirmed through further probing (Booker et al., 2014).

Common assessment techniques

A variety of assessment techniques is available to secondary mathematics teachers. Commonly used techniques include: tests, quizzes, examinations, homework and investigations. Teachers can also measure student learning through the administration of alternative assessments, including: student self-assessment, teacher observation, student journalling, oral tasks or practical tasks. When deciding precisely how student progress will be measured, Leder, Brew and Rowley (1999) suggest that teachers avoid an over-reliance on one form of assessment type. These scholars argue that differences in learner characteristics imply that a variety of assessment techniques should be used so that all students may display their mathematical knowledge, skills and abilities effectively (Leder, Brew & Rowley, 1999).

Tests

Tests will always be a part of assessment and evaluation (Van de Walle, Karp & Bay-Williams, 2014). The prevalence of test administration within mathematics classrooms is well documented, with multiple researchers reporting that mathematics teachers rely heavily on traditional tests to assess student learning (Henke, Chen & Goldman, 1999; Senk, Beckmann & Thompson, 1997; Watt, 2005). Senk, Beckmann and Thomson (1997) purport that tests are a popular assessment item because they can often address 'a dozen or more objectives and that they must often be administered in 40–50 minute periods' (p. 210). Moreover, such tests effectively assess aspects of mathematics in an unambiguous and straightforward way, including students' performance on routine skills and algorithms (Clarke & Lovitt, 1987; Grimison, 1992; Stephens, 1988). Research in an Australian secondary school context found that surveyed teachers described how 'the written test was a good measure of mathematical abilities for Years 11–12, less good for Years 9–10, and least good for Years 7–8' (Watt, 2005, p. 37). In some Australian states and territories, secondary school tests and examinations contain two sections; a section that permits the use of handwritten notes and a calculator, and a section that does not permit any resources.

A cautionary note on tests

Although testing is viewed as an efficient and popular assessment technique, multiple commentators advise teachers to exercise caution during the design and administration of tests. For instance, Senk, Beckmann and Thompson (1997) expressed concern regarding the over-reliance on textbook-generated chapter tests for mathematics students. These authors found that such tests were comprised generally of low-level test questions that seldom required students to justify a conclusion, contained few questions set in realistic contexts, and lacked any technological component. To address these research findings, 'greater efforts need to be made to incorporate more reasoning and multi-step problem solving, as well as more substantive use of both numerical and graphical technology on tests' (Senk, Beckmann & Thompson, 1997, p. 211). Below are two test items that require learners to demonstrate their knowledge of circular equations. Although both items could be used within a mathematics assessment, Item 2 requires a higher degree of reasoning and demonstration of further multi-step thinking than Item 1.

> **Item 1**
> Find the centre and the radius of the circular equation given by $(x+2)^2 + (y+3)^2 = 9$
>
> **Item 2**
> AB is the diameter of a circle whose equation is If A is the point (–2, –3), find the coordinates of B.

In addition, these characteristics suggested that teachers were not assessing according to nationally prescribed guidelines which 'reflect the mathematics that all students need to know and be able to do' (NCTM, 1995, p. 11). Clarke and Clarke (2002) highlighted how for certain students 'the pressure of the test leads to performance that is not representative of their knowledge and understanding' (p. 2). While scholars acknowledge that test administration can be justified on the grounds of increasing reliability and ensuring comparability, **validity** should not be compromised (Clarke, 1997; Lacey & Lawton,

1981). To enhance validity, Van de Walle, Karp and Bay-Williams (2014) recommend that tests should not be 'a collection of low-level skill exercises that are simple for teachers to mark', but designed to 'match the goals of instruction' and to determine 'what concepts students understand and how their ideas are connected' (p. 91). To accomplish this, tests of **procedural knowledge** should consist of items requiring students to demonstrate a conceptual basis for processes used (Van de Walle, Karp & Bay-Williams, 2014). Some examples of the types of questions teachers can use in enhancing students' mathematical procedural knowledge are to 'find the odd one out', to 'provide an example of', and to include questions that can be solved in different ways.

Quizzes

Quizzes can serve as an efficient assessment method allowing teachers to evaluate student learning at a specific point in the learning process (Webb, 2001). At a specific point in time, quizzes can provide teachers with important and timely feedback regarding student learning (Ohlsen, 2007). In particular, teachers can use the results of quizzes to 'determine if remediation or researching of key concepts or processes is needed for improved student outcomes' (Ohlsen, 2007, p. 10). In this way, Black and William (1998) contend that quizzes can be utilised as both a formative assessment method that informs and guides instruction and an evaluation component that can be used in the calculation of student summative grades.

Examinations

Major examinations are often a summative assessment tool that teachers use as a cumulative evaluation of student learning at the end of a chapter or unit (Ohlsen, 2007, p. 10). As such, examinations are designed to last for several hours and may even be divided into sections where resources (e.g. notes, formula sheets, calculators) may or may not be permitted for use.

Homework

A common and expected activity within Australian secondary schools is the allocation of homework tasks. According to Huetinck and Munshin (2008), determining the extent to which homework is used in formative assessment can be a tricky decision for teachers. Following a synthesis of research on homework tasks, Cooper, Robinson and Pattall (2006) outlined that homework positively influences both immediate and long-term achievement and learning, and has a variety of non-academic (e.g. greater self-direction, self-discipline) and familial (e.g. awareness of the school-home connection) benefits. On the other hand, potential negative effects of homework can include: satiation, denial of access to leisure time and community activities, parental interference, an increase in the onset of cheating and increased differences between high and low achievers (Cooper, Robinson & Patall, 2006). Furthermore, these authors found that a typical secondary school student who completes homework will outperform classmates who do not complete homework by 69 per cent. For middle school students this figure is 35 per cent. Alongside this research, Cooper, Robinson and Patall (2006) offer some guidelines regarding the issuing of homework:

- Homework is never given as punishment.
- The purpose of homework should be to diagnose individual learning problems.
- All students in the same class will be responsible for the same assignment.
- Homework assignments should not be used to teach complex skills.
- Homework should be a combination of material already covered and simple introduction to material about to be covered.
- Include a mixture of both mandatory and voluntary homework, with the caveat that voluntary should be intrinsically interesting.

Investigations

Alongside quizzes and tests, investigations are another popular method of formative assessment in the secondary mathematics classroom. A mathematical investigation requires students to pose their own problems after an initial exploration of a given mathematical situation (Ronda, 2010; Yeo, 2008). Such exploration – together with the formulation of problems and their corresponding solutions – provide an 'opportunity for the development of independent mathematical thinking' and to engage in 'mathematical processes such as organising and recording data, pattern searching, conjecturing, inferring, justifying and explaining conjectures and generalisations' (Ronda, 2010). Scholars have pointed out that problem-solving tasks and investigations are different activities; Ernest (1991) described problem solving as 'trail-blazing to a desired location' (p. 285) and Pirie (1987) defined investigation as the exploration of an unknown land where 'the journey, not the destination, is the goal' (p. 2). To summarise, problem solving is a convergent activity focused on a well-defined goal and solution, and investigation is a divergent activity with an open goal and solution (Evans, 1987). The open-ended nature of investigative tasks assists students in developing mathematical habits of mind (Lampert, 1990; Ronda, 2010), problem-posing skills (Brown & Walter, 2005), and conjectures and generalisations (Calder et al., 2006). In doing so, students are able to engage in a 'variety of rich mathematical activities that parallel what academic mathematicians do' (Yeo, 2008). As Civil (2002) noted, mathematicians investigate and solve mathematical problems.

Alternative assessments

There is a need for teachers to explore alternative assessment methods to assess other instructional goals (Watt, 2005). Some of these methods include:

- student self-assessment
- teacher observation
- student journalling
- oral tasks
- practical tasks.

Student self-assessment

Student self-assessment can provide a greater degree of ownership of the assessment process to students (Clarke & Clarke, 2002). Stenmark (1989) asserted that 'the capacity and willingness to assess their own progress and learning is one of the greatest gifts our students can develop … Mathematical power comes with knowing how much we know and what to do to learn more' (p. 26). Van de Walle, Karp and Bay-Williams (2014) argued that student self-assessment should not be a teacher's only measure of student learning or disposition, but rather 'a record of how they perceive their strengths and weaknesses as they begin to take responsibility for their learning' (p. 90). As well as assessing their own growth in knowledge and skills, Clarke and Clarke (2002) stated that 'students also have much to offer about their preferred learning styles and ways in which the teaching and learning process can be enhanced' (p. 4). Furthermore, Clarke and Clarke (2002) outlined how some teachers advocated the use of student-constructed tests as part of a review of content prior to formal assessment. According to these commentators, 'students are invited to create assessment tasks that they believe would assess fairly the key ideas in the topic under study. The students then work on each other's tasks' (p. 4). From their research, Clarke and Clarke (2002) concluded that this is considered excellent revision practice, and is particularly well received when the teacher makes a commitment to use at least some of the student-created tasks in the final assessment.

Other alternative assessment methods

Other alternative assessment methods are considered valuable in the instructional process. These are tabulated in Table 7.1 with brief description of the method (Watt, 2005, p. 26).

Table 7.1 Alternative assessment methods

Assessment method	Description of assessment method
Teacher observation	Teachers observe students in structured or unstructured activities and evaluate the quality of student task engagement.
Student journals	Students keep reflective accounts of their mathematics learning and processes of understanding, from which the quality of their task engagement and development may be explored by the assessor.
Oral tasks	Students give short answers, demonstrations, seminar presentations and debates.
Practical tasks	Students use instruments to apply or deduce mathematical principles.

Source: Watt, 2005, p. 26.

National Assessment Program in Literacy and Numeracy (NAPLAN)

The National Assessment Program in Literacy and Numeracy tests are given to students in Years 3, 5, 7 and 9 in all Australian states and territories annually. This standardised testing commenced in 2008 and is overseen by the Australian Curriculum and Reporting Authority (ACARA). NAPLAN tests are comprised of multiple-choice and short-answer items – and in more recent versions of the assessment there are two sections for students to complete: one that requires the use of a calculator and one that does not. Since its inception, NAPLAN testing has been met with healthy scepticism and criticism. It can be argued that high-stakes standardised testing can yield valuable data for individual schools, regions and states, as they 'measure all [mathematical] achievement against a common framework of content and questions' (Booker et al., 2014, p. 35). On the other hand, the broad scope of such testing limits the possibility for teachers to probe appropriately mathematical understanding. Additionally, these tests do not 'allow for the original thinking involved with problem solving, generalisations of ideas, and alternative ways to achieve likely answers' (Booker, et al., 2014, p. 35).

Scholars acknowledge that discussing test items with students can be used strategically to enhance mathematical thinking, promote learner confidence and resilience (Anderson, 2009; Martin, 2003). In particular, Anderson (2009) states that using NAPLAN test items within lessons can 'develop students' competence in reading mathematical text, to promote thinking strategies including estimation, and to evaluate alternative solutions for errors and misconceptions' (p. 22). To illustrate, Anderson (2009) outlines how a NAPLAN item can be used with a class of students (see Figure 7.6). To begin with, teachers should mention to students that many errors occur in items about percentages, and that 'multiple-choice items typically include common errors and misconceptions as alternative solutions' (Anderson, 2009, p. 20). In addition to highlighting the need for students to take time to think carefully about the problem, its solution and approaches to verify this solution, Anderson offers some questions to guide student thought:

- How much is the decrease in electricity bill?
- Is the decrease less or greater than 50% of the first electricity bill?
- What are the fraction equivalents for 25% and 33%? How much would the discount be for each of these?

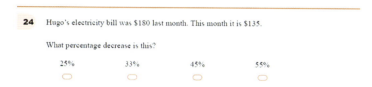

24   Hugo's electricity bill was $180 last month. This month it is $135.

What percentage decrease is this?

| 25% | 33% | 45% | 55% |

Figure 7.6 2008 Year 7 Numeracy calculator allowed test, question 24
Source: MCEETYA, 2008b.

ACTIVITY 7.3

Log in to HOTmaths and select the 'HOTmaths Global' Course list and then choose the 'Middle Secondary' Course. From the Topic list choose 'Index laws', then the 'Topic quiz' Lesson. From this page, select the 'Challenge' tab. Note that this content is generated randomly from a bank of questions each time it is accessed.

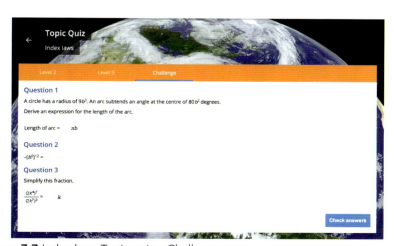

Figure 7.7 Index laws Topic quiz – Challenge

Locate the Year 7 Content Descriptions for the Australian Curriculum (Mathematics) Number and Algebra (ACMNA) strand, found at www .australiancurriculum.edu.au.

Working with a colleague, determine the extent to which this formative assessment meets the following ACMNA Content Descriptions:

- Number and place value: 149, 150, 151, 280
- Real Numbers: 152, 153, 154, 155, 156, 157, 158, 173.

Decide to what extent:

- the test sufficiently assesses the Content Descriptions
- the test sufficiently addresses the Level Description and the Proficiencies
- included questions could be modified or removed
- any questions could be included for all Content Descriptions to be assessed.

ACTIVITY 7.4

Log in to HOTmaths and select the 'HOTmaths Global' Course list and then choose the 'Middle Secondary' Course. From the Topic list choose 'Simultaneous linear equations', then the 'Topic quiz' Lesson. From this page, select the 'Challenge' tab. Note that this content is generated randomly from a bank of questions each time it is accessed.

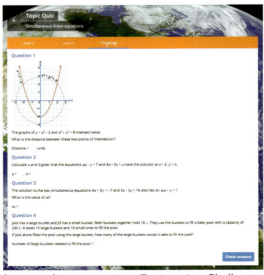

Figure 7.8 Simultaneous linear equations Topic quiz – Challenge

Working with a colleague, determine the extent to which this formative assessment meets the following ACMGM Content Descriptions:

* 038, 039, 040, 041, 042, 043, 044, 045, 046, 047

Decide to what extent:

* the test sufficiently assesses the Content Descriptions
* the test sufficiently addresses the Learning Outcomes
* included questions could be modified or removed
* any questions could be included for all Content Descriptions to be assessed.

REFLECTIVE QUESTIONS

Up until this point in time you would have completed many mathematics assessments. Recall the assessment item that had the most positive influence on you as a learner. What were the key characteristics of this assessment item? When you have a compiled a list of characteristics, compare these with a colleague. What did your comparison reveal?

Guidelines for creating assessments in secondary mathematics

Whatever model of instruction is selected to teach secondary mathematics, teachers inevitably must select and create appropriate assessments for students to demonstrate learning. During the first few years of teaching – or even teaching a new course or year level – creating completely original and valuable assessment tasks can be daunting. In addition to meeting the external curricular requirements prescribed by ACARA and Departments of Education, assessments should undergo an internal review process (via mathematics department colleagues) before administration to students. Such reviews can not only help detect simple typographical errors, but provide useful feedback regarding the procedural complexity, time constraints, feel of comprehension, balance of content – and depending on the class – accessibility for diverse learners. In this section some recommendations and guidelines are provided for secondary mathematics teachers to consider when creating assessment items. Such professional advice is drawn from educational authorities, professional associations and scholarly commentators who commonly underscore the importance of assessment within the mathematics classroom.

Principles for assessment

ACARA

ACARA has recommended assessment practices for teachers in Australian schools to follow. These practices are outlined as a series of steps, using the Australian Curriculum content and achievement standards (ACARA, 2015d):

1 Identify current learning and achievement levels.
2 Select appropriate content to teach.
3 Judge the quality of learning as demonstrated by students.
4 Review teaching programs for further instructional delivery.

Specifically, these practices are outlined as:

> Teachers use the Australian Curriculum content and achievement standards first to identify current levels of learning and achievement and then to select the most appropriate content (possibly from across several year levels) to teach individual students and/or groups of students. This takes into account that in each class there may be students with a range of prior achievement (below, at, and above the year level expectations) and that teachers plan to build on current learning.
>
> Teachers also use the achievement standards, at the end of a period of teaching, to make on-balance judgments about the quality of learning demonstrated by the students – that is whether they have achieved below, at, or above the standard. To make these judgments, teachers draw on assessment data that they have collected as evidence during the course of the teaching period. These judgments about the quality of learning are one source of feedback to students and their parents and inform formal reporting processes.
>
> If a teacher judges that a student's achievement is below the expected standard, this suggests that the teaching programs and practice should be reviewed to better assist individual students in their learning in the future. It also suggests that additional support and targeted teaching will be needed to ensure that the student does not fall behind. (ACARA, 2015d)

These practices align with the standard Plan – Teach – Evaluate model espoused by Barry and King (1998, p. 328) and with those practices offered in many teacher education programs.

The Australian Association of Mathematics Teachers (AAMT)

The Australian Association of Mathematics Teachers (AAMT) (2008) described how highly accomplished teachers use a range of assessment strategies that are appropriate, fair and inclusive, and which inform learning and action. First, the AAMT recommended that teachers should assess appropriately by matching the purpose to the information required, and by assessing the full range of learning goals via a range of strategies. Second, teachers should assess student learning in a fair and inclusive manner by:

- involving students in the processes for assessing their learning
- using assessment strategies and tasks that are as fair as possible to boys and girls, and inclusive of students from a variety of backgrounds
- assessing in ways that are clear and transparent
- making fair and inclusive judgments
- assessing through planned means and opportunities that arise in their work with students
- ensuring students are familiar with the genres of items used in their own assessment programs and in those of education authorities.

Third, the AAMT recommends that teachers use assessment of students' mathematics to inform learning and action. This can be achieved by reflecting on assessment information and subsequently planning student learning experiences, providing purposeful feedback to students about learning, and sharing assessment information with relevant key stakeholders.

Rich assessment tasks

According to Wiliam (2007), assessment should become a means of fostering growth in mathematics. Teachers must exercise care when designing assessments so that 'there is a balance in the questions that are asked, and that the range of problems reflects due importance given to each aspect of the topic under review' (Booker et al., 2014, p. 31). These authors offer broad recommendations for teachers designing assessments, including: ensuring that students can develop concepts in depth, and emphasising a wide range of problem-solving processes in reaching a solution. Additionally, practical activities (e.g. specific topics within measurement, geometry, statistics and probability) need 'to be seen to have an equal weighting with the written tasks associated with computation and problem solving' (Booker et al., 2014, p. 31). It is with the above comments in mind that the topic of *rich tasks* in mathematics can be discussed.

Rich tasks encourage students to think creatively, work logically, and communicate their ideas within a mathematical context (Piggott, 2015). Such tasks also provide opportunities for students to 'synthesise their results, analyse different viewpoints, look for commonalities, and evaluate findings' (Piggott, 2015). Various scholars suggest that rich assessment tasks contain certain characteristics (Ahmed, 1987; Clarke & Clarke, 2002; DFES, 2007; Piggott, 2015) including:

- be accessible to all learners
- engage the learner through different levels of challenge
- allow the learner to make decisions about which methods or approaches to use
- encourage students to disclose their own understanding of what they have learned
- allow students to show connections they are able to make between the concepts they have learned
- provide a range of student responses, including a chance for students to show all they know about the relevant content
- provide an opportunity for students to transfer knowledge from a known context to a less familiar one.

Figure 7.9 is an example of a rich mathematical task (Clarke & Clarke, 2002) that could be used with secondary students.

2. Area = Perimeter

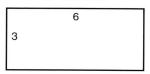

Consider the rectangle with dimensions 6 units by 3 units. We can calculate easily that the area is 18 square units and the perimeter is 18 units. So, if we ignore the units, *the magnitude of the area and perimeter are the same* for this shape.

Investigate the following questions, reporting what you find:

- Are there other rectangles that have this property? Please explain.

- Are there any circles for which area = perimeter? (ignoring units) Please explain.

- Are there any squares for which area = perimeter? (ignoring units) Please explain.

- (Extension task) Select another kind of shape (e.g., triangle, hexagon, etc.), and explore the situations in which the area is the same as the perimeter, ignoring units.

Figure 7.9 Rich assessment task
Source: Clarke & Clarke, 2002, p. 7.

Guidelines for teachers

Cangelosi (2003) offers some guidelines for teachers designing mathematics assessments. His systematic approach is offered in the way of four phases:

- clarifying the learning goal
- designing relevant mini-experiments and storing and accessing them via a computerised folder structure
- developing a test blueprint
- synthesising the test.

Clarifying the learning goal

Before designing an achievement test, teachers should answer the question 'achievement of what?' (Cangelosi, 2003). In doing so, the relevant learning goals for which the test is designed and developed must be clarified. After the learning goals are developed, allocate a weight for each learning goal according to its relative importance to the goal of the unit, chapter or syllabus.

Designing relevant mini-experiments and storing and accessing them via a computerised folder structure

Cangelosi (2003) recommends that for each unit or chapter, teachers create an electronic folder containing mini-experiments. These mini-experiments are questions and activities that link specific mathematical content to the unit learning objectives. Over time, this 'bank' of mini-experiments will become larger and creating tests will not only assist teachers in creating valid and useable tests, but to save them valuable time. For instance, such an electronic folder may take the hierarchical format of: Year 10 Mathematics > Number & Algebra > Chapter 1 > Learning Objectives > Mini-experiments. Additional folders would be created for the other two mathematical strands, chapters, learning objectives and mini-experiments.

Developing a test blueprint

According to Cangelosi (2003), a test blueprint is 'an outline specifying the features you want to build into a unit test you plan to develop from the prompts stored in your mini-experiments files' (p. 310). Such a test blueprint contains the following features:

- the title of the unit
- anticipated administration dates and times
- provisions for accommodating students with special needs

- approximate number and types of mini-experiments to be included
- an approximation of the maximum possible score for the measurement
- how points should be distributed among the objectives that define the goal (based on the weights of the objectives)
- the overall structure of the measurement
- for summative evaluations, the method for converting scores to grades.

Synthesising the test

The final stage of creating a test is to synthesise the measurement according to the variety of question formats and overall question complexity. Cangelosi (2003) recommends that teachers organise tests so that 'students have less-consuming formats (e.g. multiple choice) before responding to those that are more time consuming (e.g. essay)'. To counter this, grouping prompts together with the same format 'simplifies the directions and prevents students from having to reorient their thinking frequently due to changes in format' (Cangelosi, 2003, p. 314). Additionally, prompts using the same format should be sequenced from less difficult to more difficult. Sequencing prompts this way enables students to attempt a greater proportion of the test, rather than becoming 'stuck' on a hard prompt at any early point of the assessment.

Rubrics

A rubric (or scoring framework) is an assessment tool designed to assist planning, learning and evaluation processes. Rubrics provide students and teachers with established criteria about the intentions of a learning task which are linked to identified curriculum outcomes (Churchill et al., 2013; Siemon et al., 2015). Specifically, rubrics are comprised of two elements: the criteria about the learning task, and a scale or gradations of quality for each criterion (Lang & Evans, 2006). These elements are illustrated in the example below. Following the idea that rubrics communicate explicit pathways for developmental learning or graded performance, Churchill et al. (2013) argue that the effectiveness of rubrics rests in the specified 'content', or the development of performance statements. For instance, if a rubric is well written it 'can inform planning, strategy selection, communication and reporting, and many other aspects of the assessment process' (Churchill et al., 2013, p. 436). When designing performance assessments, Huetinck and Munshin (2008) recommend that teachers develop holistic rubrics immediately after writing the task, and well before giving the assessment to students. During the rubric-drafting phase, these authors stress the importance of using an even number of assessment points (e.g. 4 or 6). According to relevant literature of rubric development (Danielson; 1997; Stenmark, 1991), scorers are less discriminatory using a rubric with an even number of assessment points. Acknowledging that the use of an odd-numbered (e.g. 5-point) rubric may appear a more logical process to inevitably standardise grades, Huetinck and Munshin (2008) herein amplify:

> The natural tendency is to place many students in the middle scoring category (a 3 on a 5-point rubric) without examining the nuances of differences that indicate when more work is needed (1 or 2 on a 4-point rubric) versus communicating to students that they have shown satisfactory achievement of the learning goal (3 or 4 on a 4-point rubric). (2008, p. 367)

Other authors advise against the development of rubrics becoming a solitary task for teachers; rather, these should be created by teacher teams and in partnership with students (Lang & Evans, 2006).

Below is an example of a Statistics and Probability learning task that can be used for Year 10 students. The task is accompanied by a rubric that has been developed in accordance with the Australian Curriculum: Mathematics for the Year 10/Year 10A Data Representation and Interpretation content description. The task requires students to use their knowledge of statistical calculations (mean, median, mode, range, standard deviation), representations (stem and leaf diagrams, boxplots) and interpretations of data before making an informed judgment. The accompanying scoring rubric uses a four-point assessment scale with the performance criteria.

> **The lightbulb problem**: The maintenance staff of Happee High School wish to make a bulk purchase of 1000 light bulbs. To assist with this decision, the maintenance staff have obtained the results of longevity tests of three lightbulb brands, Brand A, Brand B and Brand C. The number of hours each brand of lightbulb will last (before blowing out) are displayed in this table.

Brand A	Brand B	Brand C
100	112	120
121	104	115
107	118	110
132	98	118
124	140	122
105	112	115
111	92	113
95	109	114

(*Continued*)

Brand A	Brand B	Brand C
135	111	116
126	96	112
112	115	114
120	145	115

It is also known that the prices per lightbulb are $0.38, $0.32 and $0.34 for Brands A, B and C respectively. Using the information from the table determine which brand of lightbulb should be purchased, justifying your answer with appropriate statistical calculations and representations.

Mathematical task: The lightbulb problem scoring rubric

Category	Score	Description
No response	0	The problem has not been attempted, or it has been restated without attempting any calculations.
Minimal	1	The response demonstrates a minimal knowledge of univariate statistics (mean, median, mode, range, standard deviation) with few of these measures of central tendency calculated correctly for each brand. The response indicates a graphical representation (e.g. a boxplot) containing several errors for each brand. A recommendation to purchase Brand C is justified as the cheapest option (by approximately $20) solely using calculations (no comparison of data from the three brands).
Emerging	2	The response demonstrates an emerging knowledge of univariate statistics (mean, median, mode, range, standard deviation) with some of these measures of central tendency calculated correctly for each brand. The response indicates a graphical representation (e.g. a boxplot) containing few errors for each brand. A recommendation to purchase Brand C is justified using a comparison of data from one or two brands, with calculations showing that it is approximately $20 cheaper to do so.
Satisfactory	3	The response demonstrates a satisfactory knowledge of univariate statistics (mean, median, mode, range, standard deviation) with most of these measures of central tendency calculated correctly for each brand. The response indicates a graphical representation (e.g. a boxplot) for each brand. A recommendation to purchase Brand C is justified using a comparison of data from one or two brands (standard deviation and range), with calculations showing that it is approximately $20 cheaper to do so.

Category	Score	Description
Excellent	4	The response demonstrates a thorough knowledge of univariate statistics (mean, median, mode, range, standard deviation) with all these measures of central tendency calculated correctly for each brand. The response indicates a graphical representation (e.g. a boxplot) for each brand. A recommendation to purchase Brand C is justified using a comparison of data from two or more brands (standard deviation and range), with calculations showing that it is approximately $20 cheaper to do so.

ACTIVITY 7.5

Log in to HOTmaths and select the 'HOTmaths Global' Course list and then choose the 'Early Secondary' Course. From the Topic list choose 'Length, perimeter & area', then the 'Areas of triangles' Lesson.

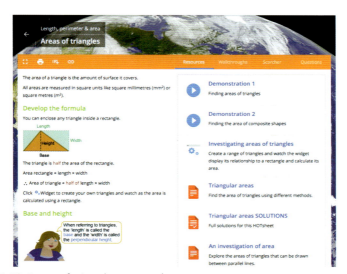

Figure 7.10 Areas of triangles screenshot

Use this lesson and its associated resources to develop a rich assessment task for upper secondary students. There are some HOTsheets located in the Resources which may be useful in creatively deciding how to find the areas of common and not-so-common shapes using triangles. You may also want to refer back to the key characteristics of rich assessment tasks before you get started.

ACTIVITY 7.6

Log in to HOTmaths and select the 'HOTmaths Global' Course list and then choose the 'Middle Secondary' Course. From the Topic list choose 'Polynomials', then the 'The remainder & factor theorems' Lesson.

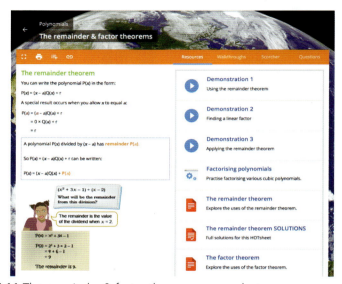

Figure 7.11 The remainder & factor theorems screenshot

Use this lesson and its associated resources to develop a rich assessment task for upper secondary students. There are some HOTsheets located in the Resources which may be useful in deciding how the remainder and factor theorems can be used to factorise polynomial functions. You may also want to refer back to the key characteristics of rich assessment tasks before you get started.

REFLECTIVE QUESTIONS

In this chapter you have looked at guidelines and principles for creating assessments for the secondary mathematics classroom. From what has been offered, together with your own professional learning and experience in education, devise a set of guidelines that will guide the way you plan, implement and evaluate assessments.

Providing feedback to students

Clarke (2001) argued that wherever possible, forms of assessment should be used that raise students' self-esteem, enable students to create success criteria and to organise students' individual targets. As such, the provision of carefully considered and delivered feedback can be instrumental in this regard. Feedback refers to 'the information gained from formal and informal assessment, provided to students, by teachers, about what the student has learnt, needs to learn, mistaken ideas and directions for improvement' (Hosking & Shield, 2001, p. 289). The key aim of feedback should be to bridge the gap between an actual level of performance and a desired learning goal (Ramprasad, 1983; Sadler, 1989). According to Latham et al. (2011), feedback on paper, spoken or conveyed through facial expressions can have a 'branding' effect. These authors suggest that verbal, gestural or written feedback has to be thoughtfully provided and it should be specific to the individual. Before providing feedback to a student, Latham et al. (2011) recommend that teachers ask themselves a number of questions about the comments that could be made (Latham et al., 2011). A list of hypothetical questions is included below, with corresponding examples of teacher-generated feedback. For instance:

- What in particular can I comment on? *Be careful to multiply first and then add in number sentences.*
- What in particular do I hope to clarify? *Between line three and line four of working out I had difficulty in following the steps you took.*
- What do I hope to extend? *Well done – could you have arrived at the same answer another way?*
- How would I feel about receiving that comment? *Your use of algebra here is great! With continued work in rearranging equations this result will continue to improve.*
- How can I help students improve what they have done? *Remember to multiply the indices of all bases (numbers and letters) within the parentheses by the index outside the parentheses.*

ACTIVITY 7.7

Log in to HOTmaths and select the 'HOTmaths Global' Course list and then choose the 'Middle Secondary' Course. From the Topic list choose 'Index laws', then the 'Topic quiz' Lesson. Here you will find tabs which will allow you to take a quiz that is divided into three levels: Level 2, Level 3 and

Challenge. Note that this content is generated randomly from a bank of questions each time it is accessed.

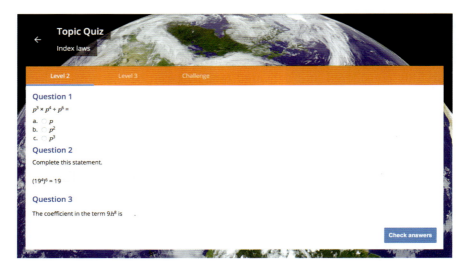

Figure 7.12 Index laws Topic quiz – Level 2

1 Complete the generated test, showing as much working out for each question as possible.
2 Critically examine the components of each question, looking for common errors that students could make with this topic (e.g. incorrect substitution, confusing x and y coordinates, rearranging an equation improperly).
3 For each 'common error' devise a feedback statement to provide to a student (if they made that particular error). Remember, this feedback statement must be succinct, focused on the error and how to correct it, and written in a positive and affirming tone.

ACTIVITY 7.8

Log in to HOTmaths and select the 'HOTmaths Global' Course list and then choose the 'Middle Secondary' Course. From the Topic list choose 'Logarithms', then the 'Topic quiz' Lesson. Here you will find tabs which will allow you to take a quiz that is divided into three levels: Level 2, Level 3 and

Challenge. Note that this content is generated randomly from a bank of questions each time it is accessed.

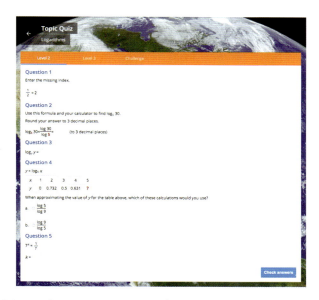

Figure 7.13 Logarithms Topic quiz – Level 2

1 Complete the generated test, showing as much working out for each question as possible.
2 Critically examine the components of each question, looking for common errors that students could make with this topic (e.g. incorrect substitution, confusing sides and vertices, rearranging an equation improperly).
3 For each 'common error' devise a feedback statement to provide to a student (if they made that particular error). Remember, this feedback statement must be succinct, focused on the error and how to correct it, and written in a positive and affirming tone.

REFLECTIVE QUESTIONS

The theory included in this chapter indicated that feedback must be specifically targeted to a student's learning – for instance, what the student has achieved, areas for improvement, and ways in which this improvement can be reached. Additionally, this feedback must raise students' self-esteem, which can initially seem difficult if mathematics teachers rely on written feedback.

Think back to when you were a student in a secondary mathematics class-room. Try to remember an assessment item where the teacher provided you with some constructive, specific and useful feedback. In recalling this feed-back, respond to the following questions and prompts:

1   How was the feedback was given to you (e.g. written or oral)?
2   For what assessment task was the feedback given?
3   Describe the teacher's feedback in terms of content.
4   How did this feedback make you feel about your efforts in secondary mathematics?
5   How did you respond to this feedback?
6   What advice would you give this teacher, based upon the feedback received?

Summary

Define the term 'assessment' and explain the function of assessment in learning

Assessment comprises the intentional and unintentional activities under-taken by teachers to gather information about student learning (knowledge, skills, dispositions). Assessment practices assist teachers with monitoring student learning and making instructional decisions, provide valuable insight to teachers for evaluating learning programs and generate valuable information for feedback and reporting purposes.

List and justify assessment types that can be used in mathematics learning

Common assessment techniques used in mathematical learning include: tests, quizzes, examinations, homework and investigations. Teachers can also measure student learning through the administration of alternative assess-ments, such as: student self-assessment, teacher observation, student jour-nalling, oral tasks or practical tasks. When deciding precisely how student progress will be measured, Leder, Brew and Rowley (1999) suggest teachers avoid an over-reliance on one form of assessment type.

Delineate and rationalise key principles for creating assessment tasks in mathematics learning

ACARA has recommended assessment practices for teachers in Australian schools to follow. These practices are outlined as a series of steps, using the Australian Curriculum content and achievement standards. In a position paper, the Australian Association of Mathematics Teachers (AAMT) described how highly accomplished teachers use a range of assessment strategies that are appropriate, fair and inclusive, and which inform learning and action. Various authors provide key principles for the development of rich mathematical tasks, which encourage students to think creatively, work logically and communicate their ideas within a mathematical context. These tasks are accessible and engaging to all learners who are involved through different levels of challenge and allow the learners to make decisions about which methods or approaches to use. Cangelosi (2003) offers some guidelines for teachers designing mathematics assessments: clarifying the learning goal, designing relevant mini-experiments and storing and accessing them via a computerised folder structure, developing a test blueprint, synthesising the test.

Recall appropriate methods of providing feedback to students

Wherever possible, forms of assessment should be used that raise students' self-esteem, enable students to create success criteria and to organise students' individual targets. Feedback refers to the information gained from formal and informal assessment, provided to students by teachers, about what the student has learnt, needs to learn, mistaken ideas and directions for improvement. To provide effective feedback to students, teachers should consider the following questions: What in particular can I comment on?, What in particular do I hope to clarify?, What do I hope to extend?, How would I feel about receiving that comment?, How can I help students improve what they have done?

Part 2

Learning and Teaching Key Mathematics Content

CHAPTER 8

Number and algebra

Learning outcomes

After studying this chapter, you should be able to:

- understand the algebraic concepts of the unknown, pronumeral and variable
- teach algebraic expression problems via analogical reasoning
- unpack the complexity of linear equations
- teach linear equations in a hierarchical level of complexity
- appreciate algebraic problem-solving in real-life contexts
- transfer algebraic problem-solving skills to the science curriculum.

Introduction

Many children and adults have reservations about algebra. Why? The abstractness of algebra partly lies in its use of variables and pronumerals. Nonetheless, algebra expressed in a mathematical principle to solve a range of problems is what makes it powerful (Kieran, 1992). Indeed, algebra is a topical theme in mathematics that requires an extensive use of problem-solving skills. Mathematics education researchers regard algebra skills as a 'gatekeeper' to higher-order mathematical thinking skills in advanced mathematics (Carpenter, Franke & Levi, 2003). Algebra skills are useful not only for solving real-life problems (e.g. 'If your father wants to increase your weekly allowance of $20 by 5%, what is your new allowance?') (Ngu, Yeung & Tobias, 2014), but are also transferrable to other curriculum domains such as physics and chemistry (e.g. 'A solution contains 1.1 g of sodium nitrate $NaNO_3$ in 250 ml, what is the molarity of this solution?') (Ngu & Yeung, 2012, 2013; Ngu, Yeung & Phan, 2015).

Despite the prominent role of algebra in the secondary mathematics curriculum, there is limited evidence of an efficient use of algebra, particularly in middle school students (Stacey & MacGregor, 1999). Such findings suggest that secondary students perceive algebra as a challenging topic to learn and master. To assist secondary students in building a foundation in algebra knowledge, this chapter will highlight several aspects of teaching and learning algebra based on the Australian Curriculum: Mathematics.

It is important to make a connection between various mathematical topics and algebra. For example, highlighting the link between fractions, percentages and decimals enhances the learning of structurally similar linear equations. Similarly, the development of prior knowledge (e.g. collect like terms, expand the bracket) impacts upon the learning of complex linear equations involving multiple solution steps. Proficiency in equation-solving skills paves the way for adopting an algebraic approach in solving real-life problems. The power of algebraic problem solving is also demonstrated in its transferability to the science curriculum.

KEY TERMS

- **Balance concept:** the operational line depicting a mathematical operation applied to both sides of the equation to preserve its equality (e.g. +2 on both sides).
- **Element interactivity:** the interaction between elements in the learning material.
- **Equal sign:** mathematical equivalence in the context of equation solving.
- **Generalisation:** a mathematical rule reflecting organised mathematical knowledge structure.
- **Inverse concept:** the operational line depicting an inverse application of a mathematical operation to preserve its equality (e.g. -2 becomes +2).
- **Learning by analogy:** mapping of the structural elements of a new problem (target) with a learned problem (source) is likely to result in analogical transfer.
- **Long-term memory:** unlimited capacity that stores a huge number of schemata.
- **Molarity:** number of moles of solute/volume in one litre of solution.

- **Operational line:** the application of a mathematical operation to change the state of the equation, but at the same time to maintain the equality of the equation.
- **Percentage change problems:** real-life problems involving an increase or decrease in percentage.
- **Percentage quantity:** a specific percentage of a quantity.
- **Pronumeral:** use of a letter to stand for a number in the context of equation solving.
- **Relational line:** the quantitative relation between elements where the left side of the equation equals the right side.
- **Schema:** a cognitive construct that permits us to treat multiple elements of information as a single element categorised according to the way in which it will be used.
- **Unknown:** an unknown is something whose value needs to be found.
- **Variable:** use of a letter to stand for different numbers.
- **Working memory:** limited capacity to process information.

The unknown, pronumeral and variable

Central to arithmetic is the concept of number sense, whereas central to algebra is an understanding of the concept of variable. A good starting point to learn algebra is to learn the meaning of the **unknown**, **pronumeral** and **variable.**

In algebraic thinking and learning, the unknown is represented by a symbol such as a letter (e.g. x). In the context of equation-solving, x stands for a number. For example, $x - 4 = 9$, where x stands for 13 so that the left side of the equation equals to the right side. Nonetheless, mathematics education researchers and educators tend to use the pronumeral instead of the unknown in equation solving. The word 'pronumeral' is derived from Latin word in which 'pro' means 'for' and 'numeral' means 'number'. In solving a simple linear equation such as $n + 2 = 8$, mathematics educators would regard n as a pronumeral because it stands for a number. More specifically, n stands for a specific number, which, in this case is 6 and not any other number. What about a variable? A variable is a letter that can stand for different numbers. For example, if y is a variable, then it can be any number (e.g. 1, 0, 5 or 7). Let us consider a formula such as $s = \frac{d}{t}$, where s stands for speed, d stands for distance, and t stands for time. Each of these letters, s, d and t, is a variable. The value of s will vary depending on the values of d and t. The concept of a variable whereby a letter stands for any number is central to algebraic thinking and learning.

Teaching the concept of variable

How shall we teach the concept of variable? Consider $2 + ? = 6$ in which '?' represents an unknown number. The sum of 2 and the unknown number will give a value of 6. Having learned the concept of an unknown number, the next step is to introduce a letter such as n to present the unknown number, $2 + n = 6$. In the context of equation solving, n is called a pronumeral. It is important to emphasise that we can use any letter to present the unknown number or the pronumeral, and the letter stands for a number or quantity. To further expand the idea of variable, we can assign students to work on activities involving growing patterns. The purpose is to highlight how a mathematical rule based on the concept of variable represents a powerful tool to solve problems in real-life contexts.

Consider the following problem situated in a real-life context (Cai, 2005, p. 340):

Sally is having a party.
The first time the doorbell rings, 1 guest enters.
The second time the doorbell rings, 3 guests enter.
The third time the doorbell rings, 5 guests enter.
The fourth time the doorbell rings, 7 guests enter.
The guests continue to arrive, with two more people than the previous time.

(i) How many guests will enter on the 10th ring? Explain how you found your answer.
(ii) Assuming the pattern continues, describe in words or write a rule to determine which doorbell ring had 99 guests arrive.
(iii) What is the benefit of having this rule?

Using g to represent the number of guests, and n to indicate when the doorbell rings (i.e. which doorbell), we can draw a table to show the growing pattern (Table 8.1). The number of guests that corresponds to the 10th ring is 19. Obviously, it will be a laborious process and therefore an inefficient strategy to continue the growing pattern in Table 8.1 in order to *determine which doorbell ring had 99 guests arrive.* In contrast, based on the relation portrayed in the growing pattern, we can generate a mathematical rule such as $g = 2n - 1$ to determine the value of g when the value of n varies or vice versa. This would be a far superior strategy than the use of the table to solve the problem. For example, by substituting $g = 99$ in the rule $g = 2n - 1$, we can easily obtain the value for n, which is 50. The benefit of the mathematical rule is obvious. It highlights the power of generalised information expressed as a mathematical rule to solve real-life problems. The number of guests (g) will vary depending on when the doorbell rings (n).

Table 8.1 The growing pattern depicting the relation between Doorbell (n) and guest (g)

n (doorbell)	1st	2nd	3rd	4th	5th	6th	7th	8th	9th	10th
g (guest)	1	3	5	7	9	11	13	15	17	19

Misconception about variables

It is not uncommon for students to regard a variable as an object rather than a number or quantity. To assist students in differentiating between a variable and an object, we can ask them to formulate an equation to solve the following problem:

> The cost of an avocado is $3, and the cost of a pear is $1. If I buy 2 avocados and 4 pears, what is the total cost?

Solution:

> Let a = number of avocados, and p = number of pears
> Total cost = $3a + $1p$
> = $3 × 2 + $1 × 4
> = $10

It is important to emphasise that a and p stand for variables and not for objects such as avocado and pear. Obviously, students can choose to use other methods to solve the problem instead of the algebra approach. Nonetheless, there are two reasons why it is important to emphasise algebraic problem solving. Firstly, the algebra approach allows the calculation of the total cost easily where a and p can stand for any number. Secondly, it is important to expose middle school students to algebraic problem solving wherever appropriate so as to pave the way for them to pursue senior mathematics, in which algebraic problem solving becomes a dominant component of the curriculum.

ACTIVITY 8.1

In my garden one day I saw a tiny plant with 4 leaves (Day 1). The following day, it had grown even more leaves. Each day I noticed it continued to grow in the same way (Wilkie & Clarke, 2014).

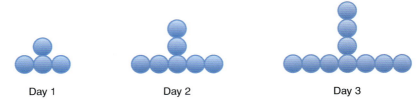

Day 1   Day 2   Day 3

Figure 8.1 Day 1, day 2, day 3
Source: Wilkie & Clarke, 2014

1 Draw a picture of what the plant will look like on Day 4.
2 Complete the table below.

Day	1	2	3	4
Number of leaves				

3 How many leaves will the plant have on Day 10?
4 Write a mathematical rule (equation) to show the relation between the Day (D) and Number of leaves (n) (i.e. what is the formula for this pattern?)
5 What is the benefit of having a mathematical rule?

ACTIVITY 8.2

Log in to HOTmaths and select the 'HOTmaths Global' Course list and then choose the 'Early Secondary' Course. From the Topic list choose 'Introducing algebra', then the 'What is algebra?' Lesson. Here you will find a range of resources which you can review to assist you in gaining insights into how you could introduce algebra to students.

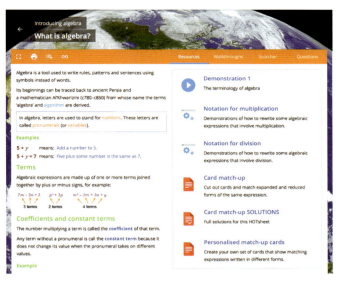

Figure 8.2 What is algebra? screenshot

REFLECTIVE QUESTIONS

1 There are six times as many students as professors at this university. Use *S* for the number of students and *P* for the number of professors. Write the equation for this problem (Clement, 1982). Most students wrote $6S = P$. Is this correct? Why?

2 Assume a correct algebraic equation is represented as $5x + 12x = 17x$. However, there may be a tendency for students to incorrectly represent the equation as $5x + 12 = 17x$ instead. Why? How can teachers help students to overcome such an error?

Algebraic expression problems

Prior to introducing simple linear equations, it is important to strengthen students' knowledge of algebraic expression problems. Structure mapping theory (Gentner, 1983) predicts that successful mapping of the structural elements of a new problem (target) with a learned problem (source) is likely to result in analogical transfer. **Learning by analogy**, underpinned by structure mapping theory, has provided a strong theoretical framework for research on mathematics education (Holyoak & Koh, 1987; Ngu & Yeung, 2012; Richland, Zur & Holyoak, 2007; Rittle-Johnson & Star, 2007, 2009).

Arithmetic manipulation and algebraic manipulation

By drawing on prior knowledge of number sense in the context of arithmetic manipulation, the learning of algebraic expression problems can be facilitated through algebraic manipulation (Table 8.2). In essence, analogical comparison between arithmetic manipulation and algebra manipulation enables the students to transfer their prior knowledge of number sense to learn algebraic expression problems.

Table 8.2 Analogical comparison between arithmetic manipulation and algebra manipulation

Arithmetic manipulation	Algebra manipulation
Addition and subtraction	
$2 + 0 = 2$	$a + 0 = a$
$2 - 0 = 2$	$a - 0 = a$
$2 + 5 = 5 + 2$	$a + b = b + a$
Multiplication and division	
$3 \times 1 = 3$	$a \times 1 = a$
$3 \div 1 = 3$	$a \div 1 = a$
$3 \times 4 = 4 \times 3$	$a \times b = b \times a$

Putting arithmetic manipulation and algebraic manipulation side-by-side reveals that algebra is analogous to the arithmetic. For example, as indicated in Table 8.2, there is one-to-one mapping of elements between $2 + 0 = 2$ and $a + 0 = a$ in terms of the underlying relational structure. The idea of analogical comparison can also facilitate the teaching and learning of the expansion of algebraic expression problems.

Arithmetic manipulation and algebraic expression manipulation

As shown in Table 8.3, there are two ways to calculate $2(5 + 1)$, and one of these is based on the expansion of the bracket. However, there is only one way to expand the algebraic expression problem, $3(a + 1)$. It is important to highlight the one-to-one mapping of elements between the expansion of 2(5 + 1) and 3(a + 1). This will help students to draw upon their prior knowledge of 2(5 + 1) to make sense of 3(a + 1).

Table 8.3 Analogical comparison between arithmetic manipulation and algebraic expression manipulation

Arithmetic manipulation		Algebraic expression manipulation
(i) 2(5 + 1) = 2 × 6 = 12	(ii) 2(5 + 1) = 2 × 5 + 2 × 1 = 12	(i) 3(a + 1) =3 × a + 3 × 1 = 3a + 3

ACTIVITY 8.3

Log in to HOTmaths and select the 'HOTmaths Global' Course list and then choose the 'Middle Secondary' Course. Next, choose 'Algebraic expressions' as your Topic. Here you will find six lessons that provide an overview of the teaching sequence of early algebra:

- Reviewing algebraic expressions
- Reviewing simplifying algebraic expressions
- Expanding & simplifying
- Factorising
- Using algebra
- A summary of working with algebraic expressions.

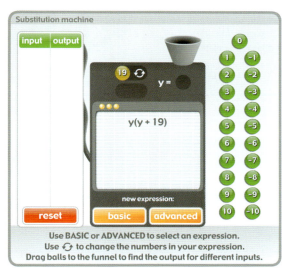

Figure 8.3 Substitution machine widget

Review each of these lessons and evaluate resources you could potentially utilise to illustrate early algebra concepts, such as the Substitution machine widget shown in Figure 8.3.

ACTIVITY 8.4

Log in to HOTmaths and select the 'HOTmaths Global' Course list and then choose the 'Middle Secondary' Course. From the Topic list choose 'Algebraic expressions', then the 'A summary of working with algebraic expressions' Lesson. This lesson contains widgets and a HOTsheet which will enable you to evaluate your general content knowledge of algebraic expressions.

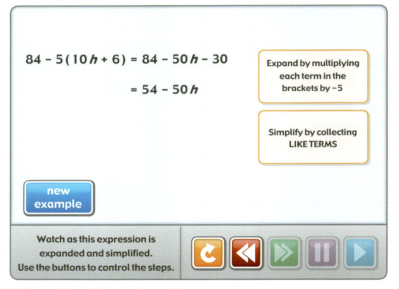

$$84 - 5(10h + 6) = 84 - 50h - 30$$
$$= 54 - 50h$$

Expand by multiplying each term in the brackets by −5

Simplify by collecting LIKE TERMS

new example

Watch as this expression is expanded and simplified. Use the buttons to control the steps.

Figure 8.4 Expanding with negatives widget

REFLECTIVE QUESTION

Draw a picture to show the expansion of both $2(5 + 1)$ and $3(a + 1)$. Highlight the similarity and difference between the two and provide reasons for this.

The classification of linear equations

Mathematics educators regard equation solving as a basic skill (Ballheim, 1999). Equation solving acts as a bridge between pre-algebra and algebra learning. The transition from arithmetic to algebraic reasoning is notoriously difficult to achieve (National Research Council, 2001). Therefore, a renewed focus on classifying linear equations provides a much-needed framework for teaching and learning linear equations.

Previous work on classifying linear equations

Mathematics education researchers have classified linear equations as one-step equations (e.g. $2x = 8$), two-step equations (e.g. $-5x + 2 = 17$), and equations with variables on both sides (e.g. $9 - x = 2x - 6$) (Matsuda et al., 2013). The classification of linear equations as one-step and two-step equations is based on the number of operations required to solve these equations. For example, performing a single operation such as $\div 2$ on both sides is required to solve the one-step equation such as $2x = 8$. Other mathematics education researchers have classified linear equations as multi-step equations when they either involve a bracket (e.g. $4(x + 1) = 20$) or variables on both sides plus brackets (e.g. $5(x - 3) = 2(x - 3) + 16$) (Rittle-Johnson & Star, 2007, 2009).

Simple linear equations have been classified as one-step (e.g. $x + 2 = 6$) and two-step equations ($2x + 4 = 10$), whereas more complex linear equations have been classified as equations with grouping symbols (e.g. $4(x + 1) = 12$) and equations with pronumerals on both sides (e.g. $5a + 7 = 2a - 1$) (McSeveny, Conway & Wilkes, 2004). Recently, mathematics educators have extended the classification of linear equations based on the number of operations required to solve equations by including three-step and four-step equations (Vincent et al., 2011). Interestingly, linear equations with a negative pronumeral (e.g. $6 - x = 15$) have been classifed as 'equations with negative numbers' (Vincent et al., 2011, p. 411) or 'one-step equations with negative integers' (McSeveny et al., 2004, p. 245).

Clearly, up to this point, neither mathematics education researchers nor mathematics educators have provided a systematic way of classifying linear equations in a hiearchical level of complexity. Based on cognitive load theory, the use of **operational** and **relational lines** to classify linear equations has captured the complexity of linear equations hierarchically (Ngu & Phan, 2016b).

Relational line and operational line

Cognitive load theory (Sweller, 2012; Sweller, Ayres & Kalyuga, 2011; Sweller, van Merrienboer & Paas, 1998) provides guidelines to design effective instructions across various discplines. Central to cognitive load theory is how people process information and learn. The interaction between elements within learning material constitutes element interactivity, and therefore the intrinsic cognitive load. An element refers to anything that needs to be learned. The complexity of learning material is proportionate to the degree of element interactivity.

To illustrate the complexity of linear equations, we use relational and operational lines to describe the solution procedure of linear equations. Figure 8.5 depicts the solution procedure of $x + 6 = 11$ via the balance method. A relational line (Lines 1 and 2) indicates the relation between elements so that the left side of the equation equals to the right side. An operational line (Line 2) refers to the performance of the same operation (e.g. –6 on both sides) to alter the problem state of the equation and yet at the same time to maintain the equality of the equation.

Line 1 $x + 6 = 11$ (–6 on both sides)
Line 2   –6 –6
Line 3 $x = 5$

Figure 8.5 Solution procedure of a simple equation

As depicted in Figure 8.5 (see also Table 8.4), Line 1 refers to a relational line. It comprises three elements (x, +6, 11) and three concepts:

1 $x + 6 = 11$ represents an algebraic sentence in which x (pronumeral) can be replaced by a number so that the left side of the equation equals to the right side.
2 The **equal sign** (=) describes a relationship between the left and right sides so that they are equal.
3 To find x, the learner needs to perform the same operation on both sides (what is done on the left side should also be done on the right side) to balance the equation.

The learner needs to manipulate the interaction between these three elements and concepts simultaneously. Line 2 refers to an operational line. It has one element (–6). The learner needs to perform an operation (–6) on both sides (cancel –6 with +6 on the left side, the same –6 must be done on the right side) in order to maintain the equivalence of the equation. Line 3 refers to a relational line and it has two elements (x, 5). If the learner can process the preceding Lines 1 and 2 successfully, then x equals 5 being the solution would seem obvious.

Table 8.4 Solution procedure: one operational line and two relational lines, two operational lines and three relational lines

One operational line and two relational lines, two operational lines and three relational lines				
Solution procedure	Number of operational line	Number of relational line	Type of element	Comment
$x + 2 = 11$ $-2 \quad -2$ $x = 9$	1	2	simple	• positive number: 2, 11
$a - 3 = -7$ $+3 \quad\quad +3$ $a = -4$	1	2	complex	• negative number: −3, −7
$\dfrac{p}{8} = 4$ $\times 8 \quad \times 8$ $p = 32$	1	2	complex	• fraction • positive number
$\dfrac{1}{3}x = 5$ $\times 3 \quad \times 3$ $x = 15$	1	2	complex	• fraction • positive number
$\dfrac{x}{0.5} = 7$ $\times 0.5 \quad \times 0.5$ $x = 3.5$	1	2	complex	• fraction • decimal number: 0.5
$\dfrac{6}{a} = 2$ $\times a \;\times a$ $6 = 3a$ $\div 3 \;\div 3$ $2 = a$	2	3	complex	• pronumeral as a denominator: $\dfrac{6}{a}$
$8 - n = 0$ $-8 \quad\quad -8$ $-n = -8$ $\div(-1) \;\div(-1)$ $n = 8$	2	3	complex	• negative pronumeral: −n

Note: when the number of operational lines and relational lines is kept constant, the presence of complex element increases its complexity. For example, $\frac{x}{0.5}$ is more complex than $\frac{p}{8} = 4$ owing to the presence of a decimal number in the former.

Table 8.4 (*cont.*)

Two operational lines and three relational lines, two operational lines and four or five relational lines				
Solution procedure	Number of operational line	Number of relational line	Type of element	Comment
$2b + 1 = 5$ $-1 \qquad -1$ $2b = 4$ $\div 2 \quad \div 2$ $b = 2$	2	3	simple	• positive number: 2, 1, 5
$15 - 7m = 1$ $-15 \qquad -15$ $-7m = -14$ $\div(-7) \quad \div(-7)$ $m = 2$	2	3	complex	• negative pronumeral: $-7m$ • manipulation results in a negative number: $1 - 15$
$\frac{y}{4} + 3 = 1$ $-3 \quad -3$ $\frac{y}{4} = -2$ $\times 4 \ \times 4$ $y = -8$	2	3	complex	• fraction: $\frac{y}{4}$ • manipulation results in a negative number: $1 - 3$
$0.3p - 1 = 5$ $3p - 10 = 50$ $+10 \qquad +10$ $3p = 60$ $\div 3 \quad \div 3$ $p = 20$	2	4	complex	• a decimal coefficient of the pronumeral: $0.3p$
$2(q + 3) = 16$ $2q + 6 = 16$ $-6 \qquad -6$ $2q = 10$ $\div 2 \ \div 2$ $q = 5$	2	4	simple	• expand the parentheses: $2(q + 3)$

$6(b + 1) - 3b = 21$ $6b + 6 - 3b = 21$ $3b + 6 = 21$ $\quad -6 \qquad -6$ $3b = 15$ $\quad \div 3 \quad \div 3$ $b = 5$	2	5	complex	• expand the parentheses: $6(b + 1)$ • simplify: $6b - 3b$

Note: The presence of a complex element increases the number of relational line ($2b + 1 = 5$ vs. $0.3p - 1 = 5$). One extra term also increases the number of relational lines ($2(q + 3) = 16$ vs. $6(b + 1) - 3b = 21$).

Three operational lines and four relational lines, three operational lines and five relational lines				
Solution procedure	**Number of operational line**	**Number of relational line**	**Type of element**	**Comment**
$\dfrac{6 - x}{3} = 4$ $\quad \times 3 \qquad \times 3$ $6 - x = 12$ $\quad -6 \qquad -6$ $-x = 6$ $\div(-1) \quad \div(-1)$ $x = -6$	3	4	complex	• negative pronumeral plus fraction: $\dfrac{6 - x}{3}$
$6p + 1 = 4p + 5$ $\quad -4p \qquad -4p$ $2p + 1 = 5$ $\quad -1 \qquad -1$ $2p = 4$ $\div 2 \quad \div 2$ $p = 2$	3	4	Complex	• operation involves pronumaral: $-4p$

(Continued)

Table 8.4 (*cont.*)

$5(k + 2) = 2k + 12$ $5k + 10 = 2k + 12$ $-2k \qquad -2k$ $3k + 10 = 12$ $-10 \qquad -10$ $3k = 2$ $\div 3 \quad \div 3$ $k = \dfrac{2}{3}$	3	5	complex	• expand parentheses: $5(k + 2)$ • operation involves pronumaral: $-2k$
$3(n + 7) = 5(n + 3)$ $3n + 21 = 5n + 15$ $-5n \qquad -5n$ $-2n + 21 = 15$ $-21 \qquad -21$ $-2n = -6$ $\div 2 \quad \div 2$ $n = 3$	3	5	complex	• expand two sets of parentheses: $3(n + 7)$, $5(n + 3)$ • operation involves pronumaral: $-5n$ • coefficient of pronumeral: $5n$ (left side) $> 3n$ (right side) • manipulation results in a negative number: $15 - 21$

Note: Expanding parentheses increases the number of relational lines ($6p + 1 = 4p + 5$ vs. $5(k + 2) = 2k + 12$)).

Naturally, the higher the number of operational and relational lines, the higher the degree of element interactivity, and thus the complexity of the linear equations (Table 8.4). For example, as shown in Table 8.4, the equation $\frac{6}{a} = 2$, involving two operational and three relational lines, is more complex than $x + 6 = 11$ (Figure 8.5), which involves one operational and two relational lines. While the number of operational and relational lines acts as a point of reference to capture the complexity of the linear equations, the complexity of linear equations is also affected by the nature of the element (simple or complex). In mathematics learning, the presence of a complex element inevitably increases its complexity. It is a known fact that operating negative numbers (Ayres, 2001; Linchevski & Williams, 1999) and fractions (Cramer & Wyberg, 2009) poses a challenge to students. Moreover, students often fail to mathematically reason and make connections between fractions, decimals and percentages (Parker &

Leinhardt, 1995). Therefore, when the number of operational and relational lines is kept constant, the presence of a complex element will increase the complexity of the linear equations.

ACTIVITY 8.5

For the three linear equations below: (1) identify the number of operational and relational lines, and (2) identify the nature of the element (simple or complex). Can you see a pattern here? What is the significance of the pattern in terms of teaching and learning linear equations?

1   $4(2 - w) = 4$
2   $2(r + 5) - 7 = 11$
3   $8(p + 2) - 6p - 4 = 18$

ACTIVITY 8.6

How would you teach $15\%x = 30$? Is it important to teach students this type of equation? Why or why not?

REFLECTIVE QUESTIONS

1   How would you classify $x - 12\%x = 72$? What is the number of operational and relational lines in the equation? How would you teach this type of equation?
2   Would this equation $5m = 2m + 1$ pose a challenge to students? Why or why not?

Teaching and learning linear equations in a hierarchical level of complexity

A previous study (Ngu, Chung & Yeung, 2015) has indicated that the inverse method is more efficient than the balance method, particularly for linear equations involving multiple solution steps (Figure 8.6). Other mathematics education researchers (Rittle-Johnson & Star, 2007, 2009) have found that the flexible method is more efficient than the distributive method for equations involving a bracket (Figure 8.7). In terms of the teaching and learning of linear equations, what are the implications of the number of operational and relational lines, the nature of the element and the availability of different equation-solving methods?

Balance method			Inverse method		
Line 1	$x + 3 = 7$	(−3) on both sides	Line 1	$x + 3 = 7$	(+3 becomes −3)
Line 2	−3 −3		Line 2	$x + 3 = 7{-}3$	
Line 3	$x = 4$		Line 3	$x = 4$	

Figure 8.6 The balance and inverse methods to solve a simple equation involving one operational and two relational lines

Flexible method			Distributive method		
Line 1	$3(x + 1) = 6$ (÷3 on both sides)		Line 1	$3(x + 1) = 6$ (expand the parentheses)	
Line 2	$(x + 1) = 2$ (−1 on both sides)		Line 2	$3x + 3 = 6$ (−3 on both sides)	
Line 3	−1 −1		Line 3	−3 −3	
Line 4	$x = 1$		Line 4	$3x = 3$ (−3 on both sides)	
			Line 5	÷3 ÷3	
			Line 6	$x = 1$	

Figure 8.7 The flexible and distributive methods in solving a complex equation involving multiple steps

Sequencing linear equations

Our human cognitive architecture comprises a working memory that can only process four or five elements at any given time (Miller, 1956), and a **long-term memory** that stores a huge number of **schemata** (learned information) in varying degrees of specification. In the mathematics domain, a schema is defined as a cognitive construct that permits us to treat multiple elements of information as a single element categorised according to the way in which it will be used (Pawley et al., 2005, p. 76). The constraint of working memory decreases when processing a schema retrieved from the long-term memory. Accordingly, the benefit of sequencing linear equations is in line with the impact of prior knowledge of a simpler linear equation upon the teaching and learning of a more complex equation (van Merrienboer, Kirschner & Kester, 2003).

A more complex linear equation requires an extra solution step (or steps) than the preceding simpler linear equation. In effect, there is a hierarchical level of schemas spanning from a lower-level schema of simple linear equations to a higher-level schema of complex linear equations. The lower-level schema of simple linear equations is embedded within the higher-level schema of complex linear equations.

Figure 8.8 depicts the solution procedure of a more complex equation (b) preceded by a simpler equation (a). For example, the equation $\frac{n}{3}+2=8$ (two operational and three relational lines) is more complex than the equation $\frac{x}{2}=7$ (one operational and two relational lines). Once the students have acquired the schema for the equation $\frac{x}{2}=7$, this schema will be stored in the long-term memory. Subsequently, when presented with the equation $\frac{n}{3}+2=8$, the students can retrieve the schema for the equation $\frac{x}{2}=7$, which is the same as $\frac{n}{3}=6$. Mathematics educators can draw students' attention to map Lines 1, 2 and 3 in the equation $\frac{x}{2}=7$, which are similar to Lines 3, 4 and 5 in the equation $\frac{n}{3}=6$. The learning of a more complex equation $\frac{n}{3}+2=8$, becomes the learning of Lines 1 and 2 only. Therefore, this manner of teaching and learning the complex equation $\frac{n}{3}+2=8$ is beneficial because it builds upon the prior knowledge of a simpler equation, equation $\frac{x}{2}=7$, thus reducing working memory load.

(a)

Line 1 $\quad \frac{x}{2}=7$ (×2 on both sides)

Line 2 $\quad \frac{x}{2} \times 2 = 7 \times 2$

Line 3 $\quad x = 14$

(b)

Line 1 $\quad \frac{n}{3}+2=8$ (−2 on both sides)

Line 2 $\quad \frac{n}{3}+2-2=8-2$

Line 4 $\quad \frac{n}{3}=6$ (×3 on both sides)

Line 5 $\quad \frac{n}{3} \times 3 = 6 \times 3$

Line 6 $\quad n = 18$

Figure 8.8 Use of the balance method to solve (a) and (b). Lines 1, 2, and 3 in (a) are similar to lines 3, 4 and 5 in (b).

Similarly, the teaching and learning of a more complex equation $5x - 8 = 3x + 4$ (three operational and four relational lines) can be built upon the prior knowledge of the simpler equation $3x - 4 = 5$ (two operational and three relational lines) (Figure 8.9), thus reducing the burden on working memory

(a)

Line 1 $\quad 3x - 4 = 5$ (+4 on both sides)

Line 2 $\quad 3x - 4 + 4 = 5 + 4$

Line 3 $\quad 3x = 9$ (−3 on both sides)

$\quad ÷3 \quad ÷3$

Line 4 $\quad x = 3$

(b)

Line 1 $\quad 5x - 8 = 3x + 4$ (−3x on both sides)

Line 2 $\quad 5x - 3x - 8 = 3x - 3x + 4$

Line 3 $\quad 2x - 8 = 4$ (+8 on both sides)

Line 4 $\quad 2x - 8 + 8 = 4 + 8$

Line 5 $\quad 2x = 12$ (÷2 on both sides)

$\quad ÷2 \quad ÷2$

Line 6 $\quad x = 6$

Figure 8.9 Use of the balance method to solve (a) and (b). Lines 1, 2, 3 and 4 in (a) are similar to lines 3, 4, 5 and 6 in (b).

load. Mapping the operational and relational lines between $5x - 8 = 3x + 4$ and $3x - 4 = 5$ reveals that Lines 1, 2, 3 and 4 in $3x - 4 = 5$ are similar to Lines 3, 4, 5 and 6 in $5x - 8 = 3x + 4$. Therefore, the learning of $5x - 8 = 3x + 4$ becomes the learning of Lines 1 and 2 only.

In summary, sequencing linear equations in a hierarchical level of complexity based on the number of operational and relational lines allows mathematics educators to map structurally similar operational and relational lines across linear equations. Ultimately, the learning of a more complex equation is preceded by a simpler equation. This would be an efficient way of teaching and learning linear equations because it will impose relatively little cognitive load.

Structurally similar linear equations

The nature of the element will affect the complexity of linear equations when the number of operational and relational lines is kept constant. Consider two equations such as $\frac{y}{3} = 2$ and $\frac{x}{0.5} = 9$. Both equations have one operational and two relational lines; and, they share a similar structure and therefore the same degree of element interactivity. Nonetheless, the presence of a complex element such as a decimal in the equation $\frac{x}{0.5} = 9$ would render it more complex than the equation $\frac{y}{3} = 2$. Because both equations are structurally similar, the equation $\frac{x}{0.5} = 9$ is analogous to the equation $\frac{y}{3} = 2$. That is, there is one-to-one match in elements (e.g., 9 matches with 2). The teaching and learning of structurally similar equations should begin with a simpler equation. In this case, it should begin with the equation $\frac{y}{3} = 2$. Once the students have acquired schema for the equation $\frac{y}{3} = 2$, they can retrieve the same schema to solve a structurally similar equation such as $\frac{x}{0.5} = 9$ via reasoning by analogy. Figure 8.10 shows pairs of structurally similar equations in which the presence of a complex element in the complex equation poses a challenge to students. To overcome this, mathematics educators could engage students in analogical reasoning to learn structurally similar equations.

Simpler equation (source)	Complex equation (target)	Complex element in the complex equation
1. $6x = 36$	$10\%x = 20$	percentage
2. $m + 8 = 12$	$p + \frac{1}{2} = 2$	fraction
3. $2b + 1 = 3$	$0.2x + 0.1 = 0.3$	decimal
4. $\dfrac{q + 6}{2} = 9$	$\dfrac{x + 2.2}{1.4} = 5$	decimal
5. $6b + 1 = 11$	$4x + \dfrac{3}{5} = 1$	fraction

Figure 8.10 Pairs of structurally similar equations

Equation-solving methods

The balance method (Figure 8.6) is a popular method for teaching equation-solving in Western countries. In contrast, some Asian countries prefer the inverse method (Figure 8.6) for equation-solving. A review by Cai et al. (2005) indicates that some Asian countries (e.g. Singapore, China, South Korea) have introduced the inverse method in their primary mathematics curriculum. What is the difference between the balance and inverse methods? As shown in Figure 8.6, the main difference between the balance and inverse methods lies in the operational line (–3 on both sides vs. +3 becomes –3). Regarding the balance method, interaction between elements occurs on both sides of the equation; however, for the inverse method, the interaction between elements occurs on one side of the equation. Thus, for each operational line, the inverse method only incurs half as many interactive elements as the balance method. In the study conducted by Ngu et al. (2015), the inverse method was better than the balance method for equations involving more than one operational and two relational lines. In particular, as shown in Figure 8.11, the inverse method was better than the balance method for equations involving a negative pronumeral. The inverse method incurs fewer operational lines (1 vs. 2) and relational lines (2 vs. 3) than the balance method. Therefore, the inverse method is more efficient than the balance method because it involves a lower degree of element interactivity.

Balance method		Inverse method	
$6 - p = 0$	(–6 on both sides)	$6 - p = 0$	(–p becomes +p)
$-p = -6$	[note: $-p$ means $(-1 \times p)$]	$6 = p$	
$\div(-6) \quad \div(-6)$	($\div(-1)$ on both sides)		
$p = 6$			

Figure 8.11 The balance and inverse methods to solve an equation involving a negative pronumeral

Nonetheless, mathematics educators in Western countries tend to advocate the balance method because the **balance concept** (e.g. –3 on both sides) highlights the 'balance' in which the same must be done on both sides of the equation to maintain the equality of the equation. They tend to view the inverse method as 'change side, change sign', which may not adequately address the equality of the equation after performing the **inverse concept** (e.g. move +3 on the left side of the equation to become –3 on the right side). Nonetheless, the inverse concept is linked to primary numeracy, for example: $3 + 5 = 8$ is the same as $3 = 8 - 5$ and $5 = 8 - 3$ (Warren & Cooper, 2005). Prior knowledge of the inverse concept is likely to help students learn the inverse concept in the context of equations. A recent study by Ngu and Phan (2016a) confirms such a prediction. Students were able to judge pairs of equations (e.g. $x + 2 = 8$, $x = 8 - 2$) as equivalent after being exposed to either the balance method or the inverse method. The ability to judge such pairs of equations as equivalent reflects the grasp of the conceptual knowledge involved in equation solving. The findings indicate that students would not be disadvantaged if they were exposed to the inverse method. Therefore, the use of the inverse method as an additional method to the balance method will likely benefit student learning of linear equations.

Researchers have provided empirical evidence in developing flexibility in solving equations involving a bracket (Rittle-Johnson & Star, 2007, 2009). For example, there are two ways to solve the equation $3(x + 1) = 6$ (Figure 8.7). The first step of the flexible method is to divide both sides of the equation by 3; whereas the first step of the distributive method is to expand the bracket. The researchers have found that the flexible method is more efficient than the distributive method. Such findings are in line with the impact of the number of operational and relational lines upon the complexity of the linear equations. The flexible method incurs fewer operational lines (1 vs. 2) and relational lines (3 vs. 4) than the distributive method. Therefore, mathematics educators should endeavour to promote the development of flexibility in equation-solving.

ACTIVITY 8.7

Log in to HOTmaths and select the 'HOTmaths Global' Course list and then choose the 'Middle Secondary' Course. From the Topic list choose 'Linear equations & inequalities', then the 'Linear equations with brackets' Lesson.

From the Resources tab, select the 'Brackets in equations' HOTsheet and compare the distributive and flexible methods in doing the tasks. Which of these methods is easier in terms of the number of operational and relational lines in the solution procedure? Why?

ACTIVITY 8.8

Log in to HOTmaths and select the 'HOTmaths Global' Course list and then choose the 'Middle Secondary' Course. From the Topic list choose 'Linear equations & inequalities', then the 'Linear equations with fractions' Lesson.

In the 'Equations with a single fraction' section of the lesson, the solution procedures for the two examples are both presented via the balance method. In the 'Equations with more than one fraction' section of the lesson, the solution procedure is also presented using the balance method.

Solve these equations with fractions using the inverse method. Which method do you prefer? Why?

Equations with a single fraction

Click ✿ Widget to work through the method of solving linear equations containing one fraction.

EXAMPLE 1

Solve $\dfrac{2k-3}{3} = 2$

SOLUTION

$\dfrac{2k-3}{3} = 2$ Multiply both sides by 3.

$2k - 3 = 6$ Add 3 to both sides.

$2k = 9$ Divide both sides by 2.

$k = 4\dfrac{1}{2}$

EXAMPLE 2

Solve $\dfrac{h-4}{2} + 6 = h$

SOLUTION

First multiply **all terms** on both sides by 2 then solve.

$$\boxed{\dfrac{h-4}{2}} + \boxed{6} = \boxed{h}$$
$$\times 2 \quad \times 2 \quad \times 2$$

$h - 4 + 12 = 2h$

$h + 8 = 2h$

$8 = h$

$h = 8$

Equations with more than one fraction

Click on ✿ Widget for a demonstration of solving linear equations containing more than one fraction.

EXAMPLE

Solve $\dfrac{3r+7}{2} = \dfrac{r}{3}$

SOLUTION

Multiply both sides of the equation by the lowest common denominator: 2 times 3.

$\dfrac{3r+7}{2} = \dfrac{r}{3}$ Multiply both sides by 6.

$6 \times \left(\dfrac{3r+7}{2}\right) = 6 \times \left(\dfrac{r}{3}\right)$

$3 \times (3r + 7) = 2 \times r$

$9r + 21 = 2r$ Subtract $2r$ from both sides.

$7r + 21 = 0$ Subtract 21 from both sides.

$7r = -21$ Divide both sides by 7.

$r = -3$

To check, substitute $r = -3$ into the equation.

LHS = $(3 \times -3 + 7) \div 2 = -1$

RHS = $-3 \div 3 = -1$

Figure 8.12 Equations with single/more than one fraction screenshot

ACTIVITY 8.9

Equations involving fractions often pose a challenge to students. Solve the following equations with fractions in two ways. Record the number of operational and relational lines for each way. Which way is more flexible and efficient? Why?

1   $\dfrac{x}{2} + \dfrac{x}{5} = 14$

2   $\dfrac{x}{6} = \dfrac{2}{3}$

REFLECTIVE QUESTION

How would you teach an equation such as 12%x = 40? Support your answer with appropriate learning theory.

Algebraic problem-solving in real life contexts

The use of a mathematical rule to solve growing pattern problems reveals the power of algebra in solving word problems in real life contexts. Indeed, in a cross-cultural comparative study, Cai (2005) found that Chinese students outperformed students from the United States on certain types of word problems because the Chinese students relied on algebra strategies. In view of the merit of algebra in problem solving, how can mathematics educators help students to acquire skills in algebraic problem solving? Mayer (1985) proposed five types of knowledge to solve word problems:

- linguistic
- factual
- schematic
- strategic
- algorithmic.

According to Mayer (1985), the main hurdle to solve word problems lies in the learners' ability to engage in problem translation and integration to gain a solution. Problem translation requires the use of linguistic and factual knowledge to translate each proposition from the problem text in light of its context. Schematic knowledge is needed to set up an equation integrating relevant information (values, variables and their relationship). The computation of a solution requires strategic and algorithmic knowledge (or equation-solving skills). Successful and unsuccessful problem solvers undertaking word problems generally differed in their ability to represent word problems as a mathematics-specific abstraction (or an equation) that could generate a problem solution (Hegarty, Mayer & Monk, 1995). In light of the findings, one of the challenges to solve word problems is the translation of the problem situation for the subsequent formulation of an equation to find the unknown variable. Because solving the equation to find the unknown variable represents part of the algebraic problem-solving process, gaining proficiency in equation-solving skills presents another challenge to learners.

There are different types of algebra word problems in real-life contexts. Mayer (1981) classified 1097 algebra word problems based on: (1) families that represent the source of formula (e.g. *distance = rate × time*), (2) categories that represent the story line (e.g. *motion*) and (3) templates that represent a specific propositional structure (e.g. *round trip*). In this chapter, we highlight the teaching and learning of percentage change problems in real-life contexts.

Teaching and learning percentage change problems

Our human cognitive architecture comprises a long-term memory which stores a large number of schemas, and a working memory which is restricted in processing novel information but not schemas that can be retrieved from the long-term memory (Sweller, 2012). In light of the respective roles of the working memory and long-term memory, the design of instructional material needs to minimise the number of interacting elements in the working memory, particularly for complex material imposing high **element interactivity**. One strategy to reduce the degree of element interactivity is by building on the learners' existing schemas (prior knowledge) that are already in the long-term memory (Ngu et al., 2014).

Consider a word problem in the real-life context of a junior secondary school: *Last semester John scored 70 marks for a mathematics test. He has improved his mathematics marks by 10% this semester. Find John's mathematics marks for this semester.* The flexibility of algebraic problem solving allows mathematics educators to use two algebraic approaches to solve the word problem.

The equation approach 1

The teaching and learning of the equation approach 1 is a two-part process. The first part focuses on revising the prior knowledge of the **percentage quantity** (e.g. 72 × 12%). The second part focuses on the translation of the problem situation and the formulation of an equation to generate a problem solution.

Part 1: review of the percentage quantity

$\frac{1}{2}$ × 20 = 10	$\frac{1}{4}$ × 12 = 3
50% × 20 = 10	25% × 12 = 3
30% × 20 = 6	

Students can use a calculator to verify the answers.

The revision of the percentage quantity capitalises on the learning by analogy theory. It draws students' attention to the relation between percentage and quantity (e.g. 25% × 12 = 3) and their prior knowledge of the relation between fraction and quantity (e.g. $\frac{1}{4}$ × 12 = 3).

Part 2: problem translation and the formulation of an equation

Figure 8.13 The equation approach 1 for the percentage change problem

The horizontal line in Figure 8.13 provides assistance in translating the problem situation. It depicts the underlying problem structure as comprising the original mark plus the increased mark. Based on the information in the diagram, the next step is to formulate an equation for John's new mark.

Let the new mark be x

Step 1: x = original mark + increased mark

Step 2: x = 70 + 70 × 10%

Step 3: x = 77

Step 1 expresses the information in the horizontal line in an equation. Step 2 replaces the original mark as well as the increased mark (percentage quantity) with numerical values. Essentially, it consists of two concepts: (1) the multiplicative relation between 70 and 10% to form 70 × 10%, and (2) the sum of 70 and 70 × 10%. The computation of Step 2 gives rise to the problem solution (Step 3).

The revision of the percentage quantity (Part 1) would help students to reinforce their schema for percentage quantity so that they can treat 70 × 10% as a single element rather than multiple interactive elements, hence reducing the burden on **working memory**. Processing the relation between the two elements in Part 2 (70 and 70 × 10%) through knowledge elaboration (Kalyuga, 2009) would constitute relatively low cognitive load, and therefore facilitate the acquisition of schema for the **percentage change problem**.

We could argue that it may be easier for students to use a new mark instead of x because of the challenge involved in using a variable. Nonetheless, it is important to provide opportunity for middle school students to engage in algebraic thinking skills wherever appropriate so as to pave the way for them to pursue algebraic problem-solving skills in solving real-life problems.

The equation approach 2

The flexibility of algebraic problem solving allows mathematics educators to explore an alternative instructional approach for percentage change problems. The equation approach 2 represents such an alternative. The main feature of the equation approach 2 is the use of a diagram to align the original mark (70) with 100%, and the new mark (x) with (110%), in which 110% represents the percentage after an increase of 10% in the mark (Figure 8.14). Essentially, the diagram would help students to visualise the relation between quantity and its corresponding percentage. Based on proportional reasoning, we can form an equation such as $\frac{70}{100} = \frac{x}{110}$, and solve for x. Hence, given appropriate preparation in establishing a schema for how to solve equations involving two fractions, the degree of element interactivity and therefore the cognitive load involved in the equation approach 2 would have been lowered. Accordingly, equation approach 2 would facilitate the teaching and learning of percentage change problems.

The diagram represents the situation

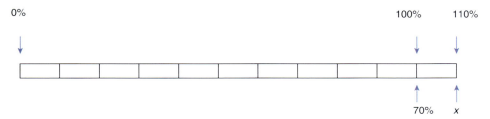

Figure 8.14 The equation approach 2 for the percentage change problem

Teaching and learning challenging percentage change problems

Students tend to make errors on challenging percentage change problems such as '*After a 12% markup, the shoes now cost $34. How much did they originally cost?*' (Parker & Leinhardt, 1995, p. 448). They tend to retrieve prior knowledge of percentage quantity $34 × 12% (wrong in this case), and rely on their intuition to subtract $34 × 12% from $34 as they may have perceived that the original cost is less than $34. Clearly, this type of challenging percentage change problem poses a challenge to students. Notwithstanding that there are other instructional approaches to teach and learn challenging percentage change problems (e.g. the unitary approach), two algebraic approaches for teaching and learning challenging percentage change problems will now be discussed.

The equation approach 1

The teaching and learning of the above challenging percentage problem requires the learners to identify the relevant information from the problem text (12%, $34), specify key words such as 'markup, original price', and construct a relation between values and a variable (the original cost) in an equation. Then, the learners need to solve for the unknown variable, which is the original cost.

Figure 8.15 The equation approach 1 for the challenging percentage change problem

Figure 8.15 shows the underlying problem structure comprising the original cost plus the increased amount (12% markup). From the information portrayed in Figure 8.15, the formulation of an equation and the subsequent solution steps to solve the problem are shown below.

Let x be the original cost.

Step 1:	Markup price = original cost + increased amount
Step 2:	$34 = x + x \times 12\%$
Step 3:	$34 = x(1 + 12\%)$
Step 4:	$x = \$34 \div (1 + 12\%)$

Answer: The original cost is $30.35

For novice learners, the solution procedure would constitute high element interactivity because of the interaction between elements on each solution step and across the solution steps. Nonetheless, the degree of element inter-activity depends on the knowledge base of the learners in the algebra domain. For example, if the learners have prior knowledge in algebraic problem solving (e.g. $x \times 25\% = \$200$, solve for x), then it is likely that they can treat $x \times 12\%$ as a single unit (Kalyuga, 2007; Kalyuga & Renkl, 2010), thus reducing the degree of element interactivity. Furthermore, prior knowledge of the factorisation skill would help learners to factorise the algebraic expression, $x + x \times 12\%$. More importantly, familiarity with how to solve $4n = 20$ may assist the learn-ers in transferring this skill to solve an equivalent equation, $\$34 = x(1 + 12\%)$. Therefore, learners' prior knowledge of basic algebra skills would affect the degree of element interactivity associated with equation approach 1. An important message for mathematics educators is to strengthen students' alge-bra skills (e.g. the variable, x, how to set up an equation, factorisation, and equation-solving skills) prior to introducing equation approach 1 for challeng-ing percentage change problems.

The equation approach 2

Figure 8.16 shows the underlying problem structure. It clearly shows that the markup price of $34 aligns with 112% (i.e. after 12% increase), and the origi-nal cost (x), which is yet to be found, aligns with 100%. In other words, the diagram will help students to visualise the relation between quantity and its corresponding percentage. Once again, the main idea is to capitalise on the proportional reasoning concept to form an equation such as $\frac{x}{100} = \frac{34}{112}$, and solve for x. Thus, given appropriate schema reinforcement regarding how to solve equations involving two fractions, the equation approach 2 would likely impose low cognitive load, facilitating the teaching and learning of challeng-ing percentage change problems.

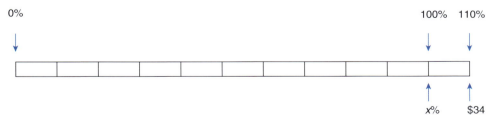

Figure 8.16 The equation approach 2 for the challenging percentage change problem

Comparing the two approaches, equation approach 2 may be easier than the equation approach 1 for challenging percentage problems because the variable (x) appears only once in equation approach 2, but twice in equation approach 1. Therefore, it is important to assess the prior knowledge of students before exposing them to either equation approach 1 or equation approach 2, or both.

In summary, when teaching percentage change problems, it is important to organise the material in a hierarchical level of complexity. Evidently, the prerequisite knowledge is the percentage quantity concept. Students not only need to have a schema for finding the part percentage quantity (e.g. 55 × 15%), they also need to have a schema for finding the whole percentage quantity (e.g. x × 25% = 56, solve for x). Empirical studies have indicated that students have great difficulty in finding whole percentage quantity (Baratta et al., 2010). Accordingly, greater attention is required to help students acquire the schemata for the percentage quantity concept before teaching them percentage change problems. More importantly, students need to have schemas for solving a range of linear equations, as equation-solving skills are another prerequisite for success in algebraic problem solving.

ACTIVITY 8.10

Log in to HOTmaths and select the 'HOTmaths Global' Course list and then choose the 'Early Secondary' Course. From the Topic list choose 'Further applications of percentages', then the 'Finding the whole amount' Lesson.

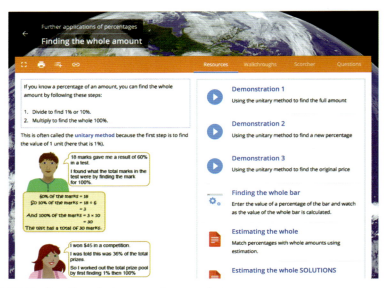

Figure 8.17 Finding the whole amount screenshot

The lesson and its associated resources are based on the unitary method. Complete the HOTsheet exercises using the algebraic approach. Which approach do you prefer? Why?

ACTIVITY 8.11

Log in to HOTmaths and select the 'HOTmaths Global' Course list and then choose the 'Early Secondary' Course. From the Topic list choose 'Using percentages', then the 'Discounts' Lesson.

Complete the HOTsheets using the algebraic approach. Would you use the algebraic approach to teach? Why or why not?

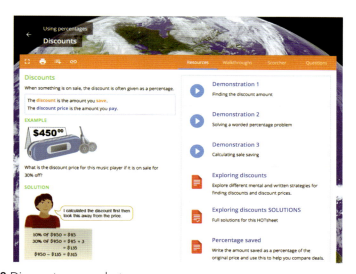

Figure 8.18 Discounts screenshot

ACTIVITY 8.12

Log in to HOTmaths and select the 'HOTmaths Global' Course list and then choose the 'Middle Secondary' Course. From the Topic list choose 'Linear equations & inequalities', then the 'Problems & formulas' Lesson. Here you will find a variety of relevant resources and activities. Explore these and outline the specific skills involved in teaching and learning algebraic problem solving.

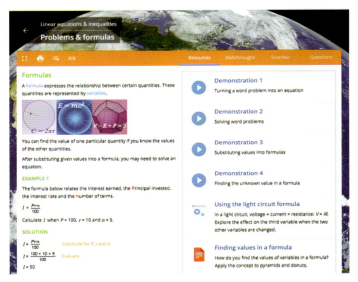

Figure 8.19 Problems & formulas screenshot

REFLECTIVE QUESTION

Tina paid $10.95 for a pizza which includes a 38% delivery charge on the price paid. What was the price of the pizza without the delivery charge?

Transferability of algebraic problem-solving to the science curriculum

Algebraic word problems represent a significant part of the secondary school curriculum. The Australian Curriculum: Mathematics highlights the need to connect mathematics across various discipline domains. Secondary school students are expected to learn how to solve both *within-domain* algebra word problems in the mathematics curriculum and *between-domain* algebra word problems, particularly in science subjects (such as in a chemistry or physics context). For example, regarding Newton's second law of motion, $F = ma$ (where F = force, m = mass, and a = acceleration), students need to understand the relation between F, m and a so that they can solve for the unknown variable (e.g. m), if the values of the other two variables (F and a) are given. Several researchers have provided empirical evidence of the transferability of prior knowledge in algebraic word problems to successfully solve physics problems requiring mathematical manipulation (Bassok, 1990; Bassok & Holyoak, 1989).

More recent research has investigated how to facilitate the teaching and learning of algebra word problems in a chemistry context (Ngu & Yeung, 2012, 2013; Ngu et al., 2015).

Molarity chemistry problems

A **molarity** problem such as the one below is an example of an algebraic word problem in a chemistry context:

A sample of vinegar is found to contain 15 grams of acetic acid, CH_3COOH, in 316 mL of the solution. Calculate the molarity of the acetic acid in vinegar (H = 1, C = 12, O = 16). Note that H = 1 means the atomic mass of hydrogen (H) is 1. The molar mass (MM) of the acetic acid is the sum of the atomic mass of all elements.

How can mathematics educators teach molarity chemistry problems? More specifically, how can mathematics educators construct an equation to solve the molarity problem?

Central to the molarity problem is the concept of the molarity. The molarity of a solution is defined as the moles of solute per litre of solution. Thus, the concept of molarity is related to a chemical substance dissolving in water to form a specific concentration. Several chemical concepts are embedded in the molarity problem:

$$Molarity = \frac{\text{number of moles of solute}}{\text{volume in one litre of solution}}$$

1 Number of moles of solute = MV where M = Molarity, V = volume in litre
2 Alternatively, the number of moles of solute = $\frac{MV}{1000}$ where V = volume in millilitres (mL)
3 Also, the number of moles of a chemical substance = $\frac{g}{MM}$ where g = mass in grams, MM = molar mass of the chemical

In effect, the problem structure of the molarity problem above comprises: (1) molarity (M), (2) volume (V in mL), and (3) number of moles of solutes (i.e. $\frac{g}{MM}$) where g and MM are mass and molar mass respectively. Each of these four chemical concepts (M, V, g and MM) represents an aspect of the problem structure. Based on the analysis of the chemical concepts, we can form an equation such as, $\frac{g}{MM} = \frac{MV}{1000}$ (where the left-hand and the right-hand sides equal to the number of moles of solutes) to solve for M, the molarity of the acetic acid in vinegar.

The solution for the vinegar problem is:

g = 15 grams, V = 360 mL, MM = molar mass of the vinegar, CH_3COOH = (2 × 12 + 1 × 4 + 2 × 16) = 60

$$\frac{g}{MM} = \frac{MV}{1000}$$

$$\frac{15}{60} = \frac{M \times 360}{1000}$$

$$M = \frac{15}{60} \times \frac{1000}{360}$$

$$M = 0.69 \text{ moles/L or } 0.69 \text{ M}$$

Once again, the power of algebra is such that we can use the equation to solve any unknown variable provided that the other chemical concepts (or variables) are given. The molarity problems pose a challenge to secondary students because the traditional method (Thickett, 2000) for teaching and learning molarity problems does not emphasise the integration of relevant chemistry concepts in a single equation to solve for the unknown variable.

ACTIVITY 8.13

An NaCl solution is 2.0 M. How many millilitres (mL) of this solution will contain 10 g of NaCl? (Na = 23, Cl = 35.5).

ACTIVITY 8.14

Phenobarbitone, $C_{12}H_{12}N_2O_3$, is commonly used in medicine as a long-acting sedative. A patient may take 5.0 mL of phenobarbitone as one dose. If the concentration of the phenobarbitone is 0.017 M, what mass of the drug is present in one dose? (C = 12, O = 16, N = 14, H = 1).

REFLECTIVE QUESTION

Underline the irrelevant information in the problem below:

20 g of sodium sulphate, Na_2SO_4, were dissolved in sufficient water to obtain 500 mL of a solution with a density of 1.11 g/mL. Calculate the molarity of the solution. (Na = 23, S = 32, O = 16).

What have you learned by underlining the irrelevant information?

Summary

Understand the algebraic concepts of the unknown, pronumeral and variable

The critical concept of algebra is underpinned by the use of an unknown or a pronumeral or a variable. In particular, the use of a variable is significant because it can stand for a range of numbers. Using variables to generate a mathematical rule in the form of an equation to solve growing pattern problems has demonstrated the power of algebra in problem solving. Nonetheless, it is important to distinguish between a variable and an object in algebraic problem solving.

Teach algebraic expression problems via analogical reasoning

Central to arithmetic manipulation is the concept of number, whereas central to algebraic manipulation is the concept of a variable. The prior knowledge of arithmetic manipulation could facilitate the learning of algebraic manipulation through reasoning by analogy. Likewise, the teaching and learning of algebraic expression can be achieved through an analogical comparison between the manipulation of an arithmetic problem and the manipulation of an algebraic expression.

Unpack the complexity of linear equations

Mathematics education researchers and educators have classified linear equations based on the number of operations (e.g. a one-step equation has one operation), grouping symbols, pronumeral on both sides of the equation, negative integers and so on. From the perspective of cognitive load theory, the classification of linear equations based on the number of relational and operational lines has captured the complexity of linear equations in a hierarchical order. Moreover, the nature of the element (simple or complex) also affects the complexity of the linear equations when the number of relational and operational lines is kept constant.

Teach linear equations in a hierarchical level of complexity

Mapping structurally similar relational and operational lines enables the teaching and learning of linear equations to occur in a hierarchical level of complexity. This is significant because the teaching and learning of a complex equation is built upon a simpler equation. Structurally similar equations share the same type and number of relational and operational lines but differ in the nature of the element (simple or complex). An effective approach to teach structurally similar equations is via the learning by analogy theory. Using the inverse method in addition to the balance method would enhance equation-solving skills, particularly for equations involving multiple solution steps.

Appreciate algebraic problem-solving in real-life contexts

Mathematics educators need to address two main challenges in teaching algebraic problem solving. Firstly, the translation of the problem situation in order to formulate an equation. Secondly, the use of equation-solving skills to solve the equation. For both equation approach 1 and equation approach 2, diagrams provide aid in translating the problem situation; and, subsequently, the formulation of an equation. The prior knowledge of percentage quantity, which is an essential concept of the percentage change problem, reduces the working memory load. More importantly, prior knowledge of equation-solving skills is critical to ensure success in using an algebraic approach in solving real-life problems.

Transfer algebraic problem-solving skills to the science curriculum

The power of algebraic problem solving has been demonstrated via molarity chemistry problems. To teach molarity problems, the first step is to identify relevant chemical concepts embedded in the chemistry word problem. The second step is to integrate the relevant chemical concepts (values, variables and their relation) in an equation. The third step is to solve the equation to obtain the answer.

CHAPTER 9

Measurement and geometry

Learning outcomes

After studying this chapter, you should be able to:

- outline the importance of measurement and geometry tasks within the secondary mathematics classroom
- develop an understanding of the concept of geometric proof as articulated in the Australian Curriculum
- develop an understanding of the concept of geometric transformation as articulated in the Australian Curriculum
- describe how technology can be used to support learning concepts in measurement and geometry.

Introduction

Within the Australian Curriculum, the topics Measurement and Geometry are presented together as one strand to emphasise their relationship to each other, and to enhance their practical relevance. Within this strand, it is required that

Students develop an increasingly sophisticated understanding of size, shape, relative position and movement of two-dimensional figures in the plane and three-dimensional objects in space. They investigate properties and apply their understanding of them to define, compare and construct figures and objects.

They learn to develop geometric arguments … make meaningful measurements of quantities, choosing appropriate metric units of measurement …
build an understanding of the connections between units and calculate derived
measures such as area, speed and density. (ACARA, 2015b)

The Measurement and Geometry sub-strands, together with the year levels for
which each sub-strand is taught, are listed in Table 9.1.

Table 9.1 Australian Curriculum: Measurement and Geometry (sub-strands)

Measurement and Geometry: sub-strands
Using units of measurement (F–10)
Shape (F–7)
Geometric reasoning (3–10)
Location and transformation (F–7)
Pythagoras and trigonometry (9–10)

At an international level, measurement and geometry questions also
enjoy a place of prominence within high-stakes mathematics testing. For
instance, the Programme for International Student Assessment (PISA)
makes explicit mention of measurement, geometry, reasoning and spatial
awareness in all of the six proficiency levels (see Thomson et al., 2013). In
the most recent PISA testing round, Australia scored 23rd out of 54 countries
with a mean score of 497. Additionally, Thomson et al. (2013, p. 62) noted
that Australia was 'among one of the countries that achieved a mean score
on the space and shape sub-scale that was lower than on the overall mathematical literacy scale, indicating students found this content area relatively more difficult'. In a similar vein to PISA, the Trends in International
Mathematics and Science Study (TIMSS) outlines explicitly how geometry
and measurement topics are tied to student achievement in the advanced,
high and intermediate international benchmarks (see Thomson et al., 2012,
p. 33). The TIMSS content domains at Year 8 level are comprised of: Number
(30%), Algebra (30%), Geometry (20%), and Data and Chance (20%). The
Geometry content domain is comprised of topic areas: Geometric Shapes,
Geometric Measurement and Location and Movement. Across Australian
schools, Thomson et al. (2012, p. 33) reported that 'Year 8 students' performance was clearly better in data and chance and number than in algebra
and geometry'. Ostensibly, it would appear that Australian schools could
benefit from increased attention to improved teaching and learning practices within this essential mathematical strand.

The importance of learning geometry

Geometry – commonly regarded as the study of space and spatial relationships – is an important and essential branch of the mathematics curriculum at all year levels (Singhal et al., 2014). According to Siemon et al. (2011), geometry is not concerned merely with learning definitions and properties, nor is it learning geometric proofs via rote memorisation. Rather, the study of geometry develops both logical and deductive thinking, which in turn helps learners expand mentally and mathematically (Singhal et al., 2014). Goldenberg et al. (2012) make two broad claims about the role of geometry within general education. The first claim is that geometry can help students connect with mathematics. Moreover, 'students connect well with properly selected geometric studies. The many hooks include no less than art, physical science, imagination, biology, curiosity, mechanical design, and play' (p. 3). The second claim is that geometry can be an ideal vehicle for building the 'habits of mind' perspective. For instance, Goldenberg et al. (2012) asserted that

> within mathematics, geometry is particularly well placed for helping people develop these ways of thinking. It is an ideal intellectual territory within which to perform experiments, develop visually based reasoning styles, learn to search for invariants, and use these and other reasoning styles to spawn constructive arguments. (2012, pp. 4–5)

Additionally, these authors contend that in a very broad sense, geometry is also ideally placed for helping students connect richly with the rest of mathematics.

The importance of learning measurement

Measurement is a key component of the mathematics curriculum, and it is considered the most practical and hands-on application of mathematics in everything from occupational tasks to day-to-day-life (Siemon et al., 2011; Van de Walle, Karp & Bay-Williams, 2014). The application and variety of measurement concepts surrounding people on a daily basis is considerable – 'from gigabytes that measure amounts of information to font size on computers, from miles per gallon to recipes for a meal' (Van de Walle, Karp & Bay-Williams, 2014, p. 397). Siemon et al. (2011, p. 624) contend that in order to extend mathematical concepts, students should be 'provided with the opportunity to apply their understandings by using number processes and geometric reasoning'.

Doing so enables students to integrate mathematics into other learning areas, to appreciate relationships existing between measurement attributes (e.g. length, area) and various units of measurement relevant to each attribute, and to fluently manage measurement tasks both in the classroom and in the outside world. Despite its importance, measurement is not an easy topic for students to understand. Data from international studies consistently indicate that US students are weaker in the area of measurement than any other topic in the mathematics curriculum (Thompson & Preston, 2004). This finding would coincide with recent high-stakes achievement data for Australian students, noted earlier in the chapter.

KEY TERMS

- **Construction of proof:** recognition of how to create deductive arguments in geometry using sufficient knowledge about definitions, assumptions, proofs, theorems and logical circularity.
- **Generality of proof:** recognition of the universality and general applicability of geometric theorems (proven statements), the roles of figures, and the difference between formal proof and experimental verification.
- **Geometry:** this important study of space and spatial relationships develops logical reasoning and deductive thinking, which in turn helps learners expand both mentally and mathematically.
- **Measurement:** the process of identifying the relationship of numbers that can be expressed in terms of length, area, volume, capacity, time, money, mass or weight.
- **Non-rigid transformation:** there are two types of transformations where an object's size or shape changes. Similar transformations keep the object the same shape but the size changes. Shear transformations keep the area the same, but the shape changes.
- **Rigid transformations:** these transformations (also known as congruence transformations) preserve size and shape, and their movement can be recorded with coordinates.
- **Transformation:** refers to the changes in positions or size of shapes, which collectively comprise the study of translations, reflections, rotations, symmetry and similarity.
- **van Hiele levels of geometric thought:** a five-level hierarchy describing the thinking processes used to understand geometric ideas and contexts.

The concept of geometric proof

The literature exhorts the concept of proof as central to the mathematics discipline and, consequently, authors contend that it should feature prominently in mathematics education (Ball et al., 2002; Siemon et al., 2011). Because of this centrality, proof is considered an essential tool for promoting mathematical understanding in students (Ball et al., 2002; Reid, 2011) and at the same time, for providing educators with insight about how students learn mathematics (Wilkerson-Jerde & Wilensky, 2011). Singh (2005, p. 21) offered insight into the idea of a classical mathematical proof as

> to begin with a series of axioms, statements which can be assumed to be true or which are self-evidently true. Then by arguing logically, step by step, it is possible to arrive at a conclusion. If the axioms are correct and the logic is flawless, then the conclusion will be undeniable. This conclusion is the theorem. Mathematical theorems rely on this logical process and once proven are true until the end of time. Mathematical proofs are absolute.

In the 1990s de Villiers suggested a framework to justify the role of proof in the mathematics classroom. The framework posits that this role can be a means of (a) verification, (b) explanation, (c) systematisation, (d) discovery and (e) communication (de Villiers, 1990). According to Yopp (2011), a majority of papers written since the advent of de Villiers' framework could be characterised by one or more of the suggested roles – although the role of proof in mathematics classrooms has yet to reach a full and exhaustive description. Following on from ideas concerning proof, geometric proof is one particular type of mathematical justification whereby students' deductive reasoning and creative thinking can be developed through systematic argumentation (Kunimune et al., 2009). According to several commentators, geometric proof once held a more prominent place in the secondary context (Ball et al., 2002; Siemon et al., 2011). Despite the centrality of deductive reasoning to successful progress in mathematics, it was found that teaching geometric proofs to lower secondary students was an extremely difficult task (Battista, 2007; Mariotti, 2007). Consequently, many teachers and students resorted to learning proofs via rote memorisation for later recitation in assessments – omitting key steps in logical reasoning and understanding along the way. It remains the case at an international level that students have great difficulty in constructing and understanding geometric proofs (Healy & Hoyles, 1998; Lin, 2000; Senk, 1985).

Approaches to teaching geometric proof

Geometric proof and reasoning in the Australian Curriculum

The concept of geometric proof and reasoning is taught implicitly and explicitly in middle year levels (Years 7–10A) of the Australian Curriculum. In Year 7, students begin by classifying triangles according to their side and angle properties, and then apply this knowledge to quadrilaterals. They also investigate various conditions for lines to be parallel and solve problems using logic and reasoning. Year 8 students investigate the concept of congruency in planar shapes using transformations, in triangles using conditions, and then solve related numerical problems using reasoning. The concept of similarity features prominently in the Year 9 curriculum, with students using principles of similarity to investigate trigonometric ratios, enlargement ratios and scale factors. In Year 10, students formulate proofs with angles, triangles and circles, and apply logical reasoning to proofs and numerical exercises. Geometric proof and reasoning is also taught explicitly and implicitly – and to varying degrees – in the senior years of mathematics courses, namely: Essential Mathematics, General Mathematics, Mathematics Methods and Specialist Mathematics. Perhaps this topic features most prominently in General Mathematics (Unit 1, Topic 3: Shape and Measurement) and in Specialist Mathematics (Unit 1, Topic 3: Geometry). Table 9.2 displays how geometric proof and reasoning is presented in the Australian Curriculum courses for middle secondary school.

Three perspectives of geometric proof and reasoning

It has already been argued in this chapter that teaching geometric proof and reasoning is very important for learning mathematics. Furthermore, Huetinck and Munshin (2008, p. 157) assert that 'the concept of proof as justification of assertions through analytic reasoning should be developed informally over time, beginning long before high school'. As such, these authors contend that verbal and written activities focused on probing mathematical questions (e.g. those requiring proof and analytic reasoning) are central within primary and middle school mathematics programs. Providing these opportunities leads students to develop a deeper understanding of and appreciation for the nature of mathematical proof at secondary school level. Huetinck and Munshin (2008) suggest that mathematics educators consider three approaches – and not least the benefits and limitations associated with each approach – concerning

Table 9.2 Geometric proof and reasoning in the Australian Curriculum (middle years)

Year 7	Classify triangles according to their side and angle properties and describe quadrilaterals (ACMMG165)	Demonstrate that the angle sum of a triangle is 180° and use this to find the angle sum of a quadrilateral (ACMMG166)	Identify corresponding, alternate and co-interior angles when two straight lines are crossed by a transversal (ACMMG163)	Investigate conditions for two lines to be parallel and solve simple numerical problems using reasoning (ACMMG164)
Year 8	Define congruence of plane shapes using transformations (ACMMG200)	Develop the conditions for congruence of triangles (ACMMG201)	Establish properties of quadrilaterals using congruent triangles and angle properties, and solve related numerical problems using reasoning (ACMMG202)	
Year 9	Use similarity to investigate the constancy of the sine, cosine and tangent ratios for a given angle in right-angled triangles (ACMMG223)	Use the enlargement transformation to explain similarity and develop the conditions for triangles to be similar (ACMMG220)	Solve problems using ratio and scale factors in similar figures (ACMMG221)	
Year 10/10A	Formulate proofs involving congruent triangles and angle properties (ACMMG243)	Apply logical reasoning, including the use of congruence and similarity, to proofs and numerical exercises involving plane shapes (ACMMG244)	Prove and apply angle and chord properties of circles (ACMMG272)	

the notion of formal proof, namely: synthetic, analytic and transformational. Each of these approaches is illustrated below through the application of the side-side-side (SSS) triangle congruence to a given situation. The situation is to prove that within a parallelogram ABCD, triangles ABC and CDA (formed by drawing a line from A to C along the 'long' diagonal) are congruent. In addition, a table outlining various ways to present the three approaches is offered in Table 9.3.

Table 9.3 Three ways to present geometric concepts

Synthetic	Analytic	Transformational
$m(length\ AB) = |B-A|$	$D = \sqrt{(x_1 - x_2)^2 + (y_1 - y_2)^2}$	
Parallel lines are two lines on the plane whose distance from each other is constant.	Parallel lines on the coordinate plane have the same gradient.	
Congruent planar figures have corresponding sides and angles equal.	Congruent figures in the coordinate plane have corresponding sides of the same measure and absolute value of corresponding gradients equal.	Congruence is a one-to-one correspondence between two geometric figures that preserves distance and angle measure (Okolica & Macrina, 1992).
Perpendicular lines are two lines on the same plane that form right angles.	Perpendicular lines on the same plane have gradients whose product equals –1.	Two lines are perpendicular if one is the reflection line of the other line in itself.
The midpoint of a line segment is equidistant from the endpoints of the segment.	The midpoint of a line segment is found by $$M = \left(\frac{x_1 + x_2}{2}, \frac{y_1 + y_2}{2} \right)$$	One way of determining the midpoint of a line segment is to see if it lies on the line of reflection of one endpoint of the segment in the other.

Adapted from Huetinck & Munshin, 2008, p. 159.

Synthetic geometry

According to Huetinck and Munshin (2008), the **synthetic approach** to geometry dates back to Euclid and the Greeks. Essentially, the synthetic approach requires the user to prepare a well-structured and formal, logical system, whereby geometric relationships are proven using rational sequences of definitions, postulates and theorems. Using this approach with middle school students enables them to 'experience the challenge and satisfaction of dealing with probing questions that ask them to justify conclusions, explain how they know a given pattern always works, or find a counterexample to show that a pattern does not always work and is therefore disproved' (Huetinck & Munshin, 2008, p. 159).

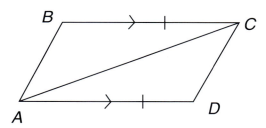

Figure 9.1 Synthetic approach

Given Figure 9.1, the task is to prove that within parallelogram ABCD, triangle ABC is congruent to triangle CDA.

The synthetic approach to proving this triangle congruence could be:

Given $\overline{BC}//\overline{AD}$ and $\overline{BC} = \overline{AD}$.
Required to Prove: $\triangle ABC \cong \triangle CDA$.

Statements	Reasons
1. **BC** ≅ **DA**	1. Given information
2. **AB** ≅ **CD**	2. Properties of a parallelogram (opposite sides are parallel and congruent)
3. **AC** ≅ **AC**	3. Identity (same line)
4. △**ABC** ≅ △**CDA**	4. Side-side-side congruence (SSS)
QED	

Analytic geometry

A second perspective of formal proof is through the use of analytic geometry. This approach was first used in the ninth century, but the concept as it is presently construed is rooted in the work of later mathematicians, notably Descartes (after whom the Cartesian plane is named) and Fermat (Huetinck & Munshin, 2008). Although a contemporary mathematics classroom would typically have graphing calculators for students to recreate and dynamically move planar figures – as well as to explore various relationships of shapes and their transformations – the mastery of skills using a manual technique is still very important for conceptual understanding.

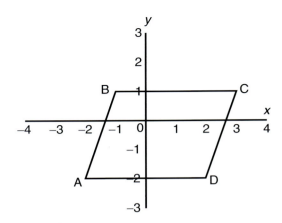

Figure 9.2 Analytic approach

To prove that triangle ABC is congruent to triangle CDA using analytic geometry, we can place the parallelogram on the Cartesian plane and use the distance formula to show corresponding sides have the same length.

$$AD : \sqrt{(-2-2)^2 + (0-0)^2} = \sqrt{16} = 4$$

$$BC : \sqrt{[(3-(-1)]^2 + (1-1)^2} = \sqrt{16} = 4$$

$$AB : \sqrt{(-2-(-1))^2 + (-2-1)^2} = \sqrt{10}$$

$$DC : \sqrt{(2-3)^2 + (-2-(-1))^2} = \sqrt{10}$$

Similar to the synthetic approach, we can easily establish that AC is a common side. Thus, with the three known side lengths we can use the side-side-side (SSS) congruence to conclude that triangle ABC is congruent to triangle CDA.

Transformational geometry

Transformational geometry techniques can be demonstrated through graphing technology (e.g. GeoGebra) or via a plane mirror. With either of these tools, students can see that rotations are compositions of reflections by doing the appropriate constructions. Hollebrands (2004) investigated how tenth-grade students in the United States constructed the images of a polygon under a reflection, rotation and translation. For the reflection activity given to the research participants, Hollebrands noted that students may have had difficulty with the task as the line of reflection was neither vertical nor horizontal, a finding consistent with the work of Schultz and Austin (1983). Another key finding was that all students seemed to think of reflection as 'flipping', which was evidenced through their explanations of attempting to draw the image congruent to the pre-image. However, the students 'did not attend to relationships between corresponding pre-image and image points and the line of reflection' (Hollebrands, 2004, p. 208). While this research focuses on transformations – which will be discussed later in this chapter – this particular finding has implications for teachers who design geometric proof lessons using a transformational approach.

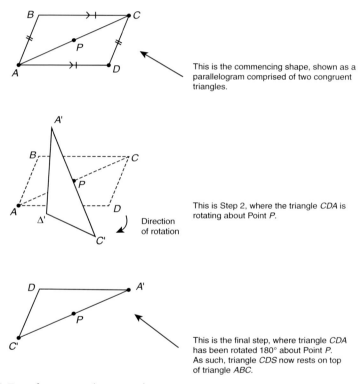

This is the commencing shape, shown as a parallelogram comprised of two congruent triangles.

This is Step 2, where the triangle *CDA* is rotating about Point *P*.

Direction of rotation

This is the final step, where triangle *CDA* has been rotated 180° about Point *P*. As such, triangle *CDS* now rests on top of triangle *ABC*.

Figure 9.3 Transformational approach

To prove that triangle *ABC* is congruent to triangle *CDA* using the side-side-side congruence, we begin by drawing diagonal *AC* and letting *P* be the midpoint of *AC*. Then we rotate triangle *CDA* 180° (in either direction) about point *P* until it rests completely on triangle *ABC*.

1 Rotate $\triangle CDA$ 180° about point *P*.
2 Now since *P* is the midpoint of *AD*, $PA \cong PC$ and both *A* and *C* rotate on to each other.

3 Since (by definition of a parallelogram) $AB//DC$ and $AD//BC$, $\angle BAC \cong \angle ACD$ and $\angle CAD \cong \angle BCA$. Therefore, the two pairs of angles ($\angle BAC$ & $\angle ACD$, and $\angle CAD$ & $\angle BCA$) rotate on to each other.

4 Because the angles $\angle CAD$ and $\angle BCA$, the lines $\overline{AC}$ and $\overline{DB}$ coincide. Similarly, the lines $\overline{AB}$ and $\overline{DC}$ coincide because the angles $\angle BAC$ and $\angle ACD$ coincide.

5 The side-side-side (SSS) congruence can now be applied, as $\triangle ADC$ completely overlaps $\triangle ABC$. Therefore, the two triangles are congruent.

Research-driven insight

Researchers in Japan were interested in discerning lower secondary students' cognitive needs in learning geometric concepts. In Japanese schools, the concept of formal proof is taught intensively to lower secondary students (Japanese grades 7–9). Following their research, Kunimune et al. (2009) suggested an analytical framework that would drive best classroom practice and satisfy student learning needs in geometry. The researchers investigated across a ten-year period how Japanese grade 8 and 9 students successfully completed geometric proof questions – specifically in terms of **construction of proof** and **generality of proof**. Considering these two aspects of geometric proof, the researchers presented their findings as a three-level analytical framework developed from the van Hiele model (1959). The framework is found in Table 9.4.

The researchers' key findings indicated that the students achieve reasonably well in terms of construction of proof, but not necessarily as well in terms of generality of proof. In other words, students demonstrated they could

Table 9.4 Student behaviours as a function of geometric proof

Level	Student behaviours
1	Students consider experimental verifications as sufficient to demonstrate that geometrical statements are true.
1a	Students do not achieve generality of proof and construction of proof.
1b	Students achieve generality of proof and construction of proof.
2	Students understand that proof is required to demonstrate geometrical statements are true.
2a	Students achieve generality of proof without understanding logical circularity.
2b	Students understand logical circularity.
3	Students understand simple logical chains between theorems.

construct a formal proof (Level 1), yet they may not appreciate the significance of such formal proof in geometry (Level 1a). According to the collected data, 80 per cent of grade 9 students remain at Level 1 in terms of understanding proof. The researchers expressed concern at this finding, as previous research had indicated that by this time in their education, students have studied formal proof in grade 8 using textbooks where 90 per cent of relevant intended lessons can be devoted to justifying and proving geometric facts (Fujita & Jones, 2003). Moreover, students 'may believe that a formal proof is a valid argument, while, at the same time, they also believe experimental verification is equally acceptable to "ensure" universality and generality of geometrical theorems' (Kunimune et al., 2009, p. 761).

Based on over 10 years of classroom-based research, Kunimune et al. (2007) proposed the following analytical framework for lower secondary school geometry (grades 7–9). This framework is designed to help students appreciate the need for formal proofs (in addition to the students being able to construct such proofs):

Table 9.5 Principles for teaching geometric proof to lower secondary school students

Year (or grade) level	Approach	Anticipated benefit
7	Commence geometric proof from problem-solving situations (e.g. consider how to draw diagonals of a cuboid).	Develops students' geometric thinking and provides experiences of mathematical processes useful in studying deductive proofs in Years 8 and 9.
8	Geometric constructions to be taught, and at the same time these constructions are to be proven.	The existing gap between geometric constructions and their proofs (according to research) to decrease.
8	Examine differences between experimental verifications and deductive proof.	Students will gain an appreciation of the differences between experimental verifications and deductive proof.
8	Lessons commence with teaching deductive geometry with a set of already learnt properties which are shared and discussed within the classroom, and used as a form of axiom.	Students are provided with known starting points for their proofs.

In a similar vein, Japanese researchers found that while geometric constructions (e.g. with ruler and compass) can be taught in grade 7, the constructions themselves are often not proved until grade 8 (Shinba, Sonoda & Kunimune, 2004). Moreover, it is in grade 8 that Japanese students learn how to prove simple geometric statements. Through a series of teaching experiments Kunimune et al. (2009) investigated the use of more complex geometric constructions (and proofs of these constructions) with grade 8 students. For instance, one lesson commenced with the direction '*Let us consider how we can trisect a given straight line AB*'. From their classroom observations, the researchers noted that students were able to first investigate theorems and properties of geometrical figures through construction activities, which led them to a consideration of why the construction worked. Following the appropriate teacher-initiated instruction, students commenced proving the geometric constructions. In addition, the active learning environment enabled students to experience some important processes which led from conjecture to proof (Kunimune et al., 2009).

ACTIVITY 9.1

Log in to HOTmaths and select the 'HOTmaths Global' Course list and then choose the 'Middle Secondary' Course. From the Topic list choose

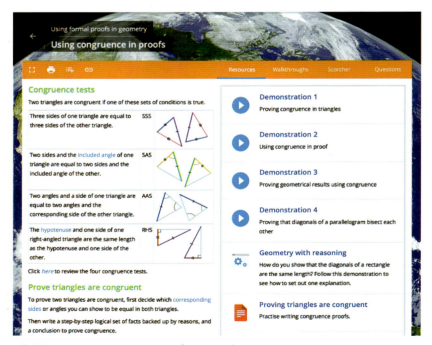

Figure 9.4 Using congruence in proofs screenshot

'Using formal proofs in geometry', then the 'Using congruence in proofs' Lesson.

Looking at the Resources, list the necessary prior learning that students require to complete these activities. Then plan a series of lessons (or a unit of work) that could be used to teach the concept of proof effectively to a class of Year 9 students.

ACTIVITY 9.2

Log in to HOTmaths and select the 'HOTmaths Global' Course list and then choose the 'Middle Secondary' Course. From the Topic list choose 'Circle geometry', then the 'Cyclic quadrilaterals' Lesson.

Looking at the Resources, list the necessary prior learning that students require to complete these activities. Then plan a series of lessons (or a unit of work) that could be used to teach the concept of proof effectively to a class of Year 12 students.

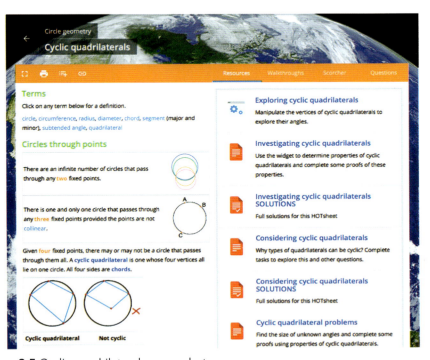

Figure 9.5 Cyclic quadrilaterals screenshot

REFLECTIVE QUESTIONS

Geometric proof is often considered by students as a difficult, inaccessible and onerous topic to master. Think back to your time spent in the classroom coming to terms with any method of geometric proof. Which method did you find the most accessible, and why? For what reasons do you feel students struggle with this important topic – and how do you plan to help them understand it well?

The concept of transformation

In mathematics the term *transformation* refers to the changes in positions or size of shapes, which collectively comprise the study of translations, reflections, rotations (slides, flips and turns), symmetry and similarity (Van de Walle, Karp & Bay-Williams, 2014). Another way of viewing transformational geometry is to consider what remains invariant (or the same) as a process happens to an object (Siemon et al., 2011). Broadly speaking, there are two categories under which all geometric transformations belong – rigid transformations and non-rigid transformations – and these will be discussed subsequently. Hollebrands (2004) asserted important reasons to teach geometric transformations within school mathematics curricula. First, transformations provide an opportunity for students to conceptualise important mathematical concepts (e.g. functions, symmetry), and to view mathematics as an interconnected discipline. Transformations also engage students in higher-level reasoning activities using a variety of representations. Moreover, Guven (2012, p. 366) argued that the study of transformation can 'lead students to exploration of the abstract mathematical concepts of congruence, symmetry, similarity, and parallelism; enrich students' geometrical experience, thought and imagination; and thereby enhance their spatial abilities'.

Rigid transformations

Transformations that do not change the size or shape of an object moved are called *rigid transformations* (also referred to as rigid motions, isometries or congruence transformations). The movement of rigid transformations can be recorded with coordinates, which enables learning activities to be placed on the Cartesian plane. Typically, three rigid-motion transformations are taught at high school level: translations (or slides), reflections (or flips) and rotations (or turns) (Van de Walle, Karp & Bay-Williams, 2014). Because rigid transformations preserve the size and shape of objects, the study of symmetry is also included under the study of transformations. A popular example of rigid transformations is the tessellation, which is a repeated shape that covers a plane without any gaps or overlaps. As Siemon et al. (2011, p. 633) note, 'for there to be no gaps or overlaps, the sum of angles of shapes meeting at each vertex must be exactly 360 degrees'. As such, there are only three regular polygons (two-dimensional shapes that have congruent sides and angles) that tessellate: triangles, quadrilaterals and hexagons.

Rigid transformations can also be applied to functions (e.g. polynomials, relations) which can allow for a richer appreciation of how modifications to a rule can lead directly to the movement of a function's graph. For instance, Figure 9.6 shows how the function $g(x) = -\dfrac{1}{x+2} - 3$ is a composition transformation of the original function $f(x) = \dfrac{1}{x}$.

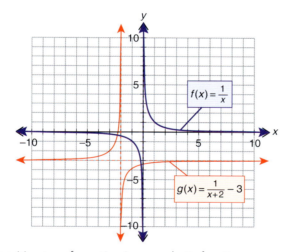

Figure 9.6 Composition transformation in a quadratic function

Composition transformations are multiple transformations that can be applied to shape, one after the other. The transformations for $f(x)$ are:

- a reflection in the x-axis
- horizontal translation 2 units left
- vertical translation 3 units down.

In senior years, the topic of matrices is taught in General Mathematics and Mathematics Methods. One of the key topics studied in Mathematics Methods concerns transformations, where objects (on the Cartesian plane – and hence which have their shape governed by coordinates) are reflected or rotated as a result of matrix multiplication. For instance, the unit square has coordinates (0, 0), (1, 0), (1, 1) and (0, 1). The two figures, Figure 9.7 and Figure 9.8, display how the unit square has undergone a reflection across the y-axis and a rotation of 90 degrees clockwise, respectively. Accompanying each figure is the transformation matrix and mathematical operations which move the object from the pre-image stage to the image.

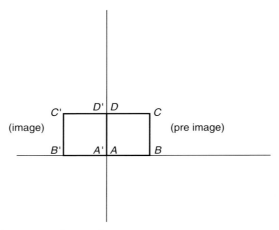

Figure 9.7 Reflection across the y-axis

$$\begin{bmatrix} -1 & 0 \\ 0 & 1 \end{bmatrix} \times \begin{bmatrix} 0 & 1 & 1 & 0 \\ 0 & 0 & 1 & 1 \end{bmatrix} = \begin{bmatrix} 0 & -1 & -1 & 0 \\ 0 & 0 & 1 & 1 \end{bmatrix}$$

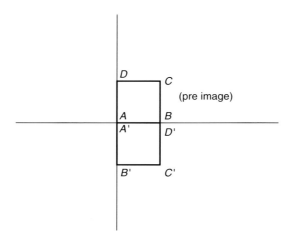

Figure 9.8 Rotation of 90° clockwise

$$\begin{bmatrix} 0 & 1 \\ -1 & 0 \end{bmatrix} \times \begin{bmatrix} 0 & 1 & 1 & 0 \\ 0 & 0 & 1 & 1 \end{bmatrix} = \begin{bmatrix} 0 & 0 & 1 & 1 \\ 0 & -1 & -1 & 0 \end{bmatrix}$$

Non-rigid transformations

Transformations that change the size or shape of an object are called *non-rigid transformations*. Chiefly, there are two types of non-rigid transformations: similar transformations and shear transformations. *Similar transformations* keep an object's shape the same but the size changes. *Shear transformations* keep an object's area the same, but the shape changes.

Similar transformations

Two shapes are said to be similar if all of their corresponding angles are congruent and the corresponding sides are proportional (Van de Walle, Karp & Bay-Williams, 2014). Using a scale factor, shapes can be enlarged or reduced – although, technically, images that are smaller than the pre-images are still considered enlargements! Applying a scale factor greater than 1 will make the image larger than pre-image; applying a scale factor less than 1 will make the image smaller. Of course, applying a scale factor of 1 will render the image congruent to the pre-image. One common method of enlarging a figure geometrically is explained by Siemon et al. (2011, p. 635):

> Many children's activity books contain examples or using a grid to enlarge a simple cartoon. A grid is drawn over the shape. A second grid is then drawn with larger or smaller squares, depending on the chosen scale factor. Points where the drawing crosses grid lines are marked on the first and copied onto a similar location on the second. Lines are then joined to construct the enlarged image.

A second method taught in high school uses a point of enlargement, which can lie inside or outside the original object. In Figure 9.9, ΔABC is enlarged by a scale factor of 3 to produce ΔDEF. Notice how the point of enlargement lies outside both tringles; the distance from the point of enlargement to each of the vertices on ΔABC is is one-third the distance to ΔDEF.

Dilations are considered to be non-rigid transformations that produce similar two-dimensional figures (Van de Walle, Karp & Bay-Williams, 2014). In Figure 9.10, we can see how the graph of $y = x^2$ has been transformed to $y = x^2$. This transformation is described as a dilation parallel to the y-axis, scale factor 4. Popular ways of explaining this transformation to students are to say the pre-image has been 'pulled or stretched' along the y-axis by a factor of 4, or that all of the output values of $f(x)$ have been multiplied by 4.

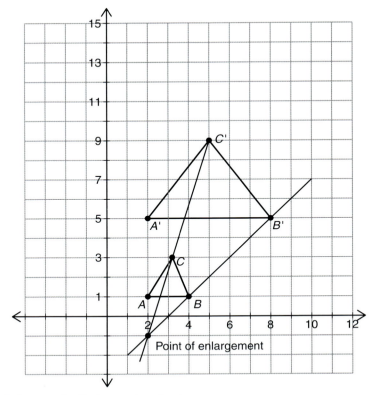

Figure 9.9 Geometric dilation: point of enlargement

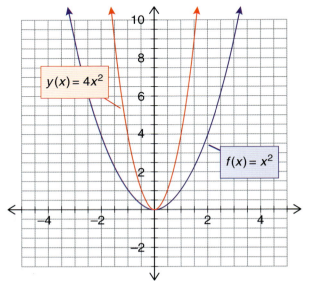

Figure 9.10 Geometric dilation: function transformation

Figure 9.11 displays how the unit square has undergone two transformations simultaneously; a dilation parallel to x-axis scale factor 3, and a dilation parallel to y-axis scale factor 2. Accompanying this figure is the transformation matrix and mathematical operations which move the object from the pre-image stage to the image.

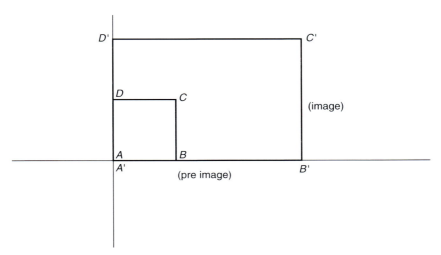

Figure 9.11 Dilation transformations

$$\begin{bmatrix} 3 & 0 \\ 0 & 2 \end{bmatrix} \times \begin{bmatrix} 0 & 1 & 1 & 0 \\ 0 & 0 & 1 & 1 \end{bmatrix} = \begin{bmatrix} 0 & 3 & 3 & 0 \\ 0 & 0 & 2 & 2 \end{bmatrix}$$

Shear transformations

Shear transformations (or shears) are transformations that do not preserve the shape of the pre-image. With shears, the area of objects remains constant but the perimeter changes. Shears have been described as the pre-image being pulled in one or more directions, or as 'slices' of the original shape sliding over each other (Siemon et al., 2011). One common application of shears is the italicisation of letters from standard letter format.

Figure 9.12 displays how the unit square has undergone two transformations simultaneously; a shear parallel to y-axis scale factor 4. Accompanying this figure is the transformation matrix and mathematical operations which move the object from the pre-image stage to the image.

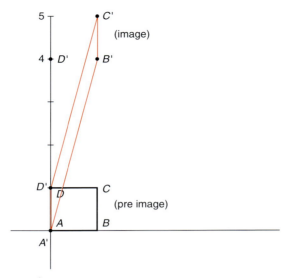

Figure 9.12 Shear transformation

$$\begin{bmatrix} 1 & 0 \\ 4 & 1 \end{bmatrix} \times \begin{bmatrix} 0 & 1 & 1 & 0 \\ 0 & 0 & 1 & 1 \end{bmatrix} = \begin{bmatrix} 0 & 1 & 1 & 0 \\ 0 & 4 & 5 & 1 \end{bmatrix}$$

Approaches to teaching transformation

Transformation in the Australian Curriculum

The topic of transformation is taught across both Number and Algebra and Measurement and Geometry strands. Although this focuses predominantly on the latter strand, we will consider in this section how transformation is presented in both strands. In Year 7, students transform planar objects according to translations, rotations and reflections – as well as identifying line and rotational symmetries. Year 8 students plot linear relationships on the Cartesian plane, and begin to examine how components of linear equations affect corresponding graphs. In Year 9, concepts of scale factor, similarity and enlargements are applied to planar figures; students also begin to graph non-linear relationships with and without technology. Year 10 students explore the connections between algebraic and graphical representations of functions, polynomials and relations. The transformation topics taught in middle secondary school are displayed in Table 9.6. In senior secondary school, students learn about transformations in all four courses: Essential Mathematics (Topic 2: Scale, Plans and Models), General Mathematics (Unit 1, Topic 3: Shape and Measurement; Unit 2, Topic 3: Linear Equations and their graphs), Mathematics Methods (Unit 1, Topic 1: Function and graphs and Topic 2: Trigonometric functions; Unit 2, Topic 1: Exponential functions) and Specialist Mathematics (Unit 2, Topic 1: Trigonometry and Topic 2: Matrices).

Table 9.6 Transformation in the Australian Curriculum (middle years)

Year 7	Describe translations, reflections in an axis, and rotations of multiples of 90° on the Cartesian plane using coordinates. Identify line and rotational symmetries (ACMMG181)		
Year 8	Plot linear relationships on the Cartesian plane with and without the use of digital technologies (ACMNA193)		
Year 9	Use the enlargement transformation to explain similarity and develop the conditions for triangles to be similar (ACMMG220)	Solve problems using ratio and scale factors in similar figures (ACMMG221)	Graph simple non-linear relations with and without the use of digital technologies and solve simple related equations (ACMNA296)
Year 10/10A	Explore the connection between algebraic and graphical representations of relations such as simple quadratics, circles and exponentials using digital technology as appropriate (ACMNA239)	Apply understanding of polynomials to sketch a range of curves and describe the features of these curves from their equation (ACMNA268)	

The van Hiele levels of geometric thought

In the 1950s Dina and Pierre van Hiele developed a hierarchical framework to suggest how students learn geometry. This sequential, five-level framework describes the thinking processes used by learners in geometric and spatial contexts, with the types of ideas referred to as 'objects of thought' (van Hiele, 1959). According to Crowley (1987), the products of thoughts (or the learning that has taken place) at any given level become the objects of thought at the next level. Assisted by appropriate instructional experiences, learners progress through the following five levels, each of which depends on successful achievement of previous levels (Fuys, Geddes & Tischler, 1988). These levels are now considered as: *visualisation, analysis, abstraction, deduction* and *rigour*.

Level 1: visualisation

Students at this level recognise and name figures based solely on the global visual characteristics of the figure, often comparing the figure to a known prototype. For instance, a square may be defined by a level 1 student as a square because 'it looks like a square' (Van de Walle, Karp & Bay-Williams, 2014). The general goal for learners is to explore 'shapes are alike and different and to use these ideas to create classes of shapes (both physically and mentally)' (Mason, 2014, p. 4).

Level 2: analysis

In level 2, learners analyse figures in terms of their components and properties, discover properties and rules of a class of shapes empirically, but do not explicitly interrelate figures or properties (Guven, 2012). To illustrate, learners are able to identify what makes a rectangle a rectangle (e.g. four sides, opposite sides parallel, opposite sides same length, four right angles, congruent diagonals) (Van de Walle, Karp & Bay-Williams, 2014).

Level 3: abstraction

Following the objects of thought from levels 1 and 2, students begin to perceive relationships between particular objects and the properties of these objects. Certain informal deductive statements (for instance, in the form of if–then reasoning) are elicited by learners when asked to create meaningful definitions and to justify their reasoning (Mason, 2014). Such a statement may be 'if all four angles are right angles, the shape must be a rectangle. If it is a square, all angles are right angles. If it is a square, all angles are right angles. If it is a square, then it must be a rectangle' (Van de Walle, Karp & Bay-Williams, 2014, p. 429). At this stage, the role and significance of formal deductive reasoning is not understood by learners.

Level 4: deduction

Level 4 learners are able to work with abstract statements about geometric properties and draw conclusions based more on logic than intuition (Van de Walle, Karp & Bay-Williams, 2014). To achieve this, learners are able to construct proofs, understand the role of axioms, definitions, theorems, corollaries and postulates, and appreciate the meaning of necessary and sufficient conditions for establishing geometric truth (Mason, 2014). At this level, students would be expected to construct proofs such as those typically found in a Year 11 Mathematics Methods class.

Level 5: rigour

At the highest level of the hierarchy, the objects of thought are axiomatic systems themselves, not merely the deductions within a system (Van de Walle, Karp & Bay-Williams, 2014). Learners – typically at the university level – can understand the use of indirect proof, contrapositive proof, non-Euclidian systems (Mason, 2014). For instance, a learner having already understood the formal aspects of deduction may establish theorems in different axiomatic systems, before analysing and comparing those systems (Guven, 2012). An example of this may be an analysis of spherical geometry, which is based on lines drawn on a sphere rather than in a plane or ordinary space (Van de Walle, Karp & Bay-Williams, 2014).

Phases of learning

Within the five levels, and supported with the appropriate instruction, learners progress through each level of thought via five phases of learning (van Hiele, 1959). These phases of learning are: *information, guided orientation, explicitation, free orientation* and *integration*.

Phase 1: information

In the initial phase, teachers engage students in conversations and activities about their prior topical knowledge. As a result of these engagements, students become oriented to the new topic and discover what direction further study will take (Mason, 2014).

Phase 2: guided orientation

Students are provided with the opportunity to explore the new topic of study through carefully structured and sequenced tasks, and learning materials. According to Crowley (1987, p. 5), the tasks should be designed to elicit specific responses that 'gradually reveal to the students the structures characteristic of this level'. For instance, a teacher may ask students to use an iPad app to create a parallelogram with the diagonals included – and then a larger and smaller version of the original parallelogram.

Phase 3: explicitation

During explicitation, students describe what they have learned about the topic in their own words (Mason, 2014). Compared with the preceding phases of learning, the teacher's role is minimal, other than to introduce relevant mathematical terms and to ensure that students use these terms accurately.

Phase 4: free orientation

In this phase, students encounter more complex tasks where they can apply the relationships they are learning to solve problems and investigate more open-ended tasks (Mason, 2014). These tasks may be open-ended in nature, they may contain multiple steps, or they could be flexibly designed for solution using various methods. Herein, Crowley (1987, p. 6) illustrates such a task: 'Fold a piece of paper in half, then in half again. Try to imagine what kind of figure you would get if you cut off the corner made by the folds on the corners at a 30 degree angle'.

Phase 5: integration

In the final phase, students review and summarise what they have learned. As a result of this review and summary, the goal is to integrate new knowledge with existing knowledge (Crowley, 1987; Mason, 2014).

Research-driven insight

The work of Hollebrands (2004) was mentioned earlier in this chapter. This researcher conducted task-based interviews with six Year 10 students who were selected from a class of 17 students. The students – none of whom had studied geometric transformations in primary or middle school – were purposely selected to reflect the range of mathematical abilities present within the class. Interviews took place before, during and after a seven-week instructional unit on geometric transformations. According to Hollebrands (2004, p. 208), the purpose of the initial interview 'was to gain insights into students' prior understandings of the four transformations that were going to be taught: translations, reflections, rotations, and dilations'. In general, students were presented with various transformative tasks on paper which comprised a picture of a polygon and a parameter (e.g. line of reflection, centre of dilation or rotation, translating vector). For instance, one type of task required students to draw the image of a polygon following a reflection, rotation, or translation. The researcher used the terms 'flip', 'slide' or 'turn' if students displayed uncertainty about the words presented. Although it was expected that students would find translations to be the easiest transformation to understand, the researchers' interviews revealed that these were the most difficult. Following an analysis of collected interview data, the key findings indicated that

> in general, students envisioned transformations as actions or motions performed on a figure; were unfamiliar with the role of parameters (lines of

reflection, translating vectors, centres of rotation); and did not focus on relationships among pre-images, images, and parameters (for example, corresponding pre-image and image points are equidistant from the centre of
rotation). (Hollebrands, 2004, p. 213).

According to the researcher, these conceptions were evidenced in students'
strategies, which appeared to rely on visual cues. Moreover, students who
seemed to rely only on visual information rather than on properties of transformations were more likely to draw images incorrectly. Again, while this
research reported on students' understandings of geometric transformations
(i.e. rotations, reflections, translations), the key findings can be considered by
teachers preparing to teach any units of work involving transformations of any
kind (e.g. functions and graphs, matrices).

ACTIVITY 9.3

Log in to HOTmaths and select the 'HOTmaths Global' Course list and then
choose the 'Middle Secondary' Course. From the Topic list choose 'Basic
non-linear graphs', then the 'Solving & graphing basic cubic equations'
Lesson.

The Resources tab contains the 'Graphs of basic cubics' widget that can be
used to facilitate learning about transformations of cubic functions. Using
this widget, develop a series of lessons for middle secondary students where

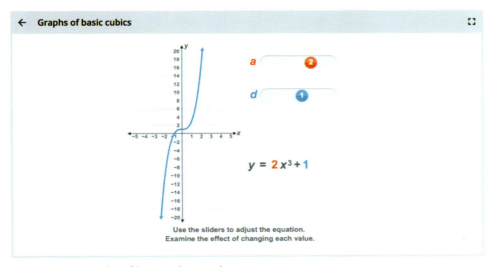

Figure 9.13 Graphs of basic cubics widget

they are able to learn about key transformations (i.e. dilations, reflections, translations) and how these can be applied to various functions. Using quadratics as a starting point, have students look at how these transformations also apply to cubic, reciprocal and exponential functions. Decide carefully on the types of questions you will use in scaffolding the learning experiences appropriate to the students' age level.

ACTIVITY 9.4

Log in to HOTmaths and select the 'HOTmaths Global' Course list and then choose the 'Middle Secondary' Course. From the Topic list choose 'Exploring functions', then the 'Inverse functions' Lesson.

Use the Resources to develop a series of lessons for senior secondary students where they are able to learn about key concepts regarding inverse functions (i.e. functions and relations, domain and range, vertical and horizontal line tests). Using linear functions as a starting point, have students determine inverses of quadratic, cubic, square root, exponential and logarithmic functions. Decide carefully on the types of questions you will use in scaffolding the learning experiences appropriate to the students' age level.

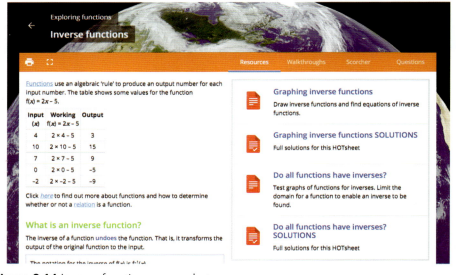

Figure 9.14 Inverse functions screenshot

REFLECTIVE QUESTIONS

Create a list of topics in secondary mathematics where transformations of any kind are used. When you have finished, check all middle and senior years courses in the Australian Curriculum to ensure that all topics have been included. Alongside each topic, write a statement indicating how students may experience conceptual difficulties in mastering the skills and concepts. This statement could be written in the form of 'common student misconceptions/errors' associated with the topic (e.g. confusing dilations parallel to the x-axis with those parallel to the y-axis). Which topic did you decide would be the most difficult for students to learn?

Using technology in measurement and geometry

The literature contains substantial evidence concerning the benefits of technology when learning mathematical concepts (Brown, 2010; Hopper, 2009; Kissane, 2007b). Researchers and teachers alike recall that several decades ago, hand-drawn diagrams were the only constructions possible for geometry students. Since the advent and increasing development of the digital age there are many dynamic geometry environments (DGE) within which to engage high school students. From a United States perspective, Goldenberg et al. (2012, p. 5) argue that geometry represents the only visually oriented mathematics offered to students, and curricula tend to present an 'otherwise visually impoverished, nearly totally linguistically mediated mathematics, a mathematics that does not use, train, or even appeal to the metaphorical right brain'. As such, many students who desire a visually rich mathematics curriculum – where visualisation and visual thinking serve not only as a potential hooks, but as the first opportunities to participate – have already dropped out before they can experience mathematics in an accessible way (Goldenberg et al., 2012). These comments represent a prevalent attitude for teachers to provide DGE for students of geometry. Doing so enables students to develop visualisation skills – which are reliant on considerable practice (Siemon et al., 2011) – and to explore geometrical problems (e.g. looking at what changes and what remains invariant after transformations). However, various factors impede the successful implementation and navigation of technology within the classroom (Allan, 2006; Goos & Bennison, 2008). In the two research summaries some of these factors are presented, concomitant with data-driven recommendations for best practice.

Research-driven insight

GeoGebra

Aventi et al. (2014) investigated the conceptual development of 25 Year 9 Australian students as they participated in a unit of work that used dynamic geometric software (DGS). Specifically, the researchers looked at (a) the characteristics of student responses when exploring linear relationships using GeoGebra, and (b) the nature of student interaction when using GeoGebra as an exploration tool. GeoGebra is free, open-source mathematics education software released in 2004. This software enables teachers and students to interact dynamically with mathematical topics such as geometry, algebra, statistics and calculus. According to researchers in mathematics education, one of GeoGebra's key strengths is that it can facilitate the mathematical learning of students at every year level from primary school to university (Hohenwarter et al., 2008; Kllogjeri & Shyti, 2010). Aventi et al. (2014) used a pre-experimental research design which involved using a pre-test, teaching intervention and end of topic test with a delayed post-test (six weeks delay). Relating to the first research question posed, the findings from this study indicate that

> GeoGebra assisted students in increasing the complexity of their responses when exploring linear relationships. At the completion of the teaching sequence, the use of correct terminology was evident with students linking the concepts of y-intercept and gradient to form the equation of a line. (Aventi, 2014, p. 84)

Regarding the second research question, students reported enjoyment at the dynamic nature of GeoGebra, as they could explore mathematical tasks quickly. Although students appeared initially reluctant to engage with the various DGS tools offered, the researchers noted that student familiarity and proficiency quickly increased as the teaching sequence progressed. Aventi et al. (2014) concluded that there are multiple benefits of using GeoGebra in the Linear Relationships sub-strand in the middle years. Besides students demonstrating an overall increase in recall knowledge from pre-test to post-test, benefits include students being assisted by GeoGebra to increase the complexity of their responses and using mathematical terminology correctly. However, the researchers underscore that when using this DGS it is essential to build familiarity with the tools so students can focus on the exploration at hand, rather than the tools they are using to explore it.

Assessment items

Lowrie et al. (2014) explored how 807 Singaporean grade 6 students (aged 11–12 years) processed spatially demanding graphic tasks using either pencil and paper or iPad. The researchers divided students into groups according to their spatial visualisation ability (low, medium and high). Then, samples of each ability group were asked to complete two tasks – namely, the Symmetry task and the Street Map task – using either pencil and paper or an iPad. The Symmetry task asked students to consider an image that required folding across a line of reflection. For the Street Map task, students had to super-impose and rotate a visual compass from its usual North position on the given graphic. For both tasks, the researchers noted that students with high spatial visualisation ability performed at a much higher level than those who pos-sessed medium or low spatial visualisation ability.

The results of the study reveal significant differences in performance across the two test modes, and there were equal differences in strategy used across the two tasks on iPad and pencil and paper. For the Symmetry task, there were performance differences in favour of those students who used iPads. The researchers hypothesised that the iPad encouraged students to reflect men-tally the given object in their mind's eye. Those students who completed this task in a pencil-and-paper mode were also more likely to use this approach, but they did so with less success. Because the students could not physically fold the object across the line of symmetry on the iPad (as they could have done on paper), the researchers speculated that the digital mode prompted them to mentally reflect the given object. By contrast, Lowrie et al. (2014, p. 434) noted

> those students who completed the Street Map task in a pencil-and-paper form scored higher than those students who solved it on iPad. The iPad students used a variety of strategies to solve the task, with the highest proportion using imag-ery to evoke a mental representation of a compass indicating the North direction.

By contrast, the pencil-and-paper mode tended to encourage students to draw a compass on the diagram. Such an encoding strategy produced a higher pro-portion of correct responses. Despite having working-out paper, iPad users appeared less likely to draw, and success rates were lower when students chose this strategy.

The results from this study hold several important educational implica-tions. First, the differences in test mode (i.e. pencil and paper, and iPad) appear to influence students' mathematical performance on graphic tasks with spa-tial demands. Given the announcement of the National Assessment Program – Literacy and Numeracy (NAPLAN) will be implemented in 2016 using a digital online environment, it may be difficult to compare the digital-based perfor-mance of students to earlier cohorts of students who used non-digital tests. Second, there may be scope for the Australian Curriculum to include visuo-spatial reasoning skills, especially given the differences in students' strategy in completing both tasks. Third, the finding that iPad users were less likely to monitor their thinking via diagram sketching or annotation positions educators wishing to use such technology with some advanced consideration.

ACTIVITY 9.5

Looking at the previous section it is apparent that technology has a role to play in middle secondary school mathematics. Log in to HOTmaths and select the 'HOTmaths Global' Course list and then choose the 'Middle Secondary' Course. From the Topic list choose 'Using formal proofs in geometry', then the 'Extending the concept of similarity' Lesson.

Your task is to plan two lessons that demonstrate the utility of technology in geometric proofs and mathematical reasoning, and which are aimed at a middle secondary audience. In the first lesson, your students are given a structured lesson to investigate similarity in triangles. For the second lesson, students must creatively design a learning experience (e.g. podcast, YouTube clip, PowerPoint presentation) where they can showcase their technological proficiency and mathematical expertise. The learning experience should be for a small group of their peers, and a marking rubric should be available to provide feedback and a mark for each student.

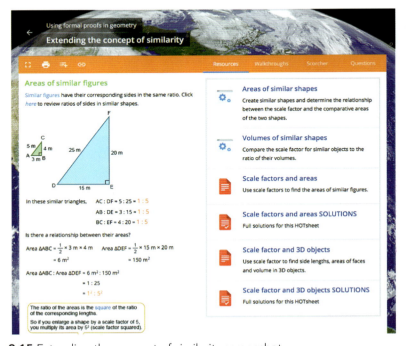

Figure 9.15 Extending the concept of similarity screenshot

ACTIVITY 9.6

Looking at the previous section it is apparent that technology has a role to play in senior secondary school mathematics. Log in to HOTmaths and select the 'HOTmaths Global' Course list and then choose the 'Middle Secondary' Course. From the Topic list choose 'Using formal proofs in geometry', then the 'Proofs & similar triangles' Lesson.

Your task is to plan two lessons for senior secondary students to demonstrate the utility of technology in geometric proofs and mathematical reasoning. In the first lesson, students are given a structured lesson to investigate similarity in triangles. For the second lesson, your senior secondary students must creatively design a learning experience (e.g. podcast, YouTube clip, PowerPoint presentation) where they can showcase their technological proficiency and mathematical expertise. The learning experience should be for a small group of their peers, and a marking rubric should be available to provide feedback and a mark for each student.

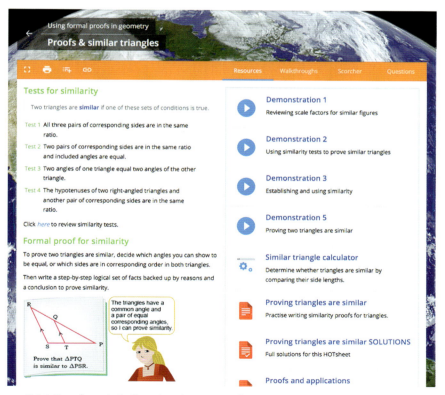

Figure 9.16 Proofs and similar triangles screenshot

REFLECTIVE QUESTIONS

The acronym ICT is generally understood to be expanded as 'Information and Communications Technology', although an interesting alternative for educators is 'It Can't Teach'. In light of this latter expansion of the ICT acronym, outline your philosophy of using technology to support student learning across Years 7–12. Also, describe what you believe to be your strengths and areas for improvement regarding the integration of technologies into a secondary mathematics classroom.

Summary

Outline the importance of measurement and geometry tasks within the secondary mathematics classroom

Geometry is an important and essential branch of the mathematics curriculum at all year levels, as it develops both logical and deductive thinking, which in turn helps learners to expand mentally and mathematically. Measurement is a key component of the mathematics curriculum, and it is considered the most practical and hands-on application of mathematics in everything from occupational tasks to day-to-day-life.

Develop an understanding of the concept of geometric proof as articulated in the Australian Curriculum

The concept of proof is considered central to the mathematics discipline and consequently it should feature prominently in mathematics education. Because of this centrality, proof is an essential tool for promoting mathematical understanding in students and at the same time, for providing educators with insight about how students learn mathematics. When teaching geometric proof, mathematics educators consider three approaches, namely: synthetic, analytic and transformational.

Develop an understanding of the concept of geometric transformation as articulated in the Australian Curriculum

In mathematics the term 'transformation' refers to the changes in positions or size of shapes, which collectively comprise the study of translations, reflections, rotations (slides, flips and turns), symmetry and similarity. Broadly speaking, there are two categories under which all geometric transformations belong – rigid transformations and non-rigid transformations. Transformations that do not change the size or shape of an object moved are called rigid transformations. Examples of rigid transformations include: translations (or slides), reflections (or flips), rotations (or turns), and symmetry. Non-rigid transformations change the size or shape of an object, and two categories of these are similar transformations and shear transformations.

Describe how technology can be used to support learning concepts in measurement and geometry

Existing literature contains substantial evidence concerning the benefits of technology when learning mathematical concepts. Researchers and teachers alike recall that several decades ago hand-drawn diagrams were the only constructions possible for geometry students. Since the advent and increasing development of the digital age, there are many dynamic geometry environments (DGE) within which to engage high school students. However, various factors impede the successful implementation and navigation of technology within the classroom – and these factors must be carefully considered when supporting any learning experiences with technology.

CHAPTER 10

Statistics and probability

Learning outcomes

After studying this chapter, you should be able to:

- describe why data are needed and how data can be produced
- describe the processes and issues involved in the calculation of descriptive statistics
- describe the processes and issues involved in producing graphical and tabular displays
- describe the principles of probability
- describe how statistical inferences are reached.

Introduction

Quantitative information is everywhere, and many decisions that we make, or are made for us, are based on statistical data. For example, whether or not a drug is allowed onto the market depends on an analysis of its efficacy and safety. In addition, quantitative data are often used to persuade us to alter our behaviour, such as to vote for a particular political party or to buy a certain brand of toothpaste (Ben-Zvi & Garfield, 2004). Yet, it is common for decisions such as these to be made on incomplete data. For example, the polling of the electorate that takes place during an election campaign surveys only a small proportion of the voting population. During this time, campaign managers make daily adjustments depending on these polls and it is just too expensive and time consuming to poll all the voters. It is the study of the collection, organisation, analysis and interpretation of such data that makes up the discipline of statistics (Jones, Langrall & Mooney, 2007).

Unfortunately, the teaching of statistics often emphasises skills, procedures and computations that do not promote understanding of the collection and interpretation of data. Many adults in our society cannot think statistically about the information they are given which affects their lives (Jones, Langrall & Mooney, 2007). Therefore, this chapter emphasises the importance of encouraging students to take a holistic view of the discipline of statistics.

It has to be acknowledged that there is keen debate as to whether the discipline of statistics is a part of mathematics at all. Some writers (e.g. Moore, 1990) claim that statistics is an independent discipline that, like science, relies heavily on mathematics. Whether or not you agree with this argument, it should be appreciated that unlike mathematics, where numbers may be used in the abstract, statistics always uses numbers in a context.

KEY TERMS

- **Biased:** the characteristic of a statistic to give consistently high or low values. A sampling method that favours one part of a population over another.
- **Bi-modal data:** a data set that has two values with the highest frequency.
- **Box plot (box-and-whisker plot):** a graph that divides the data into quartiles. A rectangle (the box) represents the middle 50% of the data. The lower edge of the box is placed at Q_1, and the upper edge is placed at Q_3. A line through the box represents the value of the median. Any data outside the middle 50% is represented by the lines (sometimes known as whiskers). If there is a value in the data set that is more than $3 \times$ IQR times lower than Q_1, or higher than Q_3, this value is deemed to be an outlier and is represented by a dot or star.
- **Complement:** the complement of a set A (consisting of the items of interest) is all the outcomes in the sample space that are not included in the set A.
- **Conditional probability:** the probability of an event 'A' given that another event 'B' has occurred. The notation is $P(A|B)$. For this conditional probability the sample space is reduced to B.
- **Event:** the subset of interest of a sample space in a random experiment.
- **Experiment:** the carrying out of several repetitions of a random process.
- **Frequency table:** a list of the values in a data set with their corresponding frequencies.

- **Histogram:** a bar graph where the data are grouped into classes of equal intervals and the bar heights represent the frequency of each class. Because histograms are used for continuous data there are no gaps between the bars.
- **Independent:** the characteristic of an outcome to have no effect on the probability of any other outcome.
- **Inference:** the process of using a sample to draw conclusions about an entire population.
- **Interquartile range:** the value that results when the value of the first quartile is subtracted from the value of the third quartile. This gives the range of the middle 50% of the data.
- **Intersection ('and'):** the subspace in the sample space where two events occur together.
- **Measures of centre:** statistics that are used to represent a data set. These include the mean, median and mode.
- **Mean:** a measure of centre. The mean is calculated by adding up the values of the data set and dividing by the number of values.
- **Measures of spread:** statistics that are used to represent the width or spread of a data set.
- **Median:** a measure of centre. The median is the value that divides the data into halves; 50% of the data are below this value, and 50% of the data are above.
- **Mode**: the value in a data set with the highest frequency.
- **Outlier:** a value in a data set that is substantially higher or lower than the rest of the data.
- **Parameter:** a descriptive statistic that applies to the entire population.
- **Percentile:** see **Quartile**.
- **Population:** the entire collection of the items of interest.
- **Probability:** the mathematical representation of the chance of an event occurring expressed as a fraction or decimal between zero and one. A probability of one indicates that the event of interest will definitely occur, and a probability of zero indicates the event will definitely not occur.
- **Probability (classical):** classical probability applies to equally likely events, where the calculation of the probability is found by dividing the number of items of interest by the total number of items in the sample space.
- **Probability (empirical):** empirical probability is that which is determined by experiment.
- **Probability (frequentist):** the frequentist idea of probability sees the probability of an event as equal to the long-term frequency of the event if a random process was repeated many times.

- **Probability (subjective):** a probability that is estimated by the knowledge a person holds about a situation.
- **Probability distribution:** the list of all possible outcomes in a random experiment with their accompanying probability of occurrence.
- **Quartile:** the first quartile (represented by Q_1) is the value that divides a data set so that 25% of the data is below it, and 75% above it. The third quartile (represented by Q_3) is the value in the data set that divides a data set so that 75% of the data is below it, and 25% of the data is above it. The second quartile is equivalent to the median. These values are also known as the 25th, 50th and 75th percentiles.
- **Random:** a process where each individual outcome cannot be determined but a long-term pattern in outcomes exists.
- **Random sampling:** a process of taking samples where every object of interest has the same probability of being selected, and the selection of one object has no influence on the probability of selecting another object.
- **Range:** the value that results from subtracting the lowest value from the highest value in a data set.
- **Representative sample:** a sample whose characteristics are in the same proportion as in the population.
- **Sample:** a set of data selected from a population often with the purpose of giving information about the population.
- **Sample space:** the list of all possible outcomes in a random experiment.
- **Scatter plot:** a graph where Cartesian coordinates are used to represent the relationship between two numerical variables.
- **Standard deviation:** the mean value of the square root of the sum of squared deviations from the mean. If the data represents the entire data set the denominator is N, the number of values in the data set. If the data is a sample then the denominator is $N - 1$, one less than the number of values in the data set.
- **Statistics (discipline of):** the study of the collection, organisation, analysis and interpretation of data.
- **Statistics (descriptive):** calculated numbers that are used to summarise data. Examples of statistics include the mean, median, mode and standard deviation.
- **Table:** a set of values arranged in rows and columns with the purpose of summarising or making sense of the data.
- **Transnumeration:** the process of transforming the representation of data so that features of the data can be recognised and described. Examples of transnumeration include the making of tables and the drawing of graphs.
- **Trial:** a single repetition in a random process.

- **Union ('or'):** the subspace in the sample space where two or more events can occur together.
- **Variable (noun):** a quantity whose value may change in a random experiment.
- **Variability:** the tendency of a quantity to have different values.
- **Variation:** the measurement of variability.

The need for data and its collection

'Many real world situations cannot be judged without the gathering and analysis of properly collected data' (Pfannkuch & Wild, 2004, p. 18). Whether or not a new school is built, or an old one is closed down depends on data. The food stocked on our supermarket shelves, the cars we buy, the latest health recommendations all depend on data. Not all of these data are properly collected, however, and it is important that students are exposed to data collection from a young age.

Data collection

Students, especially young students, are usually very interested in themselves, so early data collection usually involves the students collecting data about themselves. As they progress through school, however, they should be encouraged to broaden their basis to real data collected by others (Curcio, 1989). The development of the internet means that there are now extensive real data sets available. Such data are available from the Australian Bureau of Statistics (www.abs.gov.au), The Australian Bureau of Meteorology (www.bom.gov.au/climate/?ref=ftr) and Australian Government data (data.gov.au).

Sampling

Much of the data that are analysed are based on samples, incomplete selections from populations, and this leads to one of the main difficulties with statistical analyses. Individuals vary and as a consequence samples vary. This is one of the most important messages in **statistics**. If we wanted to know the **mean** weight and height for Year 8 students in a large city we could take a **sample** of, say, 100 students. It is very unlikely, however, that another sample of 100 students would have the same mean weight and height as the first one. In addition, neither sample is likely to have the same mean weight and height as the entire **population**. Therefore conclusions based on samples always contain an element of uncertainty.

It would be possible to avoid this uncertainty by measuring the entire population, but this rarely occurs. We noted above that the surveying of the entire voting population is too expensive and time consuming during an election campaign. There are other times when examining the entire population is not possible, such as in a study of the wasps in a forest, or a population of fish.

Sampling in the Australian Curriculum

Issues of sampling arise in the Australian Curriculum: Mathematics in Years 7, 8, and 9:

> Year 7: Identify and investigate issues involving numerical data collected from primary and secondary sources. (ACMSP169)
>
> Year 8: Investigate techniques for collecting data, including census, sampling and observation. (ACMSP284)
>
> Year 9: Investigate reports of surveys in digital media and elsewhere for information on how data were obtained to estimate population means and medians. (ACMSP227) (ACARA, 2015f)

It is apparent from these content descriptors that students should not just be collecting data from themselves, but also investigating data from other sources and considering the nature of the collection.

Understanding sampling

The use of the word 'sample' is an example of where a word in common use has a different meaning in everyday use from its use in statistics (and mathematics in general). Care must also be taken that in this context the term is not confused with **sample space**. In contrast to many of the samples taken for statistical analysis, some of the samples students may be familiar with are representative of the rest of the population. Such samples include food samples, blood samples and seed samples. In cases such as these there are procedures to follow to ensure that the mixture is homogeneous and representative of the whole. It is important that the nature of these differences in word use be made explicit to students (Watson, 2006).

When a sample is taken it is hoped that it might be *representative* of the population. A **representative sample** is one where the sample contains the same proportions of each characteristic as found in the population (Kahneman & Tversky, 1982). The larger a sample is, the more representative it is likely to be of the parent population. This is often referred to the 'law of large numbers'. It is difficult to put this idea into practice. When is a sample large enough? There comes a time when increasing a sample size becomes too expensive or time consuming to be worth the extra information obtained (Watson, 2006). In contrast, people have difficulty in realising that a relatively small sample in a poll can give very good estimates of the entire voting population and therefore may reject conclusions drawn from such polls.

It is also important that students consider what factors may result in sampling bias and be encouraged to think of this when analysing media reports or when designing their own data collection. Television stations sometimes ask a question of the audience and give telephone numbers to ring. Will such a poll be representative of the entire population of Australia? Students can also consider what could happen if the wording of a question was changed – could they write a question that will be more likely to produce the answer they want? They can also consider how they might collect data that would be more likely to be representative of their school. How should they conduct this survey if they do not have the time and resources to ask everyone? They might decide to ask members of each class. By using this process they can be introduced to stratified sampling. Newspaper polls often describe their sampling methods and they can compare these with their own plans.

ACTIVITY 10.1

Different samples

Log in to HOTmaths and select the 'Australian Curriculum' Course list and then choose the 'AC Year 8' Course. From the Topic list choose 'Data', then the 'Collecting & describing data' Lesson. From the Resources tab, select the 'Different samples' HOTsheet.

This HOTsheet illustrates how samples can vary from the parent population. It also suggests using technology (spreadsheets and calculators) to identify which members of the population will be selected; these techniques are used by practising statisticians. Try out the exercises on this HOTsheet. What can students learn from such an exercise?

HOTsheet   Collecting & describing data

DIFFERENT SAMPLES

The table below shows 400 raw, unsorted marks for a mathematics test.

The mean (average) of this data is **70.31**.

The first column in the table can be used to help you find a mark in a particular position eg 46th mark.

Position																
1 to 25	51	85	85	84	45	53	53	88	68	100	47	74	43	73	75	93 66 40 61 94	69 71 90 45 69
26 to 50	44	73	86	45	85	82	100	78	78	84	60	88	74	76	72	45 59 93 47 66	90 89 79 51 43
51 to 75	77	72	50	60	64	77	99	54	45	45	79	48	53	70	81	89 45 87 93 63	67 63 49 76 69
76 to 100	83	96	67	57	88	96	83	69	97	92	75	87	60	54	89	43 62 41 41 58	71 77 88 84 74
101 to 125	90	62	72	71	90	81	96	47	41	70	58	64	93	82	63	91 72 74 67 54	100 85 58 78 67
126 to 150	41	84	43	99	71	77	41	98	49	65	52	87	96	41	62	42 41 66 62 79	45 56 41 100 51
151 to 175	52	85	61	98	76	44	85	87	76	93	92	73	53	47	62	94 63 97 93 98	80 98 40 85 43
176 to 200	62	44	75	66	71	94	82	73	94	66	49	58	75	94	65	53 73 70 42 71	43 93 70 60 62
201 to 225	57	75	72	60	40	55	75	58	93	96	48	84	91	54	98	76 44 95 45 89	95 57 72 87 50
226 to 250	78	86	63	48	43	40	96	79	64	67	60	98	42	77	78	93 51 92 71 65	58 59 62 53 89
251 to 275	40	41	52	81	66	58	79	96	41	95	73	86	98	84	67	52 83 66 61 65	72 98 88 81 73
276 to 300	88	84	62	89	77	42	72	42	94	84	56	59	78	47	54	94 62 86 93 77	93 44 62 91 75
301 to 325	90	43	93	66	69	93	75	94	46	76	46	65	99	86	61	91 68 53 76 94	63 97 50 74 49
326 to 350	95	72	94	73	83	65	84	98	56	45	66	100	67	75	54	57 56 77 98 92	85 64 73 64 69
351 to 375	50	45	90	43	86	85	91	52	51	85	58	5	45	97	76	47 78 78 64 45	85 67 75 75 64
376 to 400	70	70	85	43	45	44	77	67	54	79	58	83	90	79	80	56 98 57 47 41	89 60 69 92 83

Figure 10.1 Different samples HOTsheet

ACTIVITY 10.2

Log in to HOTmaths and select the 'Australian Curriculum' Course list and then choose the 'AC Year 8' Course. From the Topic list choose 'Data', then the 'Collecting & describing data' Lesson. From the Resources tab, select the 'Trusting conclusions' HOTsheet.

This HOTsheet illustrates common situations involved in sampling. It is critical that students learn about the issues that relate to taking samples and not just analyse data from textbooks. What can students learn from such an exercise?

 HOTsheet   Collecting & describing data

TRUSTING CONCLUSIONS

TASK 1   **Considering bias**

People were surveyed as they left a supermarket.

List situations in which this would be an extremely biased sampling method and ones in which it would not be as biased.

_____

_____

_____

_____

Figure 10.2 Trusting conclusions HOTsheet

REFLECTIVE QUESTION

When students are asked to look at the results of polls of voting intention in newspapers they are often surprised to see how few people take part in the poll. They find it difficult to see how a reliable result could possibly be obtained. How are the participants chosen in these polls? Could you explain to your students why large numbers of participants (for example, over 2000 people) are not needed?

Descriptive statistics, tables and graphs

Once the data have been collected it is important that they are organised in a way so that the reader can make sense of them, to allow the data to tell a story. One way of getting the data to tell a story is to calculate what are known as descriptive statistics. Another way is to graph the data. The process of changing the representation of data to tell a story is known as **transnumeration** and is a vital part of statistical analysis (Pfannkuch & Wild, 2004).

Descriptive statistics

The discipline of statistics was described earlier in the chapter. Within this discipline there are numbers that are also called statistics. These are the calculated numbers designed to describe data; examples are the mean and **standard deviation**. The purpose of such calculated statistics is to find numbers that are somehow representative of all the data – thus their name *descriptive statistics*. There are two general types of descriptive statistics, those that give a measure of the centre of the data, and those that give a measure of the spread of the data. Research shows that much more teaching time is spent on the **measures of centre**; it is important that the issue of measuring **variability** is also emphasised.

Descriptive statistics in the curriculum

The commonly used measures of centre (the mean, **median** and **mode**) are reflected in the National Curriculum: Mathematics in Years 7, 8, 9, 10 and 10A.

Year 7: Calculate mean, median, mode and range for sets of data. Interpret these statistics in the context of data. (ACMSP171)

Year 8: Investigate the effect of individual data values, including outliers, on the mean and median. (ACMSP207)

Year 9: Compare data displays using mean, median and range to describe and interpret numerical data sets in terms of location (centre) and spread. (ACMSP283)

Year 10: Determine quartiles and interquartile range. (ACMSP248)

Year 10A: Calculate and interpret mean and standard deviation of data and use these to compare data sets. (ACMSP279)

These descriptors place an emphasis on interpretation. Yet, in many classrooms the concentration is on the teaching of the procedures. It is important that such calculations are carried out in the context of investigations, so that the use of these statistics is meaningful. The next sections describe commonly used measures of centre and spread.

Measures of centre

Measures of centre are used to find a number that is representative of the entire data set. The most commonly used measures of centre, or averages, are the mean, median and mode. It is essential that students should not only know how these statistics are calculated, but also know the different properties of each of these statistics, and come to appreciate why one might be used in preference to the other. The next section will describe the most commonly used statistics for these measures of centre.

The mean

The mean is calculated by adding up the values in the data set and dividing by the number of values in the data. By the time students reach secondary school they will probably be familiar with how it is calculated, but unfortunately for many students their understanding of the mean stops at knowing how to calculate it. They cannot explain why one would even go to the effort of calculating the mean (Mokros & Russell, 1995; Groth & Bergner, 2006). Because these students do not realise that a mean is somehow representative of data, they do not use it when it would be useful, such as to describe data or to compare data sets (Konold & Pollatsek, 1995; Reaburn, 2012). The mean is used to find a representative number for a data set but can be thought of in several ways. These will be explained now.

The mean as even share

The mean is also the value that you would get if all things were shared equally. Students can be led to find this out for themselves. Figure 10.3 shows five baskets, each with a different number of apples. If the apples are shared out evenly, each basket will end up with four apples. When the students calculate the mean number they will find it to be four. Therefore it can be said that the mean number of apples is 'four per basket'.

Figure 10.3 The distribution of apples between five baskets (left) and the number in each basket when they are evenly shared (right)

The mean as balance point

The mean is also the point where the distribution of all the numbers is balanced, and this too can be demonstrated physically. For example, if three coins are placed on a ruler at, for example, 4 cm, 17 cm and 24 cm, the balance point (found by balancing the ruler on a whiteboard marker or similar taped on the desk) will be at 15 cm – the mean of the three values (Figure 10.4).

Figure 10.4 Three coins balanced on a ruler: the point of balance is the mean of the three distances

The sum of the deviations from the mean is zero

Students should be introduced to the idea of examining how far each number is from the mean – the deviation from the mean. This idea becomes critically important for later understanding of the standard deviation. In the previous example students can physically measure the distance to each coin below the mean, and measure the distance to each coin above the mean, and find that the sum of the distances below the mean equals the sum of the distances above the mean. This can also be shown arithmetically with the apple example (Table 10.1).

Table 10.1 The deviation of each data value from the mean in the apple example.

Number of apples	Mean number of apples	Number of apples – mean number of apples
2	4	−2
5	4	1
6	4	2
1	4	−3
6	4	2
	Sum =	0

Common misunderstandings about the mean

The apple example illustrates one difficulty students may have with understanding the mean. The mean is somehow supposed to be representative of the data, yet no value in the data necessarily has the same value as the mean. This can be confusing for students who regard the mean as the most common value, or the most likely value. Another source of confusion comes from statements such as 'The mean number of children per family is 2.3' (Mokros & Russell, 1995). How can we have 0.3 of a child? This confusion can be overcome by a classroom activity that asks the students to distribute objects so that the mean has a fractional value. An example is shown in Figure 10.5.

The formula for the mean

It is very helpful for students if they can learn to read formulae, and the formula for the calculation of the mean is relatively simple. Students can be encouraged to see the formula as a series of instructions, a recipe even.

$$\bar{x} = \frac{\Sigma x}{N}$$

The symbol 'Σ' gives us the instruction to add up all the values in the data set. Then this number is divided by the number of data values, N.

> Here is a street with 8 houses. Can you find a way of putting children into the houses so that the street has a mean of 1.5 children per house?
> Draw the children in their gardens.

Figure 10.5 An example of an activity that builds understanding of fractional means

The median

The median is relatively simple to calculate. The values in the data set are laid out in order and, if the number of values is uneven, the middle value is chosen. If there is an even number of values, then the mean of the two middle numbers is chosen. There is a formula for finding the position of the median, but if students understand what the median represents this formula is not needed. The carrying out of calculations without formulae that need to be learned should always be encouraged and this is particularly important in statistics, as this is one area notorious for students following procedures without understanding (Garfield & Ahlgren, 1988). If students know that the median is the number that divides the data set into two equal halves (50% above, 50% below) then they should be able to calculate its value.

The question arises as to why the median may be preferred to the mean. This is because the median is resistant to changes if there is a value in the data far away from the rest of the data (an **outlier**). This is illustrated in Figure 10.6. There is close to a 500 mg/100 g difference between the mean and median. Why does this occur? Which of the mean and median is the best statistic to represent this data?

It is fruitful for students to explore such data and discuss why there may be differences between the values of the mean and median. One way of introducing students to the effects of outliers on the values of each statistic is for them to calculate the values of the mean and median for data that are reasonably close together, and then add an outlier. If a spreadsheet is used the change in the mean is automatic, and students often enjoy putting in outrageous values

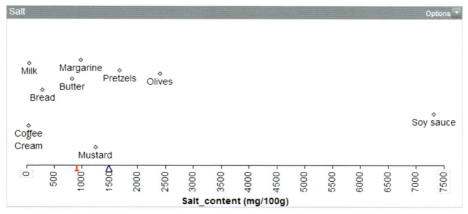

Figure 10.6 The salt content (mg/100 g) of various foods. The median value is shown by the red marker (907 mg/100 g) and the mean value is indicated by the blue triangle (1490 mg/100 g). Graph produced in Tinker Plots.

for the outlier to see what happens. Such explorations assist students to gain an understanding of these statistics in a way that merely practising the use of the formulae cannot.

The mode

The mode is the value in the data that occurs most often. If there are two values with the same highest frequency then the data are said to be bimodal. If the data have more than two numbers with the same highest frequency then the data are said to have no mode. In practice, the mode is often used to help describe the data, but is not used when comparing two or more data sets. Students can be challenged to suggest ways in which the mode could be useful, such as ordering stock into shops. There are also times when the realisation that data are bimodal can point to something important in the data, such as gender differences.

Measures of spread

Variability and variation

Without variability the discipline of statistics would not exist. If all Year 8 students had the same height, then measuring one would measure all. Without variability we would not have to be concerned that the sample we have taken might not represent the population as a whole. The words **variability** and **variation** are often used interchangeably, but in this chapter the words will be used as recommended by Shaughnessy (2007). Variability describes the tendency of a quantity to have different values (for example: the height of Year 8 students). Variation is the measure of this variability. We have already seen that data within a data set vary and that samples vary from each other. The pattern, or distribution of the data, also varies. This section will concentrate on measuring the variability; the so-called **measures of spread** variation.

Why measure variability? It is important in our continuing quest to tell the story of our data. Students should be able to see that two data sets can have the same mean but one can spread much more widely than the other. Without a description of variability this important feature will be missed, as could other important features of the data (Shaughnessy, 2007). Commonly used measures of spread include the **range** and the **interquartile range**. Of high importance, though, is the standard deviation.

The standard deviation

The standard deviation is fairly easy to calculate, and if students can read the formula for the mean they should be able to read the formula for the standard deviation. Many hours can be spent by students calculating a standard deviation, but it is not always explained why the calculations should be carried out. One way of looking at the standard deviation is to regard it as the mean of the deviations from the mean. More precisely, the standard deviation is the square root of the squared deviations from the mean, divided by the number of values in the population. If the data belong to a sample, the divisor is one less than the number of values in the sample.

An intuitive understanding of the standard deviation can be developed by asking about the mean of the differences (or more formally, *deviations*) between each value in a data set and the mean. If we subtract each value from the mean, though, the deviations add up to zero, as the apple example illustrated. One option is to think of these deviations as distances and this will allow all the measures to be positive. This should make sense even if the students are not familiar with the absolute value function. In practice, though, the effect of the negative deviations is dealt with by squaring all the deviations. This is illustrated by the apple example in Table 10.2.

Now we can find the mean of these squared deviations. This number is known as the variance. Because this is the mean of the squared deviations, we then take the square root of the answer. So to finish our calculations:

$$\text{Variance} = 22 \div 5 = 4.4$$

$$\text{Standard deviation} = \sqrt{4.4} = 2.1$$

Table 10.2 The calculation of the squared deviations from the mean in the apple example

x_i	$\bar{x}$	$x_i - \bar{x}$	$(x_i - \bar{x})^2$
2	4	−2	4
5	4	1	1
6	4	2	4
1	4	−3	9
6	4	2	4
	Sum =	0	22

If students realise that the standard deviation is the square root of the mean of the squared deviations from the mean, they will not even need to 'learn' the formula. The formula is presented here:

$$\text{Standard deviation}\,(\sigma) = \frac{\sum (x - \bar{x})^2}{N}$$

It can help students if they are encouraged to write down what the formula is telling them to do in words.

The question then arises why the squared deviations are used, and not the absolute values of these deviations in these calculations. The reason is that there are times when these deviations need to be minimised (such as in linear regression). Graphs of absolute functions have corners and flat spots and therefore they are not always differentiable, in contrast to graphs of squared functions (Shaughnessy & Chance, 2005). This may be a difficult concept for students who may not have been introduced to calculus. However, they should be able to understand that squaring is an option when one is interested in the size of deviations, and not the sign.

Do we divide by N or N − 1?

The formula for the standard deviation states that the sum of all the squared deviations should be divided by N, the population size. Depending on the brand of scientific calculator, students may observe the following two notations for the standard deviation:

$$\sigma_n \text{ and } \sigma_{n-1}$$

When the calculations are being carried out on a sample, the division is by one less than the sample size ($N - 1$) and the relevant symbol on their calculator is σ_{n-1}. When the calculations are being carried out on an entire population, then the division is by N. Student textbooks do not always make this clear. This problem is exacerbated because these symbols do not follow the correct mathematical convention. The use of the Greek letters should only be reserved for the results of calculations on entire populations (population **parameters**). This needs to be made explicit to the students. The correct notation for the mean and standard deviation for populations and samples is summarised in Table 10.3.

Table 10.3 Notation for the population and sample mean and standard deviation

	Population	Sample
Mean	μ	$\overline{x}$
Standard deviation	σ	s

Why are sample standard deviations divided by one less than the sample size? One reason samples are taken is to give an estimation of the population mean and standard deviation. It can be demonstrated that when sample standard deviations are calculated by dividing by N the standard deviation of the population gives consistently low answers. In mathematical terms, they are **biased**. When the sample standard is divided by $N - 1$ (a smaller denominator), this is to some extent corrected. This can be demonstrated by computer simulation. If students take repeated samples from a population they will find that the sample standard deviation is usually smaller than the population standard deviation (Shaughnessy & Chance, 2005).

Teaching statistics

Much of the teaching of statistics is based on calculating the measures of centre and spread. It is essential that students be given exercises that go further than practising procedures but enhance understanding of these calculations. For example, which 'average' is better for this data set? Can they make up two data sets with different means but the same standard deviation? What could a data set look like if the median is greater than the mean? How can a class have an average of 1.2 footballs each? Questions like these can enhance understanding. In the next section the other forms of transnumeration will be described: the making of graphs and tables.

ACTIVITY 10.3

Log in to HOTmaths and select the 'Australian Curriculum' Course list and then choose the 'AC Year 7' Course. From the Topic list choose 'Mean, median, mode & range', then the 'Exploring the mean' Lesson. From the Resources tab, select the 'Mean problems' HOTsheet.

This HOTsheet provides a range of examples that require both procedural and conceptual knowledge. The challenge question at the end is particularly worthwhile to increase students' interest. Try this for yourself. What do students need to know and understand to complete such an exercise?

 HOTsheet   **Exploring the mean**

MEAN PROBLEMS

TASK 1   **Changing data**

1 Calculate the mean of these four scores: {8, 5, 2, 4}

2 A score of 6 is to be included with the four scores above. Predict whether the mean of the *five* scores will be more or less than the mean of the original four. Explain your answer.

_____

_____

3 Calculate the mean of {8, 5, 2, 4, 6} to check your prediction.

Figure 10.7 Mean problems HOTsheet

ACTIVITY 10.4

Log in to HOTmaths and select the 'Australian Curriculum' Course list and then choose the 'AC Year 9' Course. From the Topic list choose 'Data & graphs', then the 'Summarising data' Lesson. From the Resources tab, select the 'Statistics from graphs and tables' HOTsheet.

This HOTsheet contains examples of calculating the mean, median, mode and range for grouped data. It also gives an example of a dot plot. There is also the challenge of creating a dot plot for data with given descriptive statistics. This exercise requires deeper understanding than just calculating descriptive statistics. Try this for yourself. At each step, note down what students will need to know and understand if they are to complete this exercise successfully.

 HOTsheet   **Summarising data**

STATISTICS FROM GRAPHS AND TABLES

TASK 1   **Starry sprinkles**

A factory makes candy stars for decorating cakes. In a quality control study, the stars in 200 randomly selected packets were counted and the results recorded. Use the data in the table to determine the mode, median, mean and range.

If there are meant to be 25 stars in each container, use the data and the summary values to determine if the factory's processes need to be changed.

No. of stars	Frequency
22	8
23	18
24	12
25	39
26	76
27	27
28	20

Figure 10.8 Statistics from graphs and tables HOTsheet

REFLECTIVE QUESTION

Why do the real estate bodies of the different regions in Australia report quarterly median house prices? Sometimes they report the mean house prices as well. Next time you see these in the media compare the means and medians. These reports can also be developed into lesson plans for students.

Tabular and graphical displays

In the quest to allow data to tell their story the data can be placed into tables and graphed – part of the process of transnumeration (Pfannkuch & Wild, 2004). In this section we will look at examples of tabular and graphical displays, and the challenges these bring to students.

Tables

Table 10.4 shows a table with some of the information taken from The Colac Otway Shire Trees data set (available from the data.gov.au website). The original

Table 10.4 Part of the data from the Colac Otway Shire Trees data set*

Red Flowering Gums (RFG)			Double Pink Flowering Gums (DPFG)		
Tree ID	Height (m)	Year planted	Tree ID	Height (m)	Year planted
6025	6	1993	10662	2	1985
6051	1	2007	10682	2	1985
6054	1	2006	10694	2	1985
6068	1	2007	12222	1	1985
6134	6	1983	12294	2	1985
6194	6	1988	12305	2	1985
6234	5	1983	12308	3	1985
6283	1	2007	12311	3	1985
.	.	.	.	.	.
.	.	.	.	.	.
.	.	.	.	.	.
6362	2	2003	8269	10	1970

*Available from data.gov.au/dataset/colac-otway-shire-trees/resource/bcf1d62b-9e72-4eca-b183-418f83dedcea. Licensed under the Creative Commons Attributes 3.0 Australia Licence (creativecommons.org/licenses/by/3.0/au/).

data set has information for 118 Red Flowering Gums and 230 Double Pink Flowering Gums among a variety of other trees. For the purposes of this section, a sample of 30 trees of each type has been taken. The list of trees with their heights is not informative on its own and some sort of transnumeration is needed.

One way of making sense of information is to use a summary table. Tables can be used as a form of data display, or they can be used as an intermediate step in data representation by graphs (Friel, Curcio & Bright, 2001). For smaller data sets the students can use tallying to assist with this summary. However, while tallying is easy, a teacher should not assume the students know how it is done and some instruction might be necessary (Friel, Curcio & Bright 2001). This data are summarised in Table 10.5.

From this table the distribution of the data can be easily observed. It is apparent that most of the trees are 4 m or less in height. Notice that the *classes* (the groups into which the trees are grouped) are equally spaced and their descriptions make it clear to which group each tree belongs. While these

Table 10.5 Frequency table

	Frequency (count)
Up to and including 2 m	28
Over 2 m and up to and including 4 m	21
Over 4 m and up to and including 6 m	9
Over 6 m and up to and including 8 m	2
Over 8 m and up to and including 10 m	1

attributes are important for a **frequency table**, care should be taken that these features are not emphasised at the expense of interpreting such tables.

Is there a difference between the Red Flowering Gums and the Double Pink Flowering Gums? This question will be answered in the next section, graphing. It is important that students are asked such questions, that their lessons do not finish at the stage of drawing the graphs.

Graphing

What is a graph? The most important feature of any graph is that it transforms numerical or other data into a form that uses spatial characteristics (e.g. height or length) to represent quantity (Curcio, 1989). Graphs may be used to describe data, summarise data, compare and contrast two or more data sets, generalise about a population or predict the next case. Graphs also allow the display of mathematical relationships that cannot be recognised in numerical form (Curcio, 1989).

Most graphs have similar structural components such as axes, scales and grids that give information about the measurements being used and the data being measured. Specifiers (for example lines on a line graph, bars on a bar graph) are used to represent data values. Knowing how to label graphs and to draw scales is important, but unfortunately, much teaching of graphing emphasises the mechanics of graphing at the expense of interpretation.

Graphs in the Australian Curriculum

Graphing appears in Years 7, 9, 10 in the Data Representation and Interpretation area of the Australian Curriculum: Mathematics (ACARA, 2015f).

Year 7: Construct and compare a range of data displays including stem-and-leaf plots and dot plots. (ACMSP170)

Interpreting graphs

Students usually find making comparisons between the lengths of bars in bar charts easy. It is more difficult, however, to make comparisons on pie charts as these require judgements of angles and area. Students also find the interpretation of **histograms** difficult. It has been proposed that line graphs are the most difficult to interpret as they deal with relationships between two **variables**. Students may also have problems with the interpretation of **box plots** because they can appear abstract (Friel, Curcio & Bright, 2001).

For all types of graphs scaling is an issue. For students to design the scales on their graphs they need to be confident in proportional reasoning (Friel, Curcio & Bright 2001; Watson, 2006). Rangecroft (cited in Friel, Curcio & Bright 2001) describes how students find it difficult to choose an appropriate scale even if they are able to draw to a given scale. In addition, changes in scale can affect the shape of the graph and this can also lead to confusion for students.

Curcio (1989) describes three kinds of behaviours that relate to graph comprehension. These are called 'reading the data, reading between the data and reading beyond the data' (pp. 5–6). *Reading the data* involves being able to describe the facts in the graph. *Reading between the data* involves interpretation and comparing quantities. *Reading beyond the data* involves making predictions and the use of knowledge that may not be explicitly stated in the graph.

It is important that teachers, when assessing their students, ask questions that require all of these behaviours. Read the data questions focus on extracting information – what was the value of …? How many books did John read? Reading between the data questions requires integration of more than one piece of information – what does the graph tell you? How has the number of passengers changed over the last week? Reading beyond the data questions require extending beyond the graph – what would happen if …? These latter forms of questions include extrapolation (Friel, Curcio & Bright, 2001).

The trees example

The curriculum for Year 9 states that students should be able to compare data displays using means, medians and ranges. They should also be using histograms. These processes will now be used to compare the two types of tree in the data set, the Red Flowering Gums and the Double Pink Flowering Gums.

The mean, median and range of the two types of tree are shown in Table 10.6. These two data sets have the same median, but it is apparent from the values of the mean and range that there are differences in the two data sets. These immediately become apparent from the histograms in Figure 10.9.

Table 10.6 Descriptive statistics for the Red Flowering Gums (RFG) and the Double Pink Flowering Gums (DPFG) data

Type of tree	RFG	DPFG
Mean (m)	3.43	2.73
Median (m)	3.0	3.0
Range (m)	7.0	9.0
Minimum	1.0	1.0
Q_1	1.0	2.0
Median	3.0	3.0
Q_3	6.0	3.0
Maximum	8.0	10.0

The use of technology

The histograms show that that most of the Red Flowering Gums are below 4 m in height, and there is a spread of values up to 8 m. The Double Pink Flowering Gums are concentrated at 4 m and below, with one tree at 10 m high. Why is there one tree that is much higher than the others in this group? The data point could be an error in transcription, or there may be something peculiar to the tree's location. Whether or not to discard such an *outlier* has no fixed answer and students can make up their own minds as to whether to include this data point or how they would investigate this data point. Problems such as these are those faced by professional statisticians and it is useful for students to have experience with the problems that occur in real data sets.

The curriculum for Year 10 suggests that students compare data sets using box plots. Box plots are based on the so-called five-figure summary: the minimum, Q_1, the median, Q_3 and the maximum. The 'box' in a box plot represents

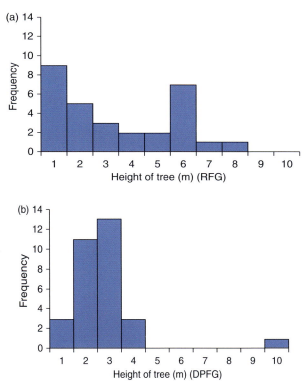

Figure 10.9 Histograms for the Red Flowering Gums (RFG) and the Double Pink Flowering Gums (DPFG) data

the middle 50% of the data. The lower edge of the box is placed at Q_1, and the upper edge is placed at Q_3. The dark line in the box is at the value of the median. Any data outside the middle 50% is represented by the lines (sometimes known as whiskers). If there is a value in the data set that is more than $3 \times$ IQR times lower or higher than Q_1 or Q_3 it is known as an outlier and is represented by a dot or star. The box plots for the tree example is found in Figure 10.10.

This figure adds to the information we have about our data. The lower 25% of the Red Flowering Gums have the minimum value. For the Double Pink Flowering Gums, the values that take up 50 to 75% of the data have the same value as the median. This figure also emphasises that the tree with 10 m height (indicated by *58) is an outlier and is somehow different from the rest of this type of tree. This value is more than three times the value of the IQR above the third **percentile**.

The curriculum for Year 10 also states that the students are able to draw scatterplots, and these are shown in Figure 10.11. These plots may answer another question about the data: does one type of tree grow faster than the other?

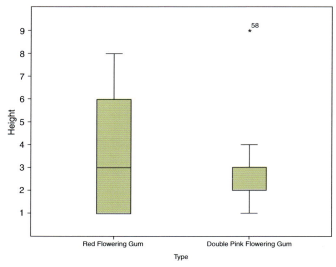

Figure 10.10 Box plots for the Red Flowering Gums (RFG) and the Double Pink Flowering Gums (DPFG) data. Box plots were produced in SPSS

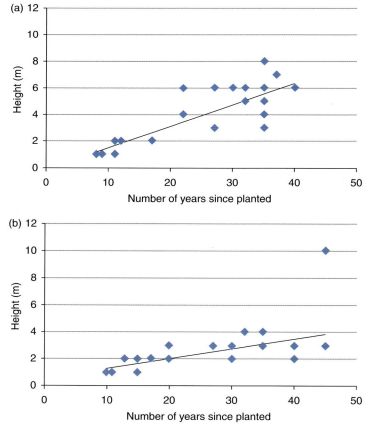

Figure 10.11 (a) Red Flowering Gum, (b) Double Pink Flowering Gum

The anomaly of the Double Pink Flowering Gum at 10 m is immediately apparent. It is also apparent that the gradient of the line for these gums (0.07 m/year) is of a lower value than that for the Red Flowering Gum (0.17 m/year). Students can then discuss what would happen to the gradient of the line for the DPFG if the suspect tree was removed from the data. They can also discuss whether the difference in the rate of growth is a real effect, or just due to the normal variation that will take place in tree growth.

Teaching graphical concepts

The ability to draw and interpret graphs requires skills that develop over time. When teaching students it is important that the techniques of graphing are not divorced from a context of real data (Watson, 2006). The trees example in this section also shows how both the descriptive statistics and the visual displays add to the interpretation of the data. Using the graphs in media reports as a source of investigation can also be very rewarding and add interest.

The use of software packages such as those used in this section can greatly enhance the ease of graphing. The trees example used in this section was based on 30 individuals of each tree. The original data contained samples of 247 Double Pink Flowering Gums and 119 Red Flowering Gums out of a total of 3400 trees. Once the techniques for using the spreadsheet package have been mastered, it is not more difficult to use all the data than it was for this reduced data set. It is still recommended that students practise graphs by hand, however, as much practise is required to develop the skills in creating scales, linking meaning to the symbols (Watson, 2006).

ACTIVITY 10.5

One important skill students need to develop is to use the descriptive statistics and their graphical skills to make judgements about the shape of the data. For example, are the data evenly spread, or are they clumped into a certain region? Do the data form the bell-shaped curve, the normal distribution?

Log in to HOTmaths and select the 'Australian Curriculum' Course list and then choose the 'AC Year 10A' Course. From the Topic list choose 'Data analysis', then the 'Standard deviation' Lesson. From the Resources tab, select the 'Match, rank and classify' HOTsheet. This HOTsheet asks students to match the descriptive statistics to the appropriate graphs for a range of data sets.

HOTsheet   **Standard deviation**

MATCH, RANK AND CLASSIFY

What information do the mean and standard deviation give about the distribution of the data?

TASK 1   **Match graph to data**

Without doing any calculations, match each graph to one of the sets of statistical values in the table. Write the letter for the matching graph in the top row of the table.

Graph letter				
Mean	4	3.45	4.5	4.65
Standard deviation	1.14	1.28	1.91	1.71
Interquartile range	2	2	3	2.5
Range	4	4	6	6

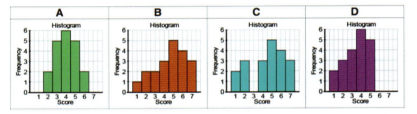

Figure 10.12 Match, rank and classify HOTsheet

ACTIVITY 10.6

Use the data from a known data set and draw graphs with different scales. Draw a histogram and then change the boundaries of the classes. What happens to the shape of these graphs?

REFLECTIVE QUESTION

Log in to HOTmaths and select the 'Australian Curriculum' Course list and then choose the 'AC Year 7' Course. From the Topic list choose 'Data displays', then the 'Different types of line graphs' Lesson. From the Resources tab, select the 'Matching graphs to data' HOTsheet. Plot the time–temperature data. As you do this, note every decision you make to draw this graph. What do students need to know and understand if they are to draw the graph successfully?

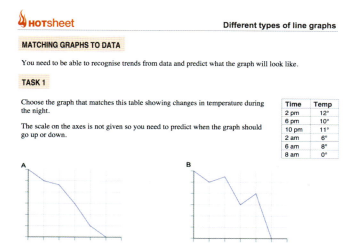

Figure 10.13 Matching graphs to data HOTsheet

Probability

Statistics and **probability** involve the mathematics involved in real-world events that cannot be predicted with certainty (Jones, Langrall & Mooney, 2007). In the section on sampling it was noted that because of the variation between individuals and samples we are often unsure of our conclusions. Therefore, a doctor will prescribe a medication knowing that the medication might produce side effects, but will not be able to predict which person will get which side effect. It is also known that smoking increases a person's chance of getting lung disease, but which smokers will be affected? Some heavy smokers live to a very old age; all that is known is that the risk of getting lung disease is increased with smoking.

If students are used to doing mathematics where there is a 'right' answer, then dealing with probability can be very frustrating. In addition, if they have a deterministic view of life, that is, they believe everything has a known cause, then they may find this difficult to comprehend.

Probability in the Australian Curriculum

From the content descriptors in the Australian Curriculum: Mathematics (ACARA, 2015f) it is noticeable that there is little mention of the use of chance experiments. In Year 6 the content descriptor suggests that students conduct chance experiments (ACMSP145) but experimentation is not mentioned in later years. Research into students' understanding in probability, however, suggests that it is essential that students compare the results of empirical experiments with the theoretical probabilities.

Year 7: Construct sample spaces for single-step experiments with equally likely outcomes. (ACMSP167)

Assign probabilities to the outcomes of events and determine probabilities of events. (ACMSP168)

Year 8: Identify complementary events and use the sum of probabilities to solve problems. (ACMSP204)

Describe events using language of 'at least', exclusive 'or' (A and B but not both), inclusive 'or' (A or B or both) and 'and'. (ACMSP205)

Year 9: List all outcomes form two-step chance experiments, both with and without replacement using tree diagrams or arrays. Assign probabilities to outcomes and determine probabilities for events. (ACMSP225)

Calculate relative frequencies from given or collected data to estimate probabilities of events involving 'and' or 'or'. (ACMSP226)

Year 10: Describe the results of two- and three-step chance experiments, both with and without replacements, assign probabilities to outcomes and determine probabilities of events. Investigate the concept of independence. (ACMSP246)

Approaches to probability

Formally, an **event** is 'a subset of the sample space for a **random** experiment' that is of interest (ACARA, 2015, p. 102). In this context, an **experiment** involves repetitions of a random process, a process where the outcomes are determined by chance. Each repetition is known as a **trial**. The probability of an event is calculated by finding the number of the events of interest and dividing by the total number of events in the sample space. This is the *classical probability* approach and will be the one with which students will be most familiar.

Other approaches to probability include the *frequentist approach*, the *empirical approach* and the *subjective approach*. The frequentist approach looks at probability as a long-term frequency. If a random experiment is repeated many times, the long-term frequency of the event will converge to the probability of the event. The empirical approach estimates the probability of an event after a large number of trials of an actual experiment. The subjective approach uses a person's knowledge of a situation to estimate the probability. For example, how is the probability of an accident at a nuclear power site evaluated? Thankfully, there are not many previous examples and the probability of an accident will depend on many factors, including the type of plant built, the safety precautions taken, and the strictness of monitoring.

Randomness

One of the seemingly simple, but poorly understood ideas in probability is that of *randomness*. In everyday conversation the word 'random' may refer to something that is disordered, irregular, patternless or unpredictable (Peterson, 1998). It may also refer to something that has no cause (Batanero, Green & Serrano, 1998). In contrast, Moore (1990) points out that in statistics random events have a long-term regular pattern of outcomes even though each individual outcome is unpredictable (compare with the frequentist approach to probability above). A simple example is that of tossing a coin. Each outcome is unpredictable, yet the long-term proportion of heads is known to be 50 per cent.

When students are shown random and non-random sequences they find it difficult to determine which one is random. In general, students expect the length of runs (repeating heads or repeating tails) to be shorter than may occur in reality. In addition, they expect the long-term frequency of an outcome (for example, 50 per cent heads) to be reflected in a short sequence.

Independence

Another important concept in probability is that of independence. Formally, two events are **independent** if the outcome of one event has no influence on the probability of the next event. Referring to the coin again, it is known that the probability of a heads or tails remains unchanged from one toss to the next; it is this stability of probability that makes the events independent.

The formal components of probability

The notation

The symbol P(A) means the probability of event A occurring.

The symbol P(A∩B) refers to the probability of A and B occurring. In set notation this is known as the intersection of the sets.

The symbol P(A∪B) refers to the probability of A or B occurring. In set notation this is known as the union of the sets.

The symbol P(A|B) is a conditional probability and refers to the probability of A, given B.

The symbol B' refers to everything in the sample space that is not in the set B. It is known as the *complement* of B.

Probability rules

These probability rules are contained in the curriculum for Mathematical Methods (ACARA, 2015). One of the first things students will need to appreciate is the difference between 'and' versus 'or'. These two words are sometimes used interchangeably in everyday English usage, but in mathematics and logic the meanings are distinct. In probability 'and' refers to two events that occur at the same time. For example, the probability of selecting King and Hearts from a pack of cards is $\frac{1}{52}$, as there is only one card that fits both criteria. In contrast, the probability of selecting a King or a Heart is $\frac{16}{52}\left(\frac{4}{13}\right)$. There are 13 Hearts and 4 Kings but as the King of Hearts belongs to both groups there are only 16 cards in total. This example illustrates the so-called *addition rule*:

$$P(A \cup B) = P(A) + P(B) - P(A \cap B)$$

If the two events A and B are mutually exclusive (cannot occur at the same time) then $P(A \cap B)$ is equal to zero. It is recommended that students be shown this formula after they have worked out such problems, as it is desirable that students gain an intuitive knowledge of these problems first.

Questions such as these also lead to the production of Venn diagrams. This can be illustrated by the example of calculating the probability of a King or a Heart. If the rectangle represents the entire pack of cards, and one circle represents the Kings, and the other circle represents the Hearts, the diagram appears as in Figure 10.14. Younger students can start with smaller sample spaces so that all the items that are not listed in the sets of interest can be listed outside of the circles.

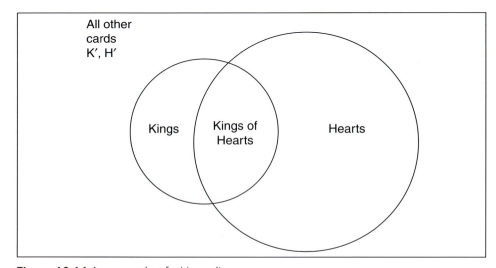

Figure 10.14 An example of a Venn diagram

For independent events the multiplication rule is used. Formally, this is presented as:

$$P(A \cap B) = P(A) \times P(B)$$

For example, if a die is thrown, and a coin is tossed, the probability of a Six and a Head is:

$$P(6 \cap H) = \frac{1}{6} \times \frac{1}{2} = \frac{1}{12}$$

Sample space

Identifying the sample space, that is, listing all the outcomes for a random process, is a deceptively simple process that holds difficulties for some students. Horvath and Lehner (1998, cited in Jones, Langrall & Mooney, 2007) suggest that the task of listing all the elements in a sample space requires the coordination of several cognitive skills. Students need to be able to recognise the different ways of obtaining an outcome and be able to systematically and exhaustively generate all the possible outcomes. If students are not comfortable with combinatorial reasoning they will find it difficult to list all the outcomes. One strategy that students find useful is the odometer strategy. With this strategy one variable is held constant and then matched with all items of the second variable until all the combinations are formed, and then repeated with further variables (English, 2005). One way of carrying out the odometer strategy is to use a tree diagram and this is illustrated in Figure 10.15.

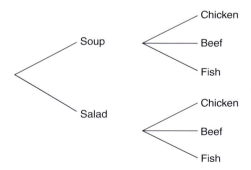

Figure 10.15 An example of a tree diagram

If students do not consider the sample space in a systematic way they can make errors in the resultant probability calculations. Jones, Langrall and Mooney (2007) have found that students may believe that if two dice are thrown that the combination 6,6 has the same probability as the combination 5,6. This error results if the sample space is not considered thoroughly. Interestingly, this error can also occur from the idea that since 'chance' is involved 'anything may happen' and therefore the events must be equally likely.

Conditional probability

A conditional probability, probability of A given B, reduces the sample space to events in B. The formula for a conditional probability A, given B, P(A|B) is:

$$P(A|B) = \frac{P(A \cap B)}{P(B)}$$

When introducing students to conditional probabilities it is useful not to always give numerical problems. An example is given in Figure 10.16.

Which probability do you think is bigger?
a. The probability that a woman is a school teacher.
OR
b. The probability that a school teacher is a woman.
c. Both (a) and (b) are equally likely.

Figure 10.16 An example of a conditional probability in words

This example was used by Watson and Kelly (2007) with students from grades 3 to 13, and by Reaburn (2013) with university students. In Watson and Kelly's study 51 per cent of the participants could not determine the difference between the meaning of statements (a) and (b). Even at university 13 per cent of the students who answered this question chose the equally likely option.

Research demonstrates that students have much more success in being able to interpret and calculate conditional probabilities when the frequencies are placed in tabular form (Watson & Kelly, 2007; Reaburn, 2013). An example is shown in Figure 10.17.

The table below shows the number of defective TVs produced every week at two factories by the day shifts and by the night shifts.

	Factory A	Factory B
Day	40	30
Night	40	60

a. How many defective TVs are produced at Factory B every week?
b. How many defective TVs are produced by a night shift every week?
c. If you were told that a defective TV was produced by Factory A, what is the probability it was produced by a night shift?

Figure 10.17 An example of a conditional probability in tabular form

Even though the school and university students had more success with this question, at the university there was still only 87 per cent who answered part (c) correctly.

Research demonstrates that some students interpret **conditional probability** as a temporal relationship. That is, in the P(A|B) they will think that B always has to precede A. another problem that may occur is that students may also see the conditional probability as cause and effect. That is, B causes A (Gras & Totohasina,1995, as cited in Batanero & Sanchez, 2005).

Implications for teaching

It has been demonstrated that students who have wide experience of carrying out probability experiments have fewer misconceptions about probability and that instructors should challenge common misconceptions with experiment. It is recommended that students' understanding is enhanced if theoretical calculations are related to the results of empirical investigation (Jones et al., 1997; Watson, 2006).

Probability distributions

In late secondary school students come to more formal probability and then have to deal with probability distributions. The first concept that needs to be dealt with is the random variable. In this context variable is a noun, and is a quantity whose value might change. A random variable (X) has values that are the outcomes of a probability experiment. A **probability distribution** consists of the list of all possible values of a random variable with their accompanying probabilities of occurrence. The distribution of a random variable may be represented by a graph, table or by a mathematical formula.

A simple example of a probability distribution that may be used to introduce students to this concept is that of rolling two dice where the variable is the sum of the two dice. The sample space needs to be determined by the use of a probability tree, or by a Table as shown in Table 10.7.

The frequency of these outcomes can be then represented in a table with their accompanying probabilities (Table 10.8) and can also be represented in a graph (Figure 10.18).

For this to represent a probability distribution students need to realise that it is important that all the probabilities add up to one. If this is not the case, an error has been made, possibly in the determination of the sample space. This is

Table 10.7 The sample space for the addition of the faces of two dice

	1	2	3	4	5	6
1	2	3	4	5	6	7
2	3	4	5	6	7	8
3	4	5	6	7	8	9
4	5	6	7	8	9	10
5	6	7	8	9	10	11
6	7	8	9	10	11	12

Table 10.8 The probability distribution for the addition of the faces of two dice

Sum (X)	Frequency	P(X)
2	1	$\frac{1}{36}$
3	2	$\frac{2}{36}$
4	3	$\frac{3}{36}$
5	4	$\frac{4}{36}$
6	5	$\frac{5}{36}$
7	6	$\frac{6}{36}$
8	5	$\frac{5}{36}$
9	4	$\frac{4}{36}$
10	3	$\frac{3}{36}$
11	2	$\frac{2}{36}$
12	1	$\frac{1}{36}$
	Sum =	1

an example of a discrete probability distribution; a distribution where the data are in distinct amounts. Examples of discrete probability distributions are the Poisson distribution and the binomial distribution.

A Poisson distribution is used when the problem refers to the number of occurrences in a given region of space, length or time. If the mean number of occurrences is represented by μ the formula is:

$$P(X;\mu) = \frac{e^{\mu} \times \mu^{X}}{X!}$$

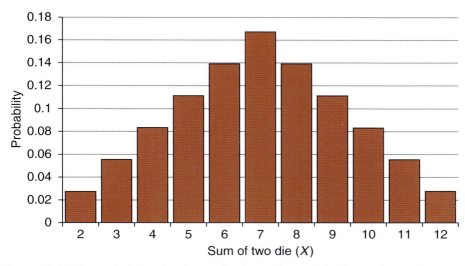

Figure 10.18 The probability distribution for the addition of the faces of two dice

A binomial distribution is used when the circumstances have the following characteristics: there are a number of trials (n), there are only two possible outcomes (designated as a success or failure), the probability of success (p) is the same for every trial, and each trial is independent. The formula is:

$$P(X;n,p) = {^nC_r}\,p^x(1 - p)^{(n - x)}$$

The answers to the questions using the Poisson and binomial distribution are not difficult but they can be tedious. It is important, though, that students are made aware of the terminology that may be used in the asking of the questions. For example, they could be asked to find the probability of 3 successes (exactly 3), greater than 3 successes (does not include 3), greater or equal to 3 successes (includes 3). The terminology is similar for 'less than' questions. It is also helpful for students to be reminded about the **complement** of the events they are interested in. For example, if asked to find the probability of 5 or less successes out of 6 trials, it is quicker to find the probability of 6 successes and then subtract this probability from one.

There is one extremely important distribution that often applies to continuous data (data that can take a range of continuous values). This is the normal distribution. This is represented by a formula.

$$P(X) = \frac{1}{\sigma\sqrt{2\pi}}\,e^{-\frac{(x - \mu)^2}{2\sigma^2}}$$

This equation gives the familiar bell curve. Probabilities are found by finding the relative areas under this curve. For example, the probability of being between the values 'a' and 'b' is found by:

$$\int_a^b f(x).dx$$

Where $f(x)$ is the normal probability distribution.

Unfortunately, this equation does not lend itself easily to integration. The solution is to transform the data using the z-score, which has a mean of zero and a standard deviation of one.

$$z = \frac{x - \mu}{\sigma}$$

This allows the use of the normal distribution tables to find the appropriate areas under the curve and hence find the necessary probabilities.

For questions using these distributions the probability of being exactly at any precise measurement (e.g. 6.000 cm) is zero. Therefore, questions about probability that apply to the normal distribution always ask for ranges: for example, what is the probability that the object is between 4 and 8 cm long? Another consequence is that it makes no difference if a question asks for being 'less than' or being 'equal to or less'. If students remember that the total area under the curve is equal to one, and that the mean is at the centre of this distribution, their calculations can be simplified.

Putting it all together: statistical inference

Statistical **inference** is the process by which conclusions are made about populations from samples. Formal methods of statistical inference include t-tests, z-tests and *chi*-squared tests. Usually these tests, such as t-tests, propose something about a population, for example the mean (the null hypothesis). A sample is taken and if the question is about the value of the population mean, the sample mean will then be compared with the value of the proposed population mean. Then the question is asked, is the sample mean consistent with a population with the given characteristic? This is answered by calculating the conditional probability that the sample mean, or a sample mean

even more extreme, could occur if the population mean should have the proposed value. If the resulting probability (known as the *p*-value) is less than a
pre-determined value (often 0.05), then it is concluded that the value population characteristic is different from the proposed value and the hypothesis is
rejected. If the resulting probability is greater than the pre-determined value,
then the hypothesis is *not rejected* and it is concluded that the sample value is
consistent with the proposed value.

Students find this hypothetical, conditional reasoning extremely difficult
(Cobb & McClain, 2004; Garfield & Ahlgren, 1988; Yilmaz, 1996). As a consequence, students in statistics often manage by blindly following the procedures. It is important, therefore, that students are asked questions that require
them to explain their reasoning.

One way that students can be introduced to this hypothetical, conditional reasoning is to use an example to which students can easily relate. One such example is provided by Shaughnessy and Chance (2005). The example is as follows:

> Hypothesis: It is warm out.
> Data: People are wearing coats.
> *p*-value: Suppose it is warm out. Then the chance that people are wearing coats
> (*p*-value) is very small.
> Conclusion: Reject the hypothesis.
>
> (Shaughnessy & Chance, 2005, p. 69)

To complicate matters, it is possible that for some reason a group of people
were wearing coats even though it was hot and the conclusion is incorrect.
Students who believe that mathematics always leads to 'correct' answers can
find this very frustrating.

Students also find that understanding the processes of finding a confidence interval (an interval in which a population parameter is likely to lie)
is very difficult. It is interesting to note that the curriculum for Specialist
Mathematics (ACARA, 2015) suggests that students should simulate repeated
random sampling to investigate the properties of the distribution of the sample mean. These simulations should be used to illustrate that for sample sizes
that are large enough ($n > 30$), sample means form normal distributions (the
Central Limit Theorem) whose standard deviations are called, most confus-
ingly, the *standard error of the mean*. The curriculum also suggests that these
simulations be used to show that most confidence intervals calculated by the
formula $(\bar{x} \pm \frac{zs}{\sqrt{n}})$ will contain the population mean. In this formula z determines the level of confidence that results, s is the sample standard deviation,
and n is the sample size. This idea of the *level of confidence* is also confusing for
students. One way of looking at it is to use the long-term frequency interpretation. In this interpretation a level of confidence of 95%, say, suggests that

if the process of sampling and calculating the confidence interval had been carried out 100 times, 95 of the resulting intervals would include the value of the population mean.

The calculation of confidence intervals is simple, but it is important that teachers do not underestimate the difficulty students have in understanding the reasoning behind these calculations. If students can understand that the process of confidence interval estimation provides a good 'guestimate' for the population mean, then their understanding can be developed from there.

ACTIVITY 10.7

Log in to HOTmaths and select the 'Australian Curriculum' Course list and then choose the 'AC Year 10' Course. From the Topic list choose 'Probability', then the 'Conditional probability' Lesson.

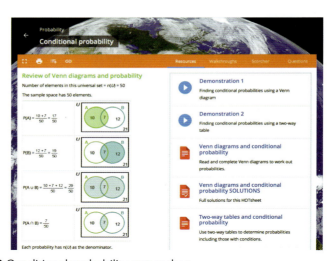

Figure 10.19 Conditional probability screenshot

There are two HOTsheets in this section: 'Venn diagrams and conditional probability' and 'Two-way tables and conditional probability'. Both these HOTsheets give experience in the use of the formal terminology used in probability.

In the 'Venn diagrams and conditional probability' HOTsheet there is a challenge question which involves a Venn diagram for three sets of data. Try it! As you do so, note each step you had to go through to get the answer. What do students have to know and understand to be able to do this task?

The 'Two-way tables and conditional probability' HOTsheet gives practice in distinguishing between conditional statements. Again, note what students need to know and understand at each step.

ACTIVITY 10.8

Log in to HOTmaths and select the 'Australian Curriculum' Course list and then choose the 'AC Year 8' Course. From the Topic list choose 'Probability', then the 'Compound events' Lesson.

In the Resources tab, there is an accompanying 'Probability cards' widget and two HOTsheets, 'Pick a card' and 'Deciphering probability'. The 'Pick a card' widget allows students to set different conditions for selecting a card from a standard deck. When the answer is given the widget lays out the cards that belong in the event (for example, all the clubs).

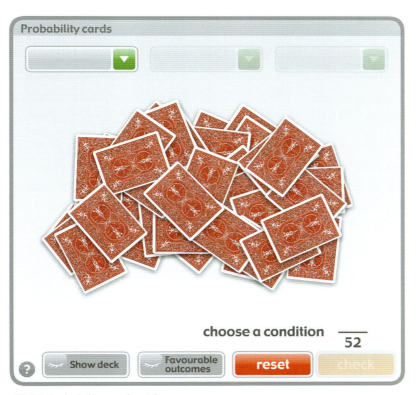

Figure 10.20 Probability cards widget

REFLECTIVE QUESTIONS

1   The 'Deciphering probability' HOTsheet, found using the pathway in Activity 10.8, contains questions where the sample space is laid out visually (for example, squares and diamonds). It also requires answers in decimal, percentage and fraction form. What are the advantages and disadvantages of using such resources?

2   This is a question with which secondary students often have difficulty: 'The weather report for yesterday stated that there was a 90 per cent probability of rain, but it did not rain'.
   Students often believe that if it did not rain then the forecast was incorrect. How can you demonstrate to students that the forecast was indeed 'correct'? Is there a simulation you can devise so that students can see how such processes work?

Summary

Describe why data are needed and how data can be produced

As a society, if good decisions are to be made data need to be collected first. Students can collect their own data (usually about themselves or their school), but as they become older they should access real data sets. It is useful for students to consider the issues around data collection and consider how data for media reports have been collected. Students also need to be aware of why samples are taken even though decisions about the parent populations are then made with uncertainty.

Describe the processes and issues involved in the calculation of descriptive statistics

Descriptive statistics are used to obtain numbers that are representative of the data. Students should be encouraged not only to calculate these statistics, but to consider the appropriateness of each statistic for each data set. In addition, these statistics should be used to investigate real data; this encourages interest and understanding in the students.

Describe the processes and issues involved in producing graphical and tabular displays

Along with the descriptive statistics, tables and graphs allow the data to tell their story. For students to use scales requires proportional reasoning, and many students find this difficult. Students need extensive practice in drawing graphs and learning to associate the symbols on the graph with their data. As students become older, it is strongly recommended that they should become familiar with commonly used technology to produce their graphs. It is also recommended that students should use different representations of the same data and explore the differences between these representations.

Describe the principles of probability

It can be difficult for students to adapt to the idea of a non-deterministic world. Not everything can be predicted with certainty, and students need to become familiar with the mathematical descriptions of probability. Students should be encouraged to see probability as a tool in making decisions.

Describe how statistical inferences are reached

When a sample is taken, any decision about the population based on the sample may be incorrect. This is because samples vary from one another and vary from the population. Therefore propositions are made about the population and the samples are assessed on how likely they are if the proposition about the population is true. Instructors need to be aware of the difficulty students have in following this hypothetical, conditional reasoning.

CHAPTER 11

Functions: a unifying concept

Learning objectives

After studying this chapter, you should be able to:

- identify content areas of the mathematics curriculum that can be learned through a function-related approach
- explain how the different settings of functions are used to develop students' understanding
- describe learning activities that develop students' understanding of functions
- identify and select appropriate technologies for developing students' understanding of functions
- describe real-life situations that can be modelled with functions.

Introduction

Scholars such as Moore (2014) have suggested that quantitative reasoning and **covariational reasoning** are both critical for success in secondary and post-secondary mathematics education. The former is the cognition required to represent (or model) an everyday situation by measurable attributes or quantities, whereas the latter is the cognition needed to understand the relationship between two or more varying quantities. Given that a function is essentially a relationship between two varying quantities, then an emphasis on functions is likely to develop covariational reasoning in our students. Moreover, applying their mathematics to everyday situations is likely to develop students' quantitative reasoning. Therefore, teachers can and should use functions, applied where possible to everyday situations, as they teach the curriculum. In doing so, teachers can use functions to develop those important connections that were outlined in Chapter 3, because functions are a unifying concept.

The chapter commences with a review of the concept of function and in particular it explores the different settings in which functions occur. It is argued that these settings are important as they allow students to encounter different representations of functions, thus building 'a robust knowledge of adaptable and transferable mathematical concepts' (ACARA, 2015, p. 6). The chapter then explores the transition to secondary school, arguing that teachers in the middle school years need to utilise the function settings that their students have already encountered, before and as they are introduced to algebra. It then explores the development of the function-related concepts through the lower secondary school, where students are progressively introduced to the different families of functions. As discussed, a number of mathematical concepts, such as rates, can and should be introduced and developed using a function-related approach because doing so will help students build important connections. The last section of the chapter explores functions in the senior secondary school, where the focus is on the algebra of functions, trigonometric functions and modelling with functions.

KEY TERMS

- **Compartmentalisation:** an inability to transfer knowledge from one setting and/or concept to another.
- **Covariational reasoning:** the cognition involved in understanding the relationship between two or more varying quantities.
- **Direct path:** a representation on the Cartesian plane of a function mapping an element from the domain to the range.
- **Function:** a one-to-one relationship between two quantities.
- **Function-related approach:** an approach to the teaching of mathematics that uses a function in one or more of its settings.
- **Function setting:** a broad area of mathematics where a function might be encountered and where the representation of the function is unique to that setting.
- **Inverse path:** a representation on the Cartesian plane of an inverse function mapping an element from the range to the domain.
- **Property-oriented approach:** a teaching approach that focuses on the property of a function irrespective of the setting involved.

Functions and their settings

A **function** is one-to-one relationship between two quantities or sets. Such a relationship occurs in a number of settings, identified by Bloch (2003) as: the formal setting, the geometric setting, the numerical setting, the graphic setting, and the algebraic setting. Typically, a given function is represented differently in each setting. In the formal setting, for example, it is represented as a mapping, whereas in the numerical setting it is represented as a table of values.

Students encounter the different settings at different stages through the mathematics curriculum. The numerical and graphic settings are introduced in late primary school, the algebraic setting in early secondary school, and the formal setting in senior secondary school. This is intentional and reflects findings from developmental psychology that suggest experience with enactive and iconic representations is necessary in order to make sense of symbolic representations (Bruner, 2006); a theme that is unashamedly repeated throughout the chapter. The remainder of this section provides a brief background to each of the **function settings.**

The formal setting

In the formal setting, encountered in senior secondary school (ACMMM022)[1], a function is represented as a one-to-one mapping from one space (called the domain) to another (called the range). Galbraith and Fitzpatrick (1977) provide the following definition:

> If for each $x \in X$, there is exactly one corresponding element $y \in Y$, we say that X is mapped into the set Y. Moreover, if the element $y \in Y$ is denoted $f(x)$, then we can write $f : X \rightarrow Y$ where $y = f(x)$.

This setting is revisited later in the chapter in the section dealing with senior secondary mathematics.

The geometric setting

Students encounter the variation between related quantities in a geometric setting: for example, they might explore the relationship between the circumference and radius of a circle. Consequently, some functions are appropriately explored in the geometric setting, though Bloch (2003) cautions that

---

1   This abbreviation references Content Descriptor 022 from the Australian Curriculum
     Mathematics: Mathematical Methods subject.

many students lack the formal geometric knowledge to work effectively in this setting. The geometric setting is discussed later in the chapter when trigonometric functions are introduced.

The other settings

The numerical setting (quantities represented in a table of values), the graphic setting (quantities represented on the Cartesian plane), and the algebraic setting (the relationship is represented as an algebraic equation), are introduced and/or developed in early secondary school. These foundational settings, however, are touched upon in late primary school where students learn to generalise, represent numbers in tabular form, and use the Cartesian plane to represent locations.

The next section explores the transition to secondary school, arguing that an introduction to algebra should come via a **function-related approach** that builds upon students' experiences with the numerical and graphic settings.

Functions at the transition

In mathematics, the transition to secondary school is usually accompanied by changes to the curriculum and the associated pedagogy. In the primary school curriculum the focus is on numeracy, whereas the transition to secondary school sees a greater emphasis on the abstract thinking associated with algebra. In addition to this, mathematics pedagogy tends to move from a hands-on approach in the early primary years to a transmission approach in the upper secondary years (Tytler et al., 2008). Reports suggest that students in lower secondary mathematics classrooms spend considerable time on repetition and drill, and are not provided with the intellectual challenges that they had anticipated (Attard, 2013; Luke et al., 2003). Moreover, their teachers are either unaware of, or ignore, their students' previous learning experiences (Skilling, 2014). It is not surprising, therefore, that studies show a hiatus in mathematics performance during the first year of secondary school (Galton, Morrison & Pell, 2000).

Given these concerns with the transition to secondary school, it is argued that more attention needs to be given to the way that algebra is introduced, and in particular that this introduction builds upon students' prior experiences with functions. According to Kieran (2007), there is a tension within the mathematics education community regarding the teaching of algebra. Many educators advocate a traditional view that draws on connections with arithmetic, introducing algebra through the generalisation of arithmetical relationships. This view sees algebraic understanding as coming from an ability to generalise and to transform, where the latter might be, for example, recognising that $x^4 - x^2$ is a difference of two squares. In contrast, mathematics educators advocating a reformist view of algebra education emphasise the importance of functions – used in problem-solving contexts – to develop algebraic thinking. A review of the Australian Curriculum: Mathematics (ACARA, 2015), suggests the writers have considered both these views, leaving it to syllabus writers and teachers to make their respective emphases. This section explores the introduction of algebra using functions, arguing that students can learn to generalise through the analysis of repeating patterns. Further, that an investment of time on pattern-related investigations allows students to make meaning of the algebraic manipulations that occupy a significant part of the secondary school syllabus.

Introducing algebra with functions

In late primary school most students encounter the concept of function in a numerical setting, through their work in describing rules used to create sequences of numbers (ACMNA133)[2]. A typical problem involves the construction of a growing pattern using matches (see Figure 11.1) and an accompanying tabular representation of the function, mapping the number of squares to the number of matches (see Table 11.1). In the primary years, the Australian Curriculum: Mathematics (ACM) suggests that students look for patterns in the way the numbers increase or decrease. In this instance a typical student response might be that the number of matches 'goes up in threes'.

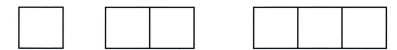

Figure 11.1 Repeating pattern of squares made from matches

---

2   As with earlier note, this references Content Descriptor 133 from the Number and Algebra strand of the Australian Curriculum Mathematics (F–10).

Table 11.1 Functional relationship between the number of matches and squares

Number of squares	1	2	3	4	5
Number of matches	4	7	10		

In the early secondary years, however, the same problem can be used to build on earlier experiences with patterns to introduce variables and linear relations. Inevitably, questions are provided to force the students to move from construction and counting processes to generalisation: 'How many matches are needed for 27 squares?' or 'Can you find a rule that predicts the number of matches for any number of squares?' Whereas some students can describe such a rule, for many this is not an easy task. Noss, Healy and Hoyles (1997) advocated that students be encouraged to move back and forth between the physical construction of the shapes (forcing them to visualise the problem) and their attempts at generalisation. For these authors this was achieved by having students build their shapes on a computer screen using FMSLogo[3] programming, though having the students use matches would serve equally well. In the previous example, Noss et al. (1997) described the efforts of two girls.

The first 'saw' the third (three-square) shape as 'three lots of three and one more' (see Figure 11.2). Consequently, the student was able to extend this logic to the four-square shape 'four lots of three and one more' and then generalise to a 27-square shape as 'twenty seven lots of three and one more'.

Figure 11. 2 First student's response to matches pattern

Students such as this should be encouraged to verbalise the relationship and eventually to write it algebraically as: $M = 3S + 1$, where M is the number of matches and S the number of squares. Such an activity can serve as an introduction to the variable, defined formally in Chapter 8, and thus the algebraic setting.

The second student 'saw' the third shape as 'two horizontal lines of three matches and four vertical matches' (see Figure 11.3). Presumably, and perhaps with some prompting, this student could extend their logic to the four-square shape 'two horizontal lines of four matches and five vertical matches' and then generalise to a 27-square shape as 'two horizontal lines of 27 matches

3   FMSLogo is a programming language suitable for young students available from fmslogo. sourceforge.net.

and 28 vertical matches'. Again, with prompting they may express this relationship using the same terminology, as: $M = 2S + S + 1$.

Figure 11.3 Second student's view of the pattern

What this example suggests is that:

- Students, even in the secondary school, benefit from being able to use concrete materials (or in the case of FMSLogo maths, the manipulation of virtual materials) as they grapple with the algebraic setting. In doing so they are able to move between a visual image of the pattern and the algebraic representations needed to describe it, in much the same way that mathematicians move between visual images of concepts and the symbolic way that they are reported.
- Allowing students to create patterns and explain them creates valuable learning opportunities. In the given example the students developed two 'rules' that look different to novices, yet are algebraically equivalent. This creates a wonderful connection with early work on algebraic manipulation and could be used as the basis for a discussion related to algebraic simplification.

With regard to the graphic setting, students are introduced to the four quadrants of the Cartesian plane in Year 6 (ACMMG143) and in early secondary school this knowledge is extended when they plot points from a table of integers and recognise linear patterns (ACMNA178). Kaput (1989) stressed the importance of the graphic setting for developing students' understanding, as it allows them to consolidate a relationship between two quantities – the number of squares and the number of matches – into a single graphical entity (see Figure 11.4). Before developing mental images of functions, students need to *see* the function and indeed the creation of a person graph (using the lines on a basketball court, and some temporary scale marks) helps provide a more tangible perspective.

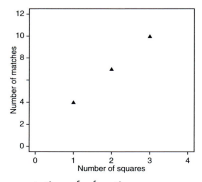

Figure 11.4 Graphical representation of a function

ACTIVITY 11.1

The painted cube investigation:

Imagine one were to assemble 27 unit cubes into a larger cube with side 3 units and then paint the surface of the larger cube (see Figure 11.5).

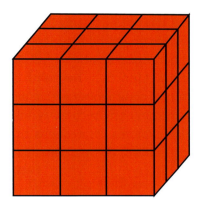

Figure 11.5 Painted cube

Students are initially asked to predict how many of the original 27 cubes are painted on: 3 faces, 2 faces, 1 face, and, no faces. The investigation then proceeds from there, because we can now ask them to predict how many of each category there would be if the original cube was now of side 4 units and comprised 64 cubes. From there they could be asked to investigate a cube of side 5 units and then generalise beyond this.

Work through this investigation yourself. There is a useful interactive version of this provided by HWB: Digital learning for Wales. Access to the resource is via their home page (hwb.wales.gov.uk), where a search for 'painted cube' is necessary.

REFLECTIVE QUESTIONS

1   When teaching this investigation, what settings of the function would you have students use and in what order would they be used? In other words, how would you teach the investigation?
2   The investigation connects to several other mathematical concepts that are quite removed from the teaching of algebra. Identify these and describe how you would explicitly show students these connections.

Developing functions through the early secondary years

The last section described how algebra can be introduced in early secondary school using a function-related approach that builds on student's primary school experiences with patterns. Given the importance of functions in developing covariational thinking, it is not surprising that the Australian Curriculum: Mathematics devotes an entire sub-strand to functions in the secondary school, which is titled 'Linear and non-linear relationships'. This section explores that sub-strand and connections that functions have with other content areas of the curriculum. More specifically, it commences with a discussion on the teaching of linear functions, introduced in early secondary school, before moving onto non-linear functions. It also explores connections that linear and non-linear functions have with other strands of the mathematics curriculum.

The linear function

As described previously, students are introduced to three settings of the function at the transition to secondary school, which are then developed through the remaining secondary years. The algebraic setting often dominates the secondary curriculum, but teachers need to provide students with opportunities to use other settings. As advocated by Bruner (2006), many learners need to move from enactive representations (physically moving objects), through iconic representations (creating pictures), before they can comfortably create symbolic representations. Thus, the numerical and graphic settings that emerge from the physical manipulation of patterns provide those iconic representations of functions that students need before they can comfortably work with algebraic representations. Yerushalmy (2000), for example, reported a longitudinal case study with two relatively low achieving students working on function-related modelling exercises. He noted that these students tended to use the numerical setting in early secondary school, and then this in conjunction with the graphic setting in middle secondary school. By late secondary school (Year 10), however, they were more inclined towards the algebraic setting and thus algebraic methods. In other words, it seems to take time for students to become familiar and competent with the algebraic setting, and teachers must therefore provide opportunities for them to utilise appropriate enactive and iconic representations.

The use of different function settings in the secondary school is not without its difficulties, because students can have difficulty connecting understandings in one setting with those in another. Several researchers (e.g. Bloch, 2003; Noss et al., 1997; Yerushalmy, 2000) have advocated a **property-oriented approach** to these settings. When solving equations, for example, students should have the opportunity to do so in different settings and be encouraged to make connections across the settings. They should be able to identify the solution of an equation in each of the settings. The introduction of linear functions through pattern-related activities, described earlier, inevitably leads to the solution of linear equations (ACMNA179) with students asked, for example, 'If we had 37 matches in the pattern (see Figure 11.1), how many squares would there be?' Before using algebraic methods, students should utilise the numerical setting, extending Table 11.1 and/or using arithmetic methods. Similarly, they should utilise the graphic setting and extend Figure 11.4, which would raise interesting questions regarding the continuity of the function. Even when using algebraic methods to solve the equation $37 = 3S + 1$, the ACM recommends that students should have access to a number of different strategies. The back-tracking method advocated by Lowe et al. (1993), for example, provides a diagrammatic approach to the solution of equations that can be used before formal algebraic manipulations occur. Though limited in scope, the method builds on students' understandings of arithmetic and requires them to present the equation as a flowchart (see Figure 11.6) and then work backwards along the flowchart to determine the unknown. The point is that a property of functions is emphasised in different settings, thus building up those important connections.

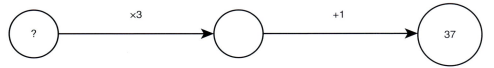

Figure 11.6 Example of simple flowchart

Developing linear functions

As students progress through the secondary school and their skills in algebraic manipulations improve, they are encouraged to explore generalisations of the linear function $y = mx + c$. Typically students are asked to plot a series of functions and to observe the effects of altering the coefficient and constant term on the shape of the plot. Ideally they should *discover* these relationships. See, for example, Activity 11.2, which explores this idea further.

Students in later secondary (Years 9 and 10) are required to solve simultaneous linear equations. Encouraging access to different function settings should lead to a deeper understanding. Consider the following example and then attempt Activity 11.3.

> **Example**: A company manufactures gadgets. The total cost of producing gadgets is broken down to a fixed component that depends on costs such as the rent of the factory, and a variable component that depends on costs such as labour and electricity. In their existing factory, fixed costs amount to $200 irrespective of the number of gadgets produced. Variable costs, however, are estimated to be $0.50 per gadget. They are considering moving to new premises, but this move will result in a higher fixed cost of $300, but a lower variable cost of $0.40 per gadget. How many gadgets would they need to sell to make this move worthwhile?

The equations $y = 200 + 0.5x$ and $y = 300 + 0.4x$ have been plotted (see Figure 11.7) using Graphmatica (Hertzer, 2014), which also provides a tabular representation of the function. Students, however, should have the opportunity to circumvent the use of equations and represent the function numerically with electronic spreadsheets (see for example, Green, 2009). Students should find a solution to these simultaneous equations in both the numerical and graphic settings before embarking on a method involving elimination or substitution. It is important, however, to stress the connections between the solutions in each of these settings, in that students should appreciate that the point of intersection (1000, $700) on the graph corresponds to the row on the table where existing and new costs align. In other words, adopting a property-oriented approach to the teaching of functions.

No.	Existing	New
0	200	300
100	250	340
200	300	380
300	350	420
400	400	460
500	450	500
600	500	540
700	550	580
800	600	620
900	650	660
1000	700	700
1100	750	740
1200	800	780
1300	850	820
1400	900	860
1500	950	900

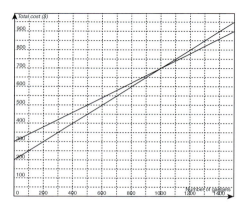

Figure 11.7 Graphical and tabular solutions to simultaneous equations

ACTIVITY 11.2

Log in to HOTmaths and select the 'HOTmaths Global' Course list and then choose the 'Early Secondary' Course. From the Topic list choose 'Graphs on the Cartesian plane', then the 'Comparing lines & equations' Lesson.

Look at the widgets 'Changing the coefficient' and 'Changing the con-stant term'. Use the first widget to graph the functions: (a) $y = x$ (b) $y = 2x$ (c) $y = 3x$ (d) $y = 100x$, without clearing successive plots. Similarly, use the second to graph the family of functions: $y = x \pm c$.

Figure 11.8 Comparing lines and equations screenshot

REFLECTIVE QUESTIONS

1   The 'Changing the coefficient' widget, found using the pathway in Activity 11.2, allows students to see the relationship between coefficient and gradient. It also demonstrates that increasingly greater coefficients produce lines that approach the vertical y-axis. Develop some 'sets' of questions that students could use with this widget. Identify the purpose of each set, in other words what feature of the linear function would this set of questions be trying to illustrate?

2   Suggest some activities that you would have students do *after* they have used these widgets to explore the influence of the coefficient and constant term.

ACTIVITY 11.3

Log in to HOTmaths and select the 'Australian Curriculum' Course list and then choose the 'AC Year 10' Course. From the Topic list choose 'Simultaneous equations', then the 'Introducing simultaneous equations' Lesson. Experiment with both the 'Guess and check simultaneous equations' and 'Discovering where lines cross' widgets, which are located in the Resources tab.

Guess and check simultaneous equations

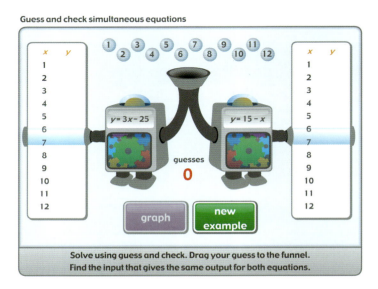

Discovering where lines cross

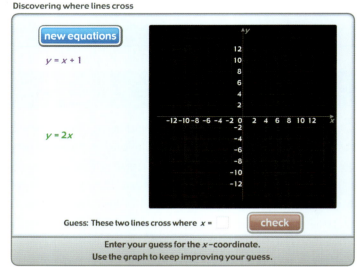

Figure 11.9 Guess and check simultaneous equations and Discovering where lines cross widgets

REFLECTIVE QUESTION

The widgets used in Activity 11.3 provide numerical and graphic settings for the solution of simultaneous equations. How would you utilise these resources and what else would you use to develop students' understanding of simultaneous equations?

Connections between linear functions and other mathematical strands

There are many instances in mathematics of relationships between two varying quantities and thus areas where teachers can utilise students' understanding of functions. In this section we explore just two, the use of rates and linear regression.

Linear functions and rates

Students are introduced to rates in Years 7 (ACMNA173) and 8 (ACMNA188). Problems involving rates connect beautifully with linear functions. Speed, for example, can be explored through time/distance comparisons in constant speed contexts. Students should have the opportunity to discover that speed is the gradient of the distance by time graph, thus laying the groundwork for later work on differential calculus that focuses on rates of change. As described in Shield (2008), cost/weight comparisons of food provide another opportunity to explore rates through a function-related approach. Students could be asked to peruse their local supermarket (or their parents' shopping trolley) and record the weight and total price of bags of apples sold at the same rate. These ordered pairs could then be presented in a tabular form, a graphic form and an algebraic form.

Linear functions and statistics: regression

Students are introduced to bivariate data analysis when they encounter scatterplots in Year 9 (ACMSP251) and lines of best fit in Year 10 (ACMSP279). Linear regression provides a good application of the linear function and opens up some wonderful opportunities for real-life investigations.

An example comes from the Australian Data and Story Library (OzDASL), which provides data describing the number of days taken for a bar of soap to completely disappear (Bar of Soap: www.statsci.org/data/oz/soap.html). This quirky investigation was undertaken by a person who suspected that the rate at which the soap disappeared would be dependent on its surface area rather than its mass. Students can be provided with these data and use technology to plot the mass of the soap against the number of days it was in use. The relationship is surprisingly very linear. Alternatively, similar investigations can be developed from the students' own experiences.

ACTIVITY 11.4

An example based on Leonardo da Vinci's Vitruvian man (leonardodavinci .stanford.edu/submissions/clabaugh/history/leonardo.html) asks students to investigate the relationship between height and arm span (believed by Vitruvius to be equal in well-proportioned people).

Obtain a sample of secondary data from CensusAtSchool (www.abs.gov .au/censusatschool), a project conducted by the Australian Bureau of Statistics that has collected data from many Australian students. Use its Random Sampler to download a sample of 200 responses from the 2008 census because it will contain heights and arm spans. Use these data to investigate the linear relationship between span and height. To what extent is it the same? What are the realistic limits (domain) of this function?

REFLECTIVE QUESTIONS

How would you introduce this investigation in such a way to engage students in Year 9 or 10? What other activities would you include with this investigation? Provide some prompting questions that would guide your students to: (a) appreciate the limited domain of this function and (b) solve a related equation.

Non-linear functions

Students will encounter non-linear functions when they sketch parabolas and circles in Year 9 (ACMNA296), if not before, and then in Year 10 when they explore links between the graphic and algebraic settings for each of: quadratics, circles, exponential functions (ACMNA239) and hyperbolas (ACMNA267). Some students in Year 10 will also encounter periodic functions (ACMMG275) as they study trigonometry.

Quadratic functions can be introduced through investigative work involving patterns (see for example, the skeleton tower, described in Activity 11.5). Exponential functions, on the other hand, can be introduced or developed through work on compound interest (ACMNA229), where students manually calculate the amount accrued in the bank after several interest periods. This information can be presented in the numerical setting, and then in the graphic setting. The algebraic setting might need to be achieved, however, through arithmetic generalisations, in that students will need to see that the amount accrued after one interest period will be $P(1+i)$, after two interest periods $P(1+i)^2$ and after n interest periods $P(1+i)^n$.

> **Example**: An amount of $10 000 was deposited for 12 years with an interest rate of 5.6% p.a. compounded annually. How much should be accrued after this period of time?

Students should initially have access to this exponential function through the numerical setting provided by a spreadsheet tool such as Excel (see first four columns of Table 11.2). Moreover, it also allows a link to the graphic setting (see Figure 11.10). In both settings students can gain an appreciation of the underlying mathematical features of exponential functions. Extending the spreadsheet, for example, reveals that whereas it takes 13 years to add $10 000 to the initial amount, it takes eight years to add a further $10 000 and then just five more years to add another $10 000.

Table 11.2 Tabular representation of an exponential function

| | Spreadsheet view | | | |
Year	Initial balance	Interest	Final balance	Later link to the algebraic setting
1	$10 000.00	$560.00	$10 560.00	=10000*(1.056)^1
2	$10 560.00	$591.36	$11 151.36	=10000*(1.056)^2
3	$11 151.36	$624.48	$11 775.84	=10000*(1.056)^3
4	$11 775.84	$659.45	$12 435.28	=10000*(1.056)^4
5	$12 435.28	$696.38	$13 131.66	=10000*(1.056)^5
6	$13 131.66	$735.37	$13 867.03	=10000*(1.056)^6
7	$13 867.03	$776.55	$14 643.59	=10000*(1.056)^7
8	$14 643.59	$820.04	$15 463.63	=10000*(1.056)^8
9	$15 463.63	$865.96	$16 329.59	=10000*(1.056)^9
10	$16 329.59	$914.46	$17 244.05	=10000*(1.056)^10
11	$17 244.05	$965.67	$18 209.71	=10000*(1.056)^11
12	$18 209.71	$1019.74	$19 229.46	=10000*(1.056)^12

Developing the link to the algebraic setting would involve establishing the formulae shown in the last column of Table 11.2, and then generalising these to $A = 10000 \times (1.056)^t$

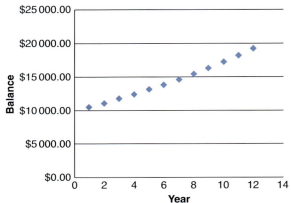

Figure 11.10 Graphic representation of exponential function

ACTIVITY 11.5

The skeleton tower (Shell Centre for Mathematical Education, n.d.) provides a pattern-based approach for introducing students to quadratic functions. Students in lower secondary school are encouraged to construct a skeleton tower with a height of six blocks. From there they are asked to establish the relationship between the height of the tower and the number of blocks. As they work through this investigation students can combine the 'wings' of the tower to form rectangular shapes and thus find a generalisation. Try this skeleton tower investigation yourself (full details are available from map.mathshell.org/index.php, search under tasks and select 'skeleton tower' under 'high school').

REFLECTIVE QUESTION

Considering the different representations of the function, describe how you could support students as they worked through this investigation. What tasks or questions might you provide?

Functions in the senior secondary school

The Australian Curriculum: Mathematics describes four mathematics subjects in the senior secondary school: Essential Mathematics, General Mathematics, Mathematical Methods and Specialist Mathematics. All of these subjects cover functions explicitly and many of them cover content, such as rates, that can be taught in a function-related way. In this section the discussion focuses on those aspects of functions that have not been discussed earlier in the context of the lower secondary school curriculum. More specifically, it examines the algebra of functions, transformations of functions, trigonometric functions and modelling with functions. The content areas covered in the section are restricted to the subjects: Mathematical Methods and Specialist Mathematics. Before covering these areas, however, a brief discussion on pedagogical consid-erations is provided.

Pedagogical considerations

In the senior secondary school students are introduced to the formal setting (at least in the more formal mathematics subjects such as Mathematical Methods) and teachers need to establish links between this setting and other function settings. Yet, as described earlier, students have difficulty connecting understandings in one function setting with those in another; they tend to **compartmentalise** their understanding (Gerson, 2010). Students, for example, might be able to provide the period of a trigonometric function in a graphic set-ting but not in an algebraic setting, thus compartmentalising within settings. On the other hand, they may be able to solve a linear equation in a graphic set-ting but be unable to solve a quadratic equation in a graphic setting, thus com-partmentalising within concepts (linear versus quadratic). Clearly, teachers need to provide opportunities for students to connect their understandings. Bloch (2003) described an experiment where students explored functions in the formal setting without resort to the algebraic setting. In what she termed the 'graphic milieu', the author described a number of tasks that allowed stu-dents to connect understandings between the formal and graphic settings. In particular, she made use of paths (see Alson, 1992) to demonstrate the link between these settings. These are described below and their use demonstrated later in the chapter.

The **direct path** (see left-most graph in Figure 11.11), shows the graphic rep-resentation of the mapping $f : X \rightarrow Y$ where an element of the domain $x_0 \in X$ is mapped onto the element $f(x_0) \in Y$. The definition of a function then fol-lows: as a mapping such that for every $x_0 \in X$, there is only one direct path. The **inverse path** (shown as the middle graph in Figure 11.11) commences with elements in the range (points on the y-axis) and shows the mapping $f^{-1} : Y \rightarrow X$. The identity paths (right-most graph in Figure 11.11) illustrate the identity function $y = x$, that maps elements onto themselves.

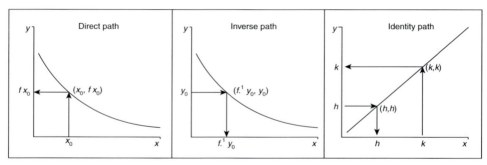

Figure 11.11 Types of paths

Bloch (2003) described how with tools such as the paths identified above, students can start to think in the formal function setting without relying necessarily on the algebraic setting. This is particularly helpful because, as noted, students tend to have difficulty working in the algebraic setting. The use of these paths in the algebra of functions is explained in the next section.

Algebra of functions

The ACM requires some students to understand the inverse of functions (ACMSM095). Using the graphic setting, students should see that a function $y = f(x)$ and its inverse (when it exists) $y = f^{-1}(x)$ are reflections of each other across the line $y = x$. Similarly, using the algebraic setting students can determine that the inverse of, for example, $f(x) = 2x+1$ is $f^{-1}(x) = \frac{x-1}{2}$. Bloch (2003) maintained that with the use of paths, students should be able to appreciate the graphic nature of the inverse without the need for algebraic methods (see Figure 11.12). The left-most diagram in Figure 11.12 shows how paths are used to find the position of the point $\left(h, f^{-1}(h)\right)$. Students can start with an abscissa $x = h$, follow a vertical path to the line $y = x$, and then a horizontal path to the function $y = f(x)$, where the point of intersection will be $\left(f^{-1}(h), h\right)$. Taking a vertical path back to the line $y = x$ at the point $\left(f^{-1}(h), f^{-1}(h)\right)$ and then a horizontal path back to the original abscissa yields the desired position of the point on the inverse function $\left(h, f^{-1}(h)\right)$. The second diagram shows the same process for the abscissa $x = k$, yielding the location of a second point $\left(k, f^{-1}(k)\right)$. Following this process for a few abscissae students should be able to sketch (or plot) the inverse function shown in the third diagram and observe the relationship between any function and its inverse as a reflection across the line $y = x$.

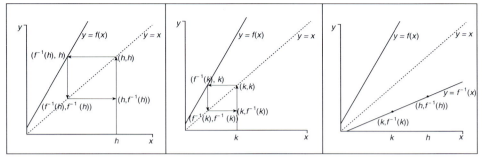

Figure 11.12 Using paths to sketch an inverse function

ACTIVITY 11.6

The ACM also requires students to understand the composition of functions (ACMMM105; ACMSM093). Given, for example, $f(x) = \frac{x^2}{2}$ and $g(x) = 2x+1$, students should understand the composition $(f \circ g)(x) = f(g(x))$. How would you use the paths described above to provide a means by which students could investigate this composite function?

Transformation of functions

The ACM suggests that students recognise features of the graphs $y = f(x)$, $y = af(x)$, $y = f(x)+c$, and $y = f(x-a)$, but separates these into concepts related to quadratic functions (ACMMM007), hyperbolic or inverse functions (ACMMM013), polynomials (ACMMM017), and trigonometric functions (ACMMM039). It is suggested that treating all of these transformations on a number of different types of functions can avoid students compartmentalising within concepts. Students need to be able to recognise, for example, that irrespective of function type, $y = f(x-a)$ is the horizontal translation of $y = f(x)$ through a units. Rather than focusing learning on a type of function (parabola, hyperbola, etc.), a focus on the type of transformation is also needed.

ACTIVITY 11.7

Use a graphing calculator or program such as GeoGebra (International GeoGebra Institute, 2015) or Graphmatica (Hertzer, 2014) to investigate the above transformations on each of the functions: $y = x^2$, $y = x^3$, $y = \frac{1}{x}$ and $y = \sin(x)$.

REFLECTIVE QUESTIONS

1 What difficulties do you think students might have as they investigate these transformations? Brown (2004), for example, highlighted the difficulties that students have in 'viewing' a function with a graphing calculator and their need for skills in determining the zoom factor (local domain and range).

2 How might you draw out the connections between the graphic setting used in the investigation and the algebraic setting? In other words, what follow-up activities and/or questions could you use to help students make these connections?

ACTIVITY 11.8

Download the projectile motion simulator available from phet.colorado. edu/en/simulation/legacy/projectile-motion (Phet Interactive Simulations, 2015). This simulation allows students to explore and apply the translations discussed above in a projectile motion context. The image shown in Figure 11.13, for example, illustrates the effects of a combined vertical and horizontal translation.

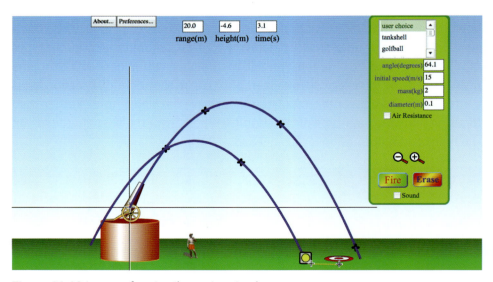

Figure 11.13 Image of projectile motion simulator

Trigonometric functions

Until they encounter trigonometric functions, students are able to apply arithmetic methods to most functions. Calculating values of the function $f(x) = 2x^2 + 3$, for example, is a relatively straightforward arithmetic process. Calculating values of the function $f(x) = \sin(x)$, on the other hand, is not as intuitive. It is not uncommon for students to incorrectly simplify an expression such as $\dfrac{\sin(x+y)}{\sin(x)}$ into $\dfrac{x+y}{x}$, possibly interpreting sin as a constant similar to π. Arguably, such errors reflect misconceptions that require a deeper understanding of the geometric basis for trigonometric functions. Moore (2014), for example, argued that students need to become familiar with the geometric interpretation of the radian as the ratio of arc length to radius and cited research suggesting that many teachers themselves tend to view the radian in terms of its degree equivalent $(180° = \pi^c)$. The introduction of trigonometric functions, therefore, needs to be situated in a geometric setting that stresses the geometric interpretation of the radian measure of angle. Gough (2010) suggested that wrapping functions are a logical means for introducing trigonometric functions that are grounded in this geometric setting. Accordingly, the following discussion focuses on wrapping functions before concluding with a description of mathematical models that are based on trigonometric functions.

Wrapping functions

Imagine a square centred on the origin with vertices (1, 0), (0, 1), (−1, 0) and (0, −1). Imagine then that we anchored a piece of string to the point (1, 0) and wrapped the string around the edges of the square in an anticlockwise direction (see Figure 11.14). We are concerned with the relationship between the length of the string (as the independent variable) and the ordinate of the point on the square, where the unit of length[4] in this example is $\sqrt{2}$. Students can

---

4   Gough (2010) recommended commencing with a simpler model for students, a square with sides parallel to the axes and vertices A(1,1), B(−1,1), C(−1,−1) and D (1,−1). The unit of length is therefore simpler.

represent this in the numerical setting (Table 11.3) and the graphic setting (Figure 11.15). In the latter, they should consider the appropriateness of connecting the points with straight line sections.

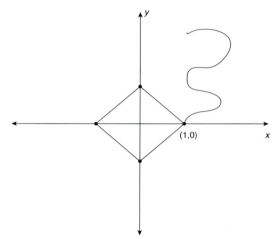

Figure 11.14 Geometric basis of a simple wrapping function

Table 11.3 Tabular representation of the wrapping function

Length of string	0	$\sqrt{2}$	$2\sqrt{2}$	$3\sqrt{2}$	$4\sqrt{2}$	$5\sqrt{2}$	$6\sqrt{2}$	$7\sqrt{2}$
Ordinate (y)	0	1	0	−1	0	1	0	−1

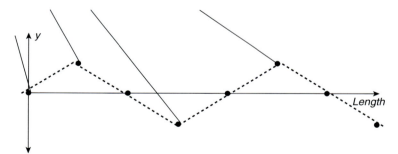

Figure 11.15 Plot of ordinate against length for wrapping function

A logical extension of the wrapping function described above is to use a unit (or indeed any circle) centred on the origin, with multiples of the arc length as the independent variable. Students could initially use length intervals of $\frac{\pi}{2}$ (see left-most graph in Figure 11.16) in much the same way as the example above, discussing the appropriateness of joining the generated points with line segments. This might inevitably lead to the recognised need for smaller length intervals of $\frac{\pi}{4}$ (see middle graph in Figure 11.16), where again discussion might focus on the nature of the function between these points. Finally, students might examine more carefully the nature of the function between these intervals (see right-most graph in Figure 11.16). Özgün-Koca, Edwards and Meagher (2013) described a hands-on approach using the unit circle and uncooked spaghetti, where students physically measured the ordinate in each case using a piece of uncooked spaghetti (broken to the correct length) and then glued that onto a plot of ordinate against length. They observed, however, that the physical manipulation (breaking and gluing) of spaghetti was time consuming for students, prompting the authors to develop an applet for this activity (see tinyurl.com/spagh1).

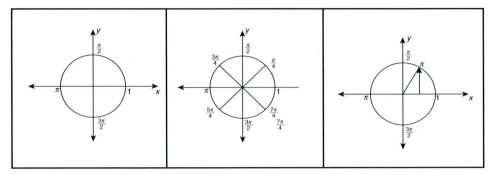

Figure 11.16 Using the unit circle as the geometric basis for a wrapping function

In this way, students should be able to appreciate the periodicity of trigonometric functions, without losing the geometric basis of the sine function or indeed the sine ratio. As with the earlier discussion, students will also need to encounter trigonometric functions in the numeric and graphic settings before mastering them in the algebraic setting. Finding the solution of the equation $\sin\left(2x + \frac{\pi}{4}\right) = 0.5$, for example, is very difficult for many students. They should have access to the function in the numerical and graphic settings. In the former, a spreadsheet would allow them to estimate the solution, whereas in the latter they should recognise the solution as being zeros to the function $y = \sin\left(2x + \frac{\pi}{4}\right) - 0.5$.

Modelling with functions

The introductory section highlighted the importance of quantitative reasoning to the success or otherwise of students in secondary school. Moreover, this reasoning was associated with the ability to model real-life situations with quantities. Galbraith (2011) identified the following steps as needed in the modelling process: understanding the real-life situation, framing a mathematical question, devising a mathematical model, analysing that model and comparing the model with the original real-life situation. In this section, we consider two situations that appear to have periodic features. The first, discussed in the example below, is messy, but many real-life situations are difficult to model. Arguably, students need to appreciate the limitations of all models. The second, covered in Activity 11.9, is much tidier and relates to the length of day in Brisbane, a context similar to that used by Wood (2003).

> **Example**: Intuitively one would expect that the distribution of births by birth date would be uniform and this statement could be posed as the basis of an investigation. The distribution of births by birth date for residents in New York during 1978 is available from www.dartmouth.edu/~chance/teaching_aids/data/birthday.txt.

Plots of these data are shown in Figure 11.17, where the first day recorded was Sunday 1 January 1978.

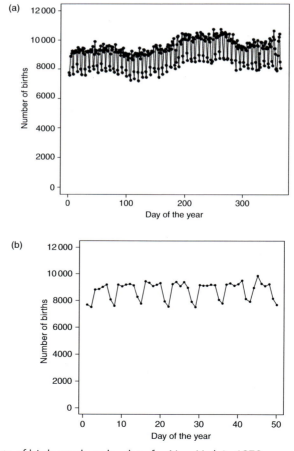

Figure 11.17 Plots of birth numbers by date for New York in 1978

Figure 11.17a shows birth numbers for each day over the year. An inspection of this suggests a seasonal variation occurs during the year. Figure 11.17b shows the number of births for each day during the first 50 days of the year and shows an approximate periodicity of seven days. At this stage, these data may merely be used to illustrate the notion of periodicity and students could be encouraged to consider the context from which they were collected to suggest why the periodicity occurs. They could also provide hypotheses to explain the apparent seasonal variation shown in the first plot. Although Galbraith (2011) cautioned against mindlessly fitting curves to data, students could explore transformations of trigonometric functions to produce curves of a similar nature to that shown in the second plot. A simple attempt is shown in Figure 11.18, which entails an understanding of periodicity, amplitude, phase shift and vertical shift; notions which are all within the reach of a senior mathematics student. Taking the absolute value of the function to obtain the final form, however, may require some prompting questions. Students could also explore the addition of periodic functions in order to model the apparent seasonal effects, thus laying the groundwork for Fourier analysis. In any case, they should be encouraged to refer back to the original context and discuss the problem: Is the distribution of birthdays uniform? More importantly, however, they could consider whether the model is generalisable to other populations.

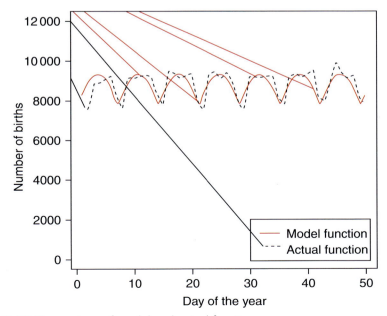

Figure 11.18 Comparison of model and actual functions

ACTIVITY 11.9

Visit the site Australian Sunrise and Sunset Times (members.iinet.net.au/
~jacob/risesetaust.html). Table 11.4 provides the predicted length of day in
Brisbane during 2015 for the 1st and 15th day of each month, though data for
every day are available from the site. Try to model these data using a trigonometric function.

Table 11.4

Month	Day	Length of day (hrs:mins)	Month	Day	Length of day (hrs:mins)
Jan	1	13:51	July	1	10:26
Jan	15	13:42	July	15	10:33
Feb	1	13:22	Aug	1	10:50
Feb	15	13:02	Aug	15	11:07
Mar	1	12:40	Sept	1	11:32
Mar	15	12:17	Sept	15	11:54
April	1	11:49	Oct	1	12:20
April	15	11:27	Oct	15	12:42
May	1	11:04	Nov	1	13:09
May	15	10:47	Nov	15	13:28
June	1	10:31	Dec	1	13:44
June	15	10:25	Dec	15	13:52

REFLECTIVE QUESTIONS

Considering the steps needed to model a real-world situation from Activity 11.9,
how would you develop the activity into a lesson or unit of work? What questions would you use to guide students through the activity? What difficulties
do you anticipate they might have with the activity? See Wood (2003), who
has used similar data with senior students.

Summary

Identify content areas of the mathematics curriculum that can be learned through a function-related approach

How many content areas of the mathematics curriculum involve one-to-one relationships between two quantities? The chapter suggests just a few examples, including rates, the compound interest formula and linear regression. Formulas used in area and volume calculations can also be taught using a function-related approach. The point is that there are many such content areas, and developing explicit links to the function helps students build those important mathematical connections. Functions are a unifying concept.

Explain how the different settings of functions are used to develop students' understanding

The function settings described in this chapter provide students with other ways of coming to know the mathematics being taught. More significantly, the numerical and graphic settings provide the iconic representations that students should encounter before and as they make sense of symbolic representations. As students are introduced to increasingly more complex mathematical topics, such as non-linear functions, families of functions and transformations of functions, many will require access to the associated functions in other (iconic) settings before mastering the algebraic setting. That said, however, teachers need to be careful that students do not compartmentalise knowledge within these settings and it is recommended that a property-orientated approach is used. The solution of equations, for example, should occur in numerical, graphic and algebraic settings. Students need to see what an equation solution looks like in each setting.

Describe learning activities that develop students' understanding of functions

The chapter has presented a number of suggested learning activities, most based on an inquiry approach that should develop students' understanding of functions. Students at the transition who are learning algebraic skills, for example, should undertake pattern-related investigations, as these build on work they have covered in their primary years. The painted cube and skeleton tower investigations, both described in this chapter, are ideal for developing early knowledge of functions. As students progress through the secondary school and become more familiar with the graphic setting, activities can be problem based. The chapter presented a problem that required simultaneous equations and recommended numerical, graphic and algebraic solutions. As students are introduced to the formal setting, activities are needed to develop connections with the other settings. The use of paths was suggested as a means for building connections between the graphic and formal settings. Similarly, activities based on wrapping functions were presented with the intent of building connections between the geometric, graphic and algebraic settings.

Identify and select appropriate technologies for developing students' understanding of functions

In early secondary school, as students work on pattern-related activities, it is recommended that students have access to physical resources and perhaps later virtual equivalents (such as the interactive painted cube described in the chapter). Electronic spreadsheets, such as Excel, and graphing programs such as Graphmatica (Herzter, 2014) and GeoGebra (International GeoGebra Institute, 2015) are ideal technologies for developing function-related concepts through the secondary school. As described, all provide access to functions in the numerical, graphic and algebraic settings.

Describe real-life situations that can be modelled with functions

As students acquire more sophisticated mathematical skills they should be able to apply their knowledge of functions to model real-life situations. The chapter presented two modeling situations appropriate for senior secondary students. The first was based on projectile motion, where students should be able to use a quadratic function to predict where the projectile lands. The second was based on periodic functions, where the length of day in a given city can be modelled using a trigonometric function.

CHAPTER 12

Calculus

Learning outcomes

After studying this chapter, you should be able to:

- describe the concepts of troublesome knowledge, threshold concepts and concept image as these apply to the teaching of calculus
- describe the skills and knowledge needed by students to study calculus and some common misconceptions that can work against students' success in calculus
- describe the alternative ways of understanding derivatives and describe the common misconceptions students may have about the derivative
- describe the alternative ways of understanding integrals and describe the common misconceptions students may have about integrals.

Introduction

How do we find the area of a region bounded by two curves? How do we find the area of a circle? How do we find the maximum and minimum values of a function? Problems such as these had been solved for centuries using methods of exhaustion by the ancients and by methods using geometry by more modern mathematicians such as Fermat, Descartes, Kepler, Cavalieri and Barrow (Katz, 1993). During the seventeenth century the solving of problems like these was revolutionised by Isaac Newton (1642–1727) in England and Gottfried Leibniz (1646–1716) in Germany. It is their work that has led to the ways that such problems are solved in science and mathematics (Katz, 1993). During their lifetimes Newton and Leibniz had a bitter argument about who had come to these ideas first; in modern times primacy is usually given to Newton, but it is Leibniz to whom we owe the notation used in modern calculus (Motz & Weaver, 1993).

In this chapter students' understanding of key concepts for the successful application of calculus will be reviewed. These concepts include rates of change, tangents, limits and the notion of infinity. Then this chapter will move onto derivatives and integrals, including their applications. Misconceptions that students may have about derivatives and integrals will also be discussed.

KEY TERMS

- **Derivative:** the derivative is a formula which allows the calculation of the gradient of the tangent at any point on a function.
- **Integration:** integration is the process by which one can get to the original function if the derivative is known.
- **Limit:** If $f(x)$ is defined on an open interval about x_o, except possibly at x_o itself, then $f(x)$ approaches the limit L as x approaches x_o, which is written as

$$\lim_{x \to x_o} f(x) = L.$$

This applies if for every $\varepsilon > 0$, there exists a corresponding number $\delta > 0$ such that for all x

$$0 < |x - x_o| < \delta \Rightarrow |f(x - L)| < \varepsilon$$

- **Rate of change:** the change in values of a function for a change in one unit of x.
- **Tangent:** the tangent of $y = f(x)$ at $x = a$ is a straight line through the point $(a, f(a))$ with the gradient of $f'(a)$. Informally, a line is a tangent if it touches a curve and has the same gradient of the curve at that point.

Troublesome knowledge, threshold concepts and concept image

Calculus, as is the case for other topics in mathematics, is notorious for the ability of students to be able to 'do it', but not to be able to explain what they are doing and why. It is also an area of mathematics where students may have ideas that are not in accord with those held by mathematicians; if their teachers do not ask for explanations this may go unnoticed. In 1981 Tall and Vinner described an individual's ideas about topics in mathematics as a *concept image*. A concept image describes the total cognitive structure that is associated with a concept. It includes all the mental pictures, properties and processes a person may hold about the concept and is built up as a result of the accumulation of all the experiences the person has had over time. Therefore, each person may have a concept image that is different from that of his or her peers (Parameswaran, 2006). This idea of the concept image is similar to the idea of the schema in constructivist learning theory. This refers to the underlying conceptual framework a person builds to make sense of the events and ideas he or she comes into contact with. When students come into contact with new information they attempt to fit this information into their current schema. If the new information does not fit into their existing schema then this schema needs to be altered; this process is known as accommodation. It has been found, however, that students can be extremely reluctant to make an accommodation, as it requires considerable mental effort. Therefore, when the students come into contact with counter-examples they may regard them as minor exceptions instead of providing reasons to change their ideas (William, 1991). They may also resort to holding simultaneously conflicting opinions, using one schema inside the classroom and another outside it. If their schema is inadequate or non-existent, in that the students do not understand the material, they will resort to using only procedural knowledge (Tall, 1992b). Therefore, it is important that teachers ask students not only to complete tasks that require procedural knowledge, but also to explain their reasoning. In this way students' understanding of the underlying concepts can be determined. Teachers also need to be aware that students may misinterpret what is being said in the classroom and this is likely when words in the technical context have different meanings from those in general conversation. In addition, students may have gained incorrect ideas early on in their learning that may persist for many years (Cornu, 1991).

Bachelard (1938, cited in Cornu, 1991) described what he called *cognitive obstacles*. One form of these obstacles is known as epistemological obstacles;

these occur because the mathematical concepts the student is dealing with are difficult. A similar idea is described by Meyer and Land (2003), who refer to *threshold concepts*. These are concepts that act as a portal, in that they open up 'new and previously inaccessible way[s] of thinking about something' (Meyer & Land, 2003, p. 1). These concepts represent a 'transformed way of understanding, or interpreting, or viewing something without which the learner cannot progress' (Meyer & Land, 2003, p. 1). These threshold concepts are transformative, in that they result in a significant shift in the perception of the learner and are probably irreversible, in that the changes that they produced are unlikely to be forgotten. These threshold concepts may also be integrative, in that previously hidden interrelations may become apparent. In this context, they may also be *troublesome*, in that they may be counter-intuitive, alien, or incoherent to the learner (Meyer & Land, 2003, p. 7). In mathematics, the **limit** is a threshold concept that may also be troublesome. This will be further discussed.

ACTIVITY 12.1

Write down what you understand the limit in mathematics is. How would you describe the limit to someone who has never seen this idea before? This is going to be a major challenge as you teach calculus.

What students need to know before beginning calculus

Students will not be successful in calculus if they do not have good algebraic skills and a good understanding of functions. In addition, students need to have a good understanding of rates of change, tangents and limits. The notion of the limit, however, is difficult for students, and therefore counts as an example of a threshold concept. This section describes the problems that students might have in these areas.

An understanding of functions

It should be obvious that to do calculus successfully students need to be competent in the manipulation of algebraic formulas and have a good understanding of exponents, including those that contain fractions and negative numbers.

Berry and Nyman (2003), however, point out that students also need to have a view of functions that does not rely on algebraic formulas alone. Ferrini-Mundy and Graham (1991) indicate that in many cases students do not regard graphical representations of functions without an accompanying formula to actually be a function. Students may also believe that functions can only be described by one formula (Dreyfus & Eisenburg, cited in Ferrini-Mundy & Graham, 1991). In addition, students may have a 'static' view of a function, in that they only consider the function by one point at a time (p. 629). If students have this static view, it is then difficult for them to understand the concept of a **derivative** when it is taught by the method of 'moving' secant lines, or to think of functions as approaching a limit.

Students also need to be familiar with the shapes of common polynomial graphs, and to have good skills in their construction and their meaning (Ubuz, 2007). Orton (1983a) has found that limited graphical understanding is one factor in low performance in differentiation. In addition, students also need to be familiar with composite functions.

Students may also fail to see the connections between the graphic and algebraic representations of a function and see these as independent from one another. Also, they may not see that functions are objects of study in themselves. Therefore, when students see a function in equation form they believe they have to 'do something to it', such as substituting in a value (Ferrini-Mundy & Graham, 1991, p. 630).

An understanding of rates of change

Quantifying rates of change is one of the main features of calculus and as such should be considered a threshold concept. Therefore, students need to have a clear idea as to the nature of rates of change in mathematical contexts. In particular, students need to have a clear idea that a **rate of change** involves a comparison of unit per unit change in quantity (Bezuidenhout, 2006). On a graph, this is obtained by finding the difference in y-values that correspond to a *unit difference* in x-values.

Students also need to have a good comprehension of the differences between straight lines and curves. Dreyfus and Eisenberg (cited by Ferrini-Mundy & Graham, 1991) have noted that students may treat other functions as linear functions. In particular, students need to appreciate that every point on a curve may lead to a different value for the rate of change. This is in contrast to straight lines, where the average rate of change is constant and is the same for the rate of change at every point (Orton, 1983a).

Orton (1983a) recommends that students should be exposed to the concept of a rate of change at the same time as students are developing understanding of ratios and graphic representations. He recommends that students should be exposed to real-life examples before more algebraic approaches are used. He recommends that students should also thoroughly explore ideas of rates of change, average rate, gradients of a line, tangents and secants to a curve before the more formal algebraic representations are introduced. During this time students should also be exploring where the functions are increasing and decreasing and connect these to positive and negative numerical measures.

Tangents

One way of viewing the derivative is that it gives a formula for finding the rate of change at any single point on a curve of a function. This rate of change is also the gradient of the **tangent** line at this point. As finding derivatives is a major task in calculus, students need to have an exact understanding of the nature of these tangent lines. A formal definition of a tangent line can be described thus:

> The tangent to $y = f(x)$ at $x = a$ is a straight line through its point $(a, f(a))$ with the gradient $f'(a)$.

This formal definition requires some understanding of the derivative and therefore, while mathematical language can have the feature of being concise and unambiguous, a definition such as that given by Barnes (1993) can be more useful for beginning calculus students:

> A tangent is a straight line through a point on the curve, going in the same direction as the curve. At the point in question, the gradient of the tangent is the same as the gradient of the curve, that is, the derivative of the function. Near the point, the tangent is closer to the curve than any other straight line through the point. (1993, p. 13)

There are other ways to describe a tangent. It is useful to use more than one way when teaching, as a method that suits one student may not benefit another. According to Artigue (1991) tangents can be thought of in these ways:

> One can think of the tangent to the curve at point A as:
> A line passing through A but not crossing the curve in the neighbourhood of A.
> A line having a double intersection with the curve at A.
> A line passing through points infinitely close to A on the curve or the line which the curve becomes when one magnifies it in the neighbourhood of A.

The limit of the secants as the intersection at one point on the curve tends towards the other intersection.

The best linear approximation or the only linear approximation of the first order to the curve in the neighbourhood of A.

The line passing through A whose slope is given by the derivative at A of the function associated with the curve (where the derivative is assumed to exist).

(Artigue, 1991, p. 174)

As the students become more flexible and experienced they tend to use the concept that best suits for each purpose.

The limit

Understanding the limit is one of the tasks required for advanced mathematics. However, owing to its difficulty, the nature of the limit is a form of knowledge that is not only a threshold concept, but is 'troublesome' (Meyer & Land, 2003). By this it is meant that not only does this concept lead to a new form of understanding, but because of its nature it may appear alien and/or counter-intuitive to the learner. In fact Davis and Vinner (1986) state that as students develop their ideas of limits, misconceptions may be unavoidable. To complicate the teaching of this concept, if teachers try to simplify the concept too much 'serious conceptual problems' may arise (Tall, 1992b, p. 502). Trying to introduce students to the thinking required for formal mathematics using a logical sequence where the concepts are introduced through definitions and logical deductions is also not necessarily successful (Cornu, 1991).

Here is one version of a formal definition:

Let $f(x)$ be defined on an open interval about x_0, except possibly at x_0 itself. We say that $f(x)$ approaches the limit L as x approaches x_0, and write:

$$\lim_{x \to x_0} f(x) = L$$

if for every $\varepsilon > 0$, there exists a corresponding number $\delta > 0$ such that for all x

$$0 < |x - x_0| < \delta \Rightarrow |f(x - L)| < \varepsilon$$

(Thomas & Finney, 1996)

In words, this can be described as: 'A limit is a number that the y-value of a function can be arbitrarily close to by restricting x-values' (William, 1991, p. 221).

Unfortunately, these formal definitions do not help students to develop an understanding of the limit and students may not, at first, relate to it (Robert & Schwarzenberger, 1991). Being able to repeat the formal definition does not guarantee understanding. For many students their introduction to the limit is the first time that students may be introduced to an idea in mathematics that does not have a finite computation with a definite answer (Cornu, 1991 cited in Cottrill et al., 1996). This difficulty can be enhanced if students have been taught in a manner that leads them to believe that mathematics always has one definite, correct answer.

One of the most common misconceptions is for students to think that a limit is never reached. One of reasons for this is the way that the word 'limit' is used in everyday language. The use of the word *limit* in things such as credit limits and speed limits can inhibit the understanding of this idea in mathematics (Cornu, 1991).

How should students be introduced to the concept of the limit?

How should students be introduced to the concept of the limit? There is no easy answer to this question, but teachers need to be aware that research shows that the context in which students are first introduced to limits dominates their thinking about limits in future work (Tall, 1992b). For example, if students are introduced to limits in the context of terms of a sequence that tend to but don't reach the limit, then students will think that limits are never reached.

If limits are being taught so that derivatives can be introduced, an example such as this may be helpful.

For the function:

$$f(x) = \frac{x^2 - 4}{x - 2}$$

What happens at $x = 2$? What is the value of the function as x gets closer to two, from below it and above it?

Students should be able to see that this function is a straight line of the form $f(x) = x + 2$ with an important difference at $x = 2$; here the denominator equals zero and the function is therefore undefined at this point. To answer the second question, a spreadsheet can be used to show what happens as the value of x approaches two and this is illustrated in Table 12.1.

It is apparent that as the x-values approach two the value of the function approaches four even though at exactly $x = 2$ the function is undefined. It has

Table 12.1 Values of f(x) as x approaches 2

x	f(x)
1.9	3.9
1.99	3.99
1.999	3.999
1.9999	3.9999
1.99999	3.99999
1.999999	3.999999
2	value undefined
2.000001	4.000001
2.00001	4.00001
2.0001	4.0001
2.001	4.001
2.01	4.01

to be remembered, however, that this is an example where the limit is not actually reached, so this may also give students the idea that a limit is never reached. An advantage of such an example, however, is that the idea of the limit is introduced where a function is undefined because the denominator is zero, which is something the students will come across if moving secants are used to introduce the derivative. Yet, if the limit is introduced with a function that is defined at every point, then the students may gain the idea that the limit of a function at a point is equal to the value of the function at that point (Przenioslo, 2004). It is, therefore, important that a variety of examples with different properties is shown to the students.

One form of understanding the limit is the *dynamic* idea of the limit. This idea of a limit emphasises motion (for example, as x moves towards a, f(x) moves towards L). Whereas this idea may lead to conflicts with the formal definition (Szydlik, 2000), this form of understanding the limit has advantages (Keene, Hall & Duca, 2014). By using language such as 'getting closer to' and 'approaching' students can be assisted in building their intuitions about the nature of limits and how they are found (Keene, Hall & Duca, 2014, p. 564). It is also important that students understand that 'approach' means that the values of the function become less and less different from a certain number and these differences become arbitrarily small (Przenioslo, 2004, p. 122).

The dynamic conception of the limit also helps covariational reasoning, which is important for the understanding of derivatives. Jones (2014) explains this as follows:

> A limit, by its nature, involves at least two quantities that are potentially changing simultaneously. For example, in the limit $x \to \infty$ $f(x) = L$ one can think of the x variable as an increasing quantity which corresponds to a tendency of the y coordinate to be close to L as x becomes large. These two quantitates can be considered as covarying. (p. 108)

An example of finding a limit that is simplified by using the dynamic view is

$$\lim_{x \to \infty} \frac{\sin(x)}{x}$$

If it is understood that as x gets very large the denominator becomes large, and that the numerator has a maximum of 1, then the limit can be easily conceived to be zero.

What is infinity: the answer to dividing by zero?

Students often think that infinity is a number, and not as a property that a set of numbers may have (Mamona-Downs, 2001). This is because students do not usually think of numbers in terms of sets. Therefore, if faced with a limit in the form of

$$\lim_{x \to \infty} \frac{\sin(x)}{x}$$

they will substitute infinity into the problem, to get

$$\lim_{x \to \infty} \frac{\sin(\infty)}{\infty}$$

Another problem that arises is that students may believe that it is possible to divide by zero, and that a number divided by zero has the answer infinity, or even zero. It is the author's experience that teachers may believe this as well. It is easy to see why someone may think that a number divided by zero is infinity. As a number is divided by a smaller and smaller number the answer becomes larger and larger. If one believes that infinity is a number, then the logical consequence is that dividing by zero gives infinity as the answer. This problem can be avoided if students understand that infinity is not a number, and that dividing by zero is not an operation that makes sense.

ACTIVITY 12.2

In her 2001 paper Mamoma-Downs lists some 'approach types' (p. 264) that students have given when attempting to define a limit, and these are listed in Table 12.2. Examine each definition and describe how a student may come to this particular view. What are the advantages and disadvantages for each approach? For those you think are incorrect, what counter-examples could you find?

Table 12.2 Approach types describing the limit

Approach type	Typical informal statement
Dynamic–theoretical	A limit describes how a function moves as x moves to a certain point.
Boundary	A limit is a number or point beyond which a function cannot go.
Formal	A limit is a number, to which the y-values of a function can be made arbitrarily close by restricting x-values.
Unreachable	A limit is a number or point the function gets close to but never reaches.
Approximation	A limit is an approximation that can be made as accurate as you wish.
Dynamic–practical	A limit is determined by plugging in numbers closer and closer to a given number until the limit is reached.

Derivatives

The derivative is a formula that can be used to find the rate of change at any single point on a curve of a function. This rate of change is also the gradient of the tangent line at this point (Barnes, 1993). Because the rate of change is for a single point, this rate of change is also known as the instantaneous rate of change. These two interpretations, that the derivative is the gradient of the tangent line at a point, and the instantaneous rates of change at a point, are requirements of the Senior Mathematics Curriculum, Mathematical Methods (ACARA, 2015). The following definitions of a derivative also apply:

- The derivative of a function is the limit of the ratio of the increase in the value of the function to the change in the independent variable as the change of the independent variable approaches zero (Kaplan, Ozturk & Ocal, 2015).
- Related to this idea is that the derivative is the following limit:

$$\lim_{h \to 0} \frac{f(x+h) - f(x)}{h}$$

- The derivative of the function is equal to the tangent of the angle between the tangent line at a point $x = a$ and the x-axis.
- The derivative is defined as the gradient of the tangent line at a point on the curve (Kaplan, Ozturk & Ocal, 2015).
- The derivative is the first order coefficient of the expansion limited to order one of the function at a point (Artigue, 1991).
- The derivative is the slope of the tangent at a point.
- The derivative is the slope of the highly magnified portion of the graph itself, where the graph is locally straight.

This latter view, that the derivative is the slope of a highly magnified portion of the graph itself has the advantage that it can be easily demonstrated on computer software by zooming in on a particular graph (Tall, 1992b).

In some European countries the derivative is introduced as first order expansion of the function. In Australia, the derivative is usually introduced as the limit of the secants. This has the advantage that it can easily be linked to finding the gradient of a linear function, with which the students should already be familiar. The students will also need to be familiar with the idea that the average rate of change between two points is approximated by the gradient of the secant. This process can be modelled by computer programs. For example, Figure 12.1 shows how the secant line (in blue) becomes closer and closer to the tangent (in red) of the line at A as the horizontal distance between A and B is narrowed until finally it becomes co-linear with the tangent (www.geogebra.org/student/m15671).

Once students have a visual understanding of the process, then the process can be introduced algebraically. In reality, finding the derivative is no more than finding the rate of change of a function per unit change in x with the addition

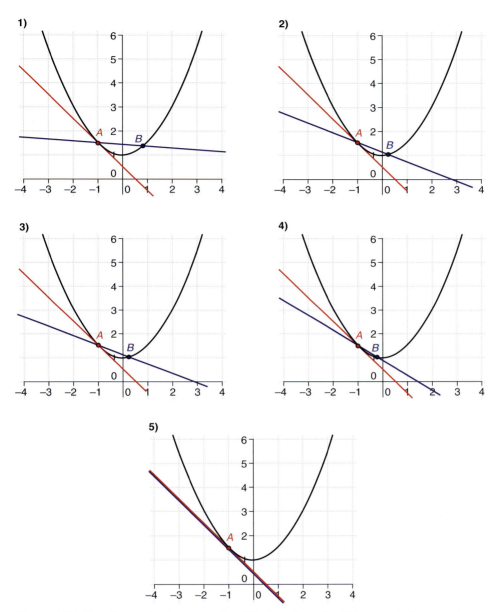

Figure 12.1 The change in the secant line (blue) as the point *B* is moved closer and closer to *A*

of having to use the concept of the limit. Therefore, the process is based on the formula for calculating the value of the gradient for any straight line.

$$m = \frac{y_2 - y_1}{x_2 - x_1} \qquad (1)$$

If students are familiar with function notation, then the formula can be written as:

$$m = \frac{f(x_2) - f(x_1)}{x_2 - x_1} \qquad (2)$$

Then, if the students can think of the difference between x_2 and x_1 as represented by the distance 'h', then with a slight modification the formula for calculating the gradient of the line can be written as.

$$m = \frac{f(x+h) - f(x)}{h} \qquad (3)$$

Once the similarity between this equation and the usual formula for finding the gradient of a straight line is appreciated, any mystery behind the process of finding the derivative 'from first principles' should be removed. This means that the process will not need to be learnt by rote. After this point has been reached the derivative is just the limit as $h \to 0$ of equation 3 to produce equation 4.

$$f'(x) = \lim_{h \to 0} \frac{f(x+h) - f(x)}{h} \qquad (4)$$

This algebraic process can also be accompanied by the use of a spreadsheet and it may even be worthwhile to use a spreadsheet before the formal algebraic process. Table 12.3 shows the gradient of the function $f(x) = x^2$ calculated at $x = 3$. In this spreadsheet the formula

$$m = \frac{y_2 - y_1}{x_2 - x_1}$$

is used where the difference between x_2 and x_1 is gradually reduced. As this difference between the values x_2 and x_1 becomes smaller the value of the gradient approaches 6, that is, $2x$. The spreadsheet formulas for Table 12.3 are shown in Table 12.4.

Students' misconceptions about derivatives

Some of the problems that students may have with derivatives can result from mistakes in simple algebra. For example, students may make errors in the

Table 12.3 Values of the gradient at $x = 3$ and with changes in the difference between x_2 and x_1

x_1	y_1	x_2	$x_2 - x_1$	y_2	gradient
3	9	3.1	0.1	9.61	6.1
3	9	3.01	0.01	9.0601	6.01
3	9	3.001	0.001	9.006001	6.001
3	9	3.0001	0.0001	9.0006	6.0001
3	9	3.00001	0.00001	9.00006	6.00001
3	9	3.000001	0.000001	9.000006	6.000001
3	9	3	1E-07	9.000001	6

Table 12.4 Formulas for Table 12.3 (in Microsoft Excel)

	A	B	C	D	E	F
1	x_1	y_1	$x_2 - x_1$	x_2	y_2	gradient
2	3	9	1	= C2 + A2	= D2^2	= (E2 - B2)/(D2 - A2)
3	3	9	= C2/10	= C3 + A3	= D3^2	= (E3-B3)/(D3 - A3)
4	3	9	= C3/10	= C4 + A4	= D4^2	= (E4 - B4)/(D4 - A4)

expansion of brackets such as $(x + h)^2$ or $(x + h)^3$. Other problems may come about because students may not develop an intuitive understanding of the concepts. If these understandings have not developed they will rely on learning the algorithmic rules and manipulation of the symbols to get through their exams (Berry & Nyman, 2003). Students may also have problems connecting the idea of the derivative, allowing the calculation of the value of the gradient of the tangent line at a point and the derivative as a generalised function (Ubuz, 2007).

Ubuz (2007) has found that students may have one or more of these beliefs about derivatives:

- The derivative at a point gives the function of a derivative.
- The tangent equation is the derivative function.
- The derivative at a point is the tangent equation.
- The derivative at a point is the value of the tangent equation at that point.
- In differentiation the slope of the secant line approaches a smaller value.

Students may also find the notation confusing. For example, they might not understand that the quotient $\frac{dy}{dx}$ is a real number that is not infinitely small (Tall, 1992a). They also need to be familiar with the f prime notation.

In his study of high school and university students, Orton (1983a) found that in general students were reasonably competent in the algebra required to calculate the derivatives of simple functions. When it came to explaining the idea of the limit, however, less than half of the students could give satisfactory explanations. For example, when questioned about what happens in a diagram similar to that found in Figure 12.1, these students could not see that as point B moved towards point A the process would result in the tangent to the curve. In addition, he found that some students could find the gradients to tangents on curves when the curves were expressed algebraically, but could not find the gradients of these tangents when graphs were presented. He also found that some students could calculate the average growth in a function between $x = a$ and $x = a + h$, but then could not see how to use this to calculate the growth at a particular point.

Uses of derivatives

Stationary points and dominant terms

One of the applications of derivatives stipulated in the curriculum for Mathematical Methods (ACARA, 2015) is to 'sketch curves associated with simple polynomials, find stationary points, and local and global maxima and minima; and examine behaviour as $x \to \infty$ and $x \to -\infty$ (ACMMM095).

When searching for stationary points students may find the algebra worrisome, but the principle is simple if students have a good understanding of how tangent lines to curved functions work. At a stationary point the gradient of a tangent line is horizontal and therefore has a gradient of zero. Therefore, in functions whose formulas are known, finding these stationary points simply requires finding the values for x where the derivative is equal to zero. Students may, however, forget that the stationary points found this way are often only local maxima or minima and are not the global maximum or minimum of the function. Students also need to be aware that the derivative may also be equal to zero at a point of inflexion.

It was mentioned in a previous section that finding limits can be effective if a dynamic view of the limit is used, which uses covariational reasoning. This is also a useful strategy when thinking about what happens to the values of functions as their x-values take on very large positive or negative values. Here, students need to think about *dominant terms*. For example, for the function $f(x) = x^2 + 4x + 1$, once the x-value is greater than two the value of x^2 will always be greater than the value of $4x$, so it is the x-squared term that needs to be thought about as the student considers what happens as $x \to \infty$ or

$x \to -\infty$. However, this becomes more complicated when rational functions are considered.

Graphing of rational functions

The National Curriculum document for Mathematics Specialised states that students should 'sketch the graph of simple rational functions where the numerator and denominator are polynomials of low degree' (ACMSM 100) (ACARA, 2015).

Rational functions are in the form of

$$f(x) = \frac{g(x)}{h(x)}$$

A function in this form is more complicated to graph than a simple polynomial as the function is undefined when $h(x) = 0$. In addition, a function in this form will have *asymptotes*. An asymptote is a line that the graph of the function $f(x)$ approaches as the x-values approach infinity. Vertical asymptotes are found at the values where $f(x)$ is undefined because $h(x)$ equals zero, but not all values of x that lead to $h(x)$ having a value of zero will have a vertical asymptote. Therefore, students need to be given examples that illustrate a variety of circumstances. Horizontal asymptotes are lines that the function approaches as the values of x approach infinity in the positive or negative direction and will occur if the degree of $g(x)$ is equal to the degree of $h(x)$. A function may have an oblique asymptote in the form of $y = ax + b$ if the degree of the numerator is one more than the degree of the numerator.

Here is an example of a rational function:

$$f(x) = \frac{x^2 - 4}{x^2 - 5x + 6}$$

In this example the degree of $g(x)$ is the same as the degree of $h(x)$. As the x values approach infinity in either the positive or negative direction, the value of the function can be approximated by the dominant terms so that

$$f(x) \sim \frac{x^2}{x^2}$$

and the function approaches one. Therefore there is a horizontal asymptote at this value.

The function will be undefined where $h(x) = 0$; these can be found by factorising $h(x)$ and equating it to zero.

$$x^2 - 5x + 6 = 0$$

$$(x - 3)(x - 2) = 0$$

Therefore the values of x for which $h(x) = 0$ are $x = 2$ and $x = 3$.

It is tempting for students to then think that vertical asymptotes will be found at both these x-values, but there is only one vertical asymptote, at $x = 3$. Why this occurs can be found by factorising both $g(x)$ and $h(x)$.

$$f(x) = \frac{(x+2)(x-2)}{(x-3)(x-2)}$$

This tells us that the function $f(x)$ is undefined at $x = 2$ in both the numerator and denominator. Therefore at this point the $f(x)$ is undefined. The final graph is shown in Figure 12.2.

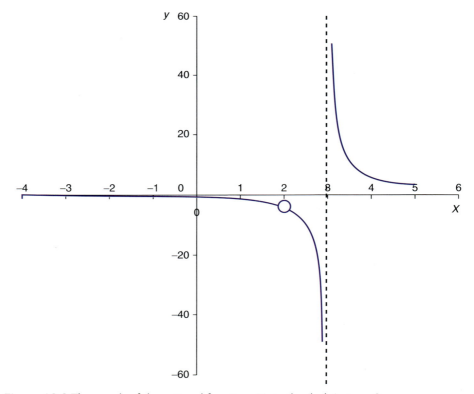

Figure 12.2 The graph of the rational function. Note the 'hole' at $x = 2$

Now that graphing calculators are in common use, a useful teaching strategy is to give students examples where they are likely to make errors and then get them to plot the graphs by hand and then compare their answers with that given by the calculator. They then can explore the differences.

Optimisation

What is the maximum volume of a box that can be made with a fixed area of cardboard? If we want to make a soft drink container that can hold 300 ml, what shape will use the least materials? Because derivatives can be used to find maxima and minima of functions, such questions can be answered and this process of *optimisation* is one of useful applications of derivatives. Optimisation is one of the requirements of Mathematical Methods (ACARA, 2015), which states: 'Solve optimisation problems arising in a variety of contexts involving simple polynomials on finite interval domains' (ACMMM096).

For example, the students could be required to find the maximum area of a box (without a lid) that could be made out of a sheet of A4 paper. This is a useful example to use to introduce students to optimisation as it can be solved in several ways. The students can use trial and error with actual paper, or use a diagram with a spreadsheet. They can also see the relationship between what they do in calculus with a physical object. Using calculus, the students need to find the equation for the volume of the box in one variable, and then use the derivative to find the maximum of the function. Before this can be carried out they need to draw a diagram (Figure 12.3) so that they can examine the relationships between the variables length, width and volume. With the diagram it also becomes apparent that the problem can be solved if the search is made for the width of each square that needs to be cut out of each corner of the paper.

Figure 12.3 The diagram to help solve the box problem

When this problem is solved using calculus, the derivative is a quadratic equation where only one of the solutions gives a suitable size of the square.

This assists students to realise that they have to align the abstract mathematics with the physical problem. It must be admitted, however, that such problems can be more complicated than this one. It can be difficult for the students to use the different formulas that describe a particular situation and then express the connections between the variables in terms of one variable. In many cases such as this one the students can be greatly helped by sketching appropriate diagrams, and this can also be difficult for students with poor spatial reasoning.

The increments formula

There are times when we know the value of a function $f(x)$ at a point a, and wish to know the value of the function at a point very close to this x-value. This can be achieved by using the increments formula and is one of the applications of differentiation listed in the curriculum for Mathematical Methods (ACMMM107).

The increments formula is given by:

$$\delta y \cong \frac{dy}{dx} \times \delta x$$

Here, δy and δx represent small changes in y and x respectively. This is yet another example where the teacher needs to consider whether or not to 'teach' the formula alone, or to allow the students to think about the principle. Once the principle is understood, the formula is not needed. The notation $\frac{dy}{dx}$ gives the derivative, which, as we have seen, allows us to find the gradient of the tangent line at any point on a curve. Once we know the gradient at a point c, then the gradient of the curve at a small distance, δx, away will be very close to that at point c. Therefore the formula will give a close approximation to the change in the overall value of the function.

Derivatives of trigonometric functions

In this section we will look at the derivatives of sine(x), cosine(x) and tan(x). For some teachers, it is sufficient just to give the students the formulas, but a more investigative approach can be taken. For example, look at the graph of sine(x) in Figure 12.4.

If the students trace out the figure from left to right they will see that the initial negative gradient becomes less steep until it becomes zero. After this the

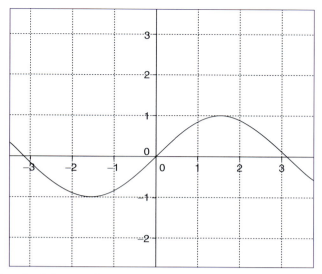

Figure 12. 4 Graph of the sine function

gradient is positive becoming steeper until it again becomes less steep until it is zero, and so on. They may see that the pattern is similar to the original function. This is confirmed with appropriate software, as shown in Figure 12.5. It is then apparent that the derivative function is out of phase by 90° or approximately 1.5 radians, that is, it is equivalent to the cosine function.

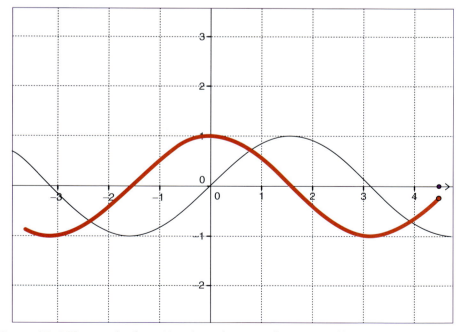

Figure 12. 5 The graph of sine(x) with its derivative function (red line)

A similar process can be followed for the tan(x) function. It should be apparent from the graph of tan(x) that the derivative must always have a positive value and this is shown in Figure 12.6. The quotient rule can be used to confirm that the derivative of tan(x) is sec²(x).

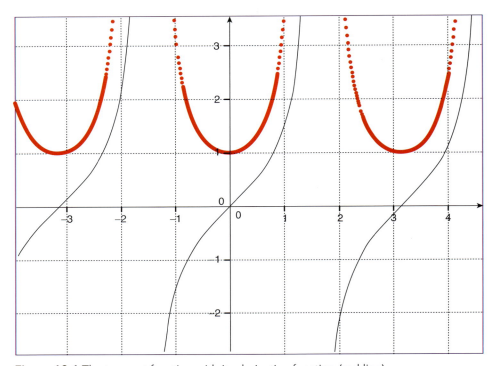

Figure 12.6 The tangent function with its derivative function (red line)

Product, quotient and chain rules

While we know that the derivative of $f(x) + g(x) = f'(x) + g'(x)$, can this rule be extended to when two functions are multiplied together? In other words, is the derivative of $f(x) \times g(x)$ equal to $f'(x) \times g'(x)$? What about the derivative of $\frac{f(x)}{g(x)}$? Is this equal to $\frac{f'(x)}{g'(x)}$? If students experiment with two polynomial functions, it should become immediately apparent that the derivative of two functions that are multiplied together is not the product of their derivatives. For example:

$$\text{If } f(x) = (x^2 + 2)(x + 3)$$

If we multiply the derivatives of both brackets together we get $f'(x) = 2x$.

But if the brackets are expanded, then the derivative is found to be

$$f'(x) = 3x^2 + 6x + 2.$$

Similarly, the derivative of one function divided by another is not the quotient of the two derivatives. Once this is established, then the rules for finding of the derivative of the product or quotient of two functions can be introduced. The more capable and curious students can investigate the proofs of these rules.

To introduce the chain rule, students can be given a function to experiment with. Here is an example.

$$f(x) = (\sin(x))^2$$

What is the derivative of this function? Because we know that the derivative of $\sin(x)$ is $\cos(x)$, and the derivative of x^2 is $2x$, then we know the derivative should have a '2' and '$\cos(x)$' in the answer. However, the derivative is not $2\cos(x)$. This would be the derivative if our original function were to be $f(x) = 2\sin(x)$. If students are prepared to accept that two different functions will not have the same derivative, then they should be ready to accept that something else needs to be done and in this case this is using the chain rule. The chain rule can be written in two different ways.

1   If $F(x) = (f \circ g)(x)$, then the derivative of $F(x)$ is, $F'(x) = f'(g(x))g'(x)$
2   If $y = f(u)$ and $u = g(x)$, then the derivative of y is:

$$\frac{dy}{dx} = \frac{dy}{du} \times \frac{du}{dx}$$

If students are familiar with both notations they can use the notation they prefer. The more curious and capable students can investigate the proof of this rule. The chain rule is commonly summarised as the 'inside outside' rule, but care must be taken if the rule is to be taught this way. Students can become so confident in this shortcut that they can be reluctant to go back to using the chain rule when more complicated derivatives arise. As a consequence, these students make errors when they are faced with more complicated examples, such as

$$y = \sqrt{\sin(x^2)}$$

Try it and see.

Connection between function graphs and graphs of their derivative functions

Students can have the idea that doing calculus only involves manipulating symbols and numbers (Hughes Hallet, 1991). Drawing the graph of a derivative function from the original function, or drawing an original function from the graph of its derivative function, however, requires higher-order thinking, in particular, a thorough understanding of rates of change.

Here again a graphing tool such as GeoGebra can be helpful. Students can watch the gradient of the tangent line change as a function is traced, and see what this gradient is when a local maximum or minimum point is reached. They can also see how the gradient changes when the function is increasing and decreasing (Figure 12.7). With the appropriate software they can also see the graph of the derivative function being traced out as the x-values change (Figure 12.8).

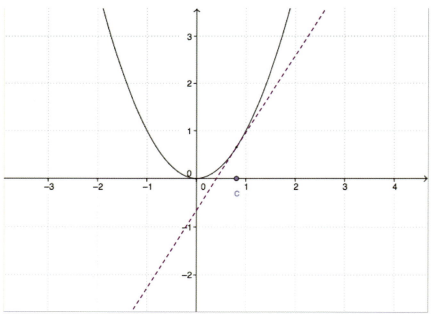

Figure 12.7 An example of the tangent line being drawn on the function $y = x^2$. This was produced using GeoGebra.

Second derivatives

The first derivative is the derivative of an original function which allows the calculation of the rate of change on any point on the function. Second

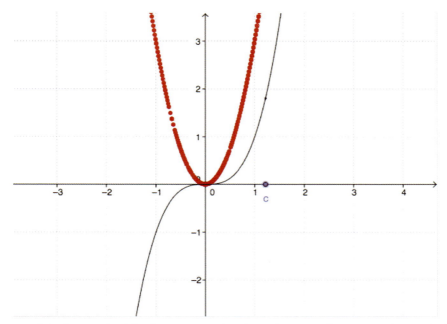

Figure 12.8 The graph of the function $y = x^3$ with its derivative function (red points). The red points are created as the point c is moved along the x-axis. This was produced using GeoGebra.

derivatives are found by finding the derivative of the first derivative, and tell us the rate of change of the derivative.

Therefore, if the second derivative is at a particular x-value is positive, that means $f'(x)$ is increasing as it passes through that x-value. Similarly, $f'(x)$ is decreasing where the second derivative is negative. This information allows a method for confirming if a particular stationary point of the original function is a local maximum or minimum. Say that a possible minimum or maximum is found at $x = c$. If at a nearby x-value to the left of c $f'(x)$ is positive, zero at $x = c$, and negative at a point to the right of c, then the original function will have a maximum at $x = c$. $f'(x)$ will be decreasing at this point and $f''(x)$ will be negative. Similarly, for a minimum point on the original function $f'(x)$ will be negative, zero and positive and $f''(x)$ will be positive.

It is well known that in general, students may try to accommodate a lack of understanding by learning the rules by rote, and then making mistakes as they do not remember the rules correctly (Chance, del Mas & Garfield, 2004). Therefore, it is important that students gain an understanding of the reasoning that leads to these rules. Figure 12.9 gives an example of how this might be done. The students are given a polynomial function to the power of three (but not in algebraic form) and asked to draw the graph of the first and second derivatives. Once it is appreciated that the first derivative is a quadratic

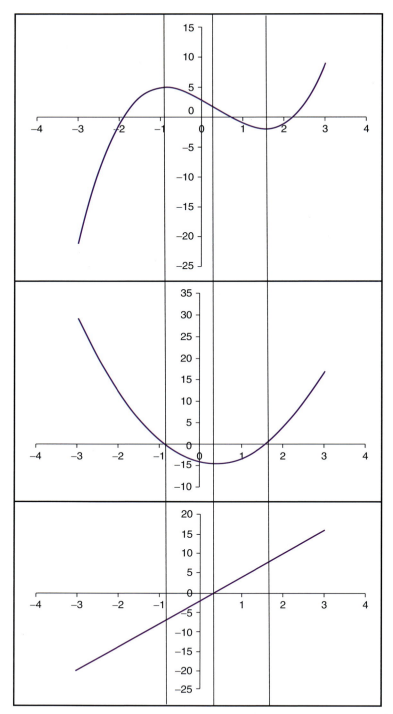

Figure 12.9 A graph of a function with the graphs of its first and second derivatives

function, they should be able to reason that the graph of the second derivative will be a straight line. What happens when $f''(x)$ equals zero? From such an exercise the ideas about inflection points and concavity can be introduced. The students can also observe what happens at the point where the second derivative is zero. This then leads naturally to a discussion about inflection points.

ACTIVITY 12.3

Commonly, an asymptote is defined as a line to which the graph 'never reaches'.

Graph the following function and look carefully to see what happens at the values $x = 8$, $x = 9$ and $x = 10$. What does this say about this common definition?

$$f(x) = \frac{(x-3)^2}{x^2 - 5x}$$

ACTIVITY 12.4

Do all points where $f''(x)$ equals zero form inflecion points? What about $f(x) = x^3$?

Reversing the process: anti-differentiation and integration

If we know the derivative function, what is the original function? For simple polynomials students should easily be able to work out how the original function can be found. With guidance, they should also be able to see that without further information, if there is a constant in the original function its value cannot be determined. Therefore the '+ C' given at the end of the process of anti-differentiation, or **integration**, is a necessity. Students should also be able to work out the resulting integrals of the sine and cosine functions.

The terminology can be confusing. If f is a function defined on an interval, I, and f has an antiderivative on this interval, the set of the antiderivatives of f is called the indefinite integral and has this notation.

$$\int f(x)\,dx$$

Finding these antiderivatives is known either as anti-differentiation or integration.

The function f is known as the *integrand* of the integral and x is the *variable of integration*. An antiderivative is one of the set of functions that can be a solution to the process of integration.

A source of further confusion to students is determining the difference between an *indefinite integral* and a *definite integral*. Indefinite integrals have no limits and the solution to finding an indefinite integral is a function. In contrast, definite integrals have limits, and while the process of integration is required in finding their solutions, the final answer is a number. For example, to find the solution to $\int_2^4 x^3 dx$ we need to first find the integral, $\frac{x^4}{4}$ and then evaluate it with substituting in the values '4' and '2' for x.

$$\int_2^4 x^3 dx = \frac{4^4}{4} - \left(\frac{2^4}{4}\right) = 60.$$

As the functions get more complicated there are endless possibilities for students to become confused with the signs, so the use of brackets is essential, and in addition the fraction buttons on their calculators also become a very useful tool. Students are often tempted to use decimals, but by doing this the errors can quickly compound which is of importance when, for example, they calculate areas under curves.

Applications of integration

Areas under curves

The curriculum for Mathematical Methods (ACARA, 2015) states that students should be able to 'use sums in the form of $\sum_i f(x_i)\delta x_i$ to estimate the area under the curve $y = f(x)$' (ACMMM124) and to 'interpret the integral $\int_a^b f(x)dx$ as the area under the curve $y = f(x)$ if $f(x) > 0$' (ACMMM125). It then continues to state that students should 'recognise the definite integral $\int_a^b f(x)dx$ as a limit of sums of the form $\sum_i f(x_i)\delta x_i$' (ACMMM126). These statements sum up the process of finding areas between curves and the x-axis.

The area between a curve and a straight line can be approximated by dividing the area into rectangles as shown in Figure 12.10.

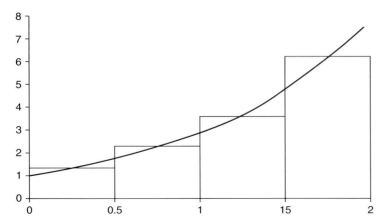

Figure 12.10 An approximation to the area under a curve using rectangles

If the height of each rectangle is found by finding the value of the function at each appropriate x-value, then the width of each rectangle can be represented by Δx. If there are k rectangles the total area is then approximated by

$$A = \sum_{i}^{k} f(x)\Delta x$$

This formula calculates what is known as a Riemann sum.

Students should be able to see that by reducing the width of each rectangle, the error that is introduced by using the rectangles and the actual area should be minimised until it approximates zero (Tarvainen, 2006). The area is then found by letting Δx→0, so that the area between a curve and the x-axis over an interval [a,b] is found by the definite integral.

$$A = \int_{a}^{b} f(x)\,dx$$

That is, the limit of the Riemann sum as the interval approaches zero is the same as the area between the curve and the x-axis. The lower letter (in this case a) is known as the lower limit of integration, and the upper letter (in this case b) is known as the upper limit of integration.

Students can see this process with the use of a spreadsheet. Students can set up the spreadsheet so that the intervals along the x-axis become smaller and smaller and can thus observe the area approaching a limiting value, which is the same as that obtained by calculating the integral. It is also important for students to discover what happens when the area is below the x-axis.

To understand the process of integration, it is useful for students to be able to imagine the process of the intervals along the x-axis becoming smaller and that these intervals represent the base of a series of rectangles. A similar process of imagination is needed for students to follow the process of finding solids of revolution. If each rectangle is rotated around the x-axis, then the result is a series of cylinders, each with a radius equal to the value of the function and with the width equal to Δx. Once this is appreciated, the actual calculations are not difficult. Instructors may need a variety of representations for students to understand how the process works.

ACTIVITY 12.5

What is the answer to this definite integral? How is the answer to this different from finding the area between these two limits and the x-axis?

$$\int_{-\pi}^{\pi} \sin(x)\,dx$$

Accumulation functions

Whereas integration can be described in the form of the limit of Riemann sums, integrals are also used to calculate total change and without this understanding students will only have a reduced understanding of integrals (Sealey, 2006). The 'accumulation' is calculated by 'integrating instantaneous or marginal rate of change' (ACMMM133) (ACARA, 2015). One example is that of water leaking out of a tank; if water leaks out of the tank at rate of $l(t)$ litres per minute, then the total water that has leaked out of the tank at a given time, t, is $\int_{0}^{t} l(t)\,dt$. Similarly, if the velocity of an object is represented by the function $v(t)$ then the displacement at a time t is represented by $\int_{0}^{t} v(t)\,dt$. However, the total distance travelled is represented by $\int_{0}^{t} |v(t)|\,dt$. In the second example, students need to be able to recognise what a 'negative' velocity represents.

Thompson and Silverman (2007) report that students can find the idea of accumulation in such contexts difficult, as it is hard to think of something accumulating if they cannot 'conceptualize the "bits" that accumulate' (p. 117). In addition, students also need a covariational understanding of the relationship between x and f, that is, to understand that as x changes the value of $f(x)$ also changes accordingly (Thompson & Silverman, 2007, p. 118).

Students' misconceptions about integration

Students can have several problems in calculating the answers to integration and comprehending the process. Some errors arise from a poor understanding of fractions, fractional indices and factorisation. On a deeper level, they can have problems understanding notions such as 'positive' and 'negative' areas (Orton, 1983b). Sometimes, however, such area problems are merely owing to a reluctance on behalf of the student to graph the function. Other problems arise if students do not use brackets. In general, however, the problems that students have can be more fundamental – a problem with their concept image.

Orton (1983b) found that some students did not understand the idea of the Riemann sum. Another problem with calculating areas is that students may treat the area under the curve as a literal representation of an area in nature (Thompson & Silverman, 2007) and get confused because a 'negative' area has no physical meaning (Tall & Rasslan cited in Hall, 2010). Orton also found that some students saw the integral symbol as code for 'do something' without thinking further of its meaning. Other problems arise from the way that the words 'definite' and 'indefinite' are used in general English. Therefore for these students a 'definite' integral is one that is clear and accurate, while an 'indefinite' integral lasts for a prolonged period of time, or is unclearly defined (Hall, 2010).

Hall (2010) and Rasslan and Tall (1997) also report that the units that result from integration to solve real problems can be confusing for students. For example, the units that result in calculating an area are squared units, but if the question deals with velocity, then the problem is in terms of unit/time.

Position, velocity and acceleration

In the curriculum for Mathematical Methods students are required to 'determine positions given acceleration and initial values of position and velocity' (ACMMM135). Because acceleration is the rate of change of velocity, and velocity is the rate of change of position, then the functions that describe position, velocity and acceleration are all linked together through the processes of integration and differentiation. McDermott, Rosenquist and van Zee (1987) have found that some students have trouble making the connections between position, velocity and acceleration in this way and therefore cannot move through from one function to another. When looking at position time graphs, they believe that an increasing height on the graph represents the object as speeding up, even if the gradient is decreasing. They also have difficulties in conceiving that uniform motion on a level surface can be represented on a position time graph with a steep gradient. They may also have problems interpreting the nature of a negative velocity. Instead of representing a change in direction back to the starting point as a negative velocity, they indicate the change in direction as a V on the graph.

Connecting it all: Part 1 of the Fundamental Theorem of Calculus

Schwalbach and Dosemagen (2000) note that the realisation that the derivative and anti-derivative are opposite operations as a very important connection for first-year calculus students. This is summarised by the Fundamental Theorem of Calculus (sometimes known as Part 1 of the Theorem).

If f is continuous on $[a,b]$, then $F(x) = \int_a^x f(t)\,dt$ has a derivative at every point of $[a,b]$ and

$$\frac{dF}{dx} = \frac{d}{dx} \int_a^x f(t)\,dt = f(x), \quad a \leq x \leq b$$

(Thomas & Finney, 1996, p. 333)

This equation states that the differential equation $\frac{dF}{dx} = f$ has a solution for every continuous function f. It also states that every continuous function f is the derivative of some other function, namely $\int_a^x f(t)\,dt$. In addition, it states that every continuous function has an antiderivative, and that integration and differentiation are the inverse of one another.

The problem of 'e' and the natural log (ln)

The Fundamental Theorem of Calculus allows us to deal with the problem of finding the indefinite integral of $\frac{1}{x}$.

If we follow the rules for polynomials:

$$f(x) = x^{-1}$$

$$\text{then } \int x^{-1}dx = \frac{x^{-1+1}}{-1+1} = \frac{x^0}{0}$$

which is undefined.

But we know from the Fundamental Theorem of Calculus that this integral must exist, and this is confirmed by looking at the graph of $f(x) = \frac{1}{x}$ (Figure 12.11).

There are many ways of describing the number 'e'. The way that is most useful for this purpose is found in Figure 12.11. If we find the area between the curve and the x-axis from one to 'e', then the area has the value of 1 squared unit.

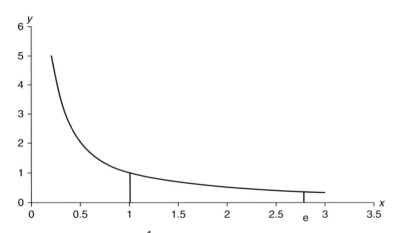

Figure 12.11 The graph of $f(x) = \frac{1}{x}$ showing the area from x = 1 to x = e

From this we can deduce that:

$$\int_{1}^{e} \frac{1}{x} dx = \ln(e) = 1$$

Therefore, in general,

$$\int \frac{1}{x} dx = \ln(x)$$

And we can also find the converse, that is, if $f(x) = \ln(x)$, then $f'(x) = \frac{1}{x}$.

Technology

It should be apparent from this chapter that to be successful in calculus students need to not only be able to follow algebraic procedures, but to imagine processes such as finding limits, imagining derivatives as the gradient of tangent lines, and imagining integrals as areas, to name just a few. Modern teachers now have the advantage of using technology that can enhance these concept images. According to Awang and Zakaria (2013, p. 205), 'A well-designed instructional approach in using technology for the teaching and learning can produce positive impacts on students' mathematical understanding'. The use of graphing calculators and programs such as GeoGebra[1], Mathematica[2], MATLAB[3] and spreadsheets can all assist in the building of the required concept images.

These programs can assist by helping students see relationships without being caught up in tedious calculations (Tall, 2000). Furthermore, students can also develop their understanding through sensorimotor activity available in the systems and be helped to 'comprehend the high-level theoretical concepts' (Awang & Zakaria, 2013, p. 205).

However, it is not wise to assume that the use of technology automatically helps students in their comprehension, and one should take note of Awang and Zakaria's mention of the phrase 'well-designed'. For example, students may get bogged down in the instructions for the technology and miss the principles they are meant to learn (Lipson, Kokonis & Francis, 2003). In addition, teachers need to be cautious so that students do not end up forgetting simple pen and paper techniques (Tall, 2000).

The most successful students are those more versatile using different representations – visual, numeric and verbal cues, as most appropriate. Unfortunately, students often restrict themselves to working with a single representation. For example, students may draw a sketch and then ignore it in the solving of the problem (Mundy, 1984 as cited in Dreyfus, 1991). Therefore, as teachers we need to encourage students to think as broadly and flexibly as we can.

1   International GeoGebra Institute
2   Wolfram Institute
3   MathWorks

Summary

Describe the concepts of troublesome knowledge, threshold concepts and concept image as these apply to the teaching of calculus

When students begin calculus they are reaching a level of mathematics where they are required to imagine processes, such as finding limits, that are inherently difficult. It may take some time and many examples from different contexts before students gain an understanding of these processes. This difficulty should not be underestimated. In addition, at this level of mathematics students require an increased flexibility in going from algebraic to graphic representations.

Describe the skills and knowledge needed by students to study calculus, and some common misconceptions that can work against students' success in calculus

To be successful at calculus students need a deep understanding of functions, rates of change, tangents and the concept of infinity. Therefore, teachers need to be aware of the misconceptions that students may have about these concepts and frequently check students' understanding. Questions that require procedural knowledge alone may not lead to the detection of these misconceptions.

Describe the alternative ways of understanding derivatives, and describe the common misconceptions students may have about the derivative

If students are aware of the alternative ideas that can be used to represent a derivative, they can choose the most suitable concept for each problem they tackle. Being familiar with these alternatives and making the connections between them also leads to understanding.

Describe the alternative ways of understanding integrals, and describe the common misconceptions students may have about integrals

The use of the Riemann sum to assist students in learning the process of integration is a successful teaching strategy. It should be remembered, however, that for a deep understanding of integration students should also be aware of accumulation functions. Teachers need to also be aware of the misconceptions that students may have so that they can pose questions to detect these should they occur.

References

ACARA (Australian Curriculum, Assessment and Reporting Authority) (2011a). *The Australian Curriculum: Mathematics*. Sydney: ACARA. Retrieved from www.australiancurriculum.edu.au/mathematics/content-structure.

—— (2011b). *The Australian Curriculum: Mathematics F–10*. Sydney: ACARA. Retrieved from www.australiancurriculum.edu.au/mathematics/content-structure.

—— (2015). *Australian Curriculum: Mathematics* (Version 7.5). Sydney: ACARA. Retrieved 10 May 2016 from www.australiancurriculum.edu.au/mathematics.

—— (2015a). *Mathematics: Aims – The Australian Curriculum*. Sydney: ACARA. Retrieved 14 May 2015 from www.australiancurriculum.edu.au/mathematics/aims.

—— (2015b). *Mathematics: Content Structure – Overview*. Sydney: ACARA. Retrieved 18 August 2015 from www.australiancurriculum.edu.au/mathematics/content-structure.

—— (2015c). *Mathematics: General Capabilities – The Australian Curriculum*. Sydney: ACARA. Retrieved 14 May 2015 from www.australiancurriculum.edu.au/mathematics/general-capabilities.

—— (2015d). *Mathematics: Implications for teaching, assessment and reporting*. Sydney: ACARA. Retrieved 4 November 2015 from www.australiancurriculum.edu.au/mathematics/implications-for-teaching-assessment-and-reporting.

—— (2015e). *Student Diversity*. Retrieved 15 May 2015 from www.australiancurriculum.edu.au/StudentDiversity/student-diversity-advice.

—— (2015f). *The Australian Curriculum: Mathematics* (Version 8.1). Sydney: ACARA. Retrieved 28 July 2015 from www.australiancurriculum.edu.au/download/f10.

Adams, T.L., Thangata, F. & King, C. (2005). 'Weigh' to go! Exploring mathematical language. *Mathematics Teaching in the Middle School*, 10(9), 444–8.

Ahmad, W.F.B.W., Shafie, A.B. & Janier, J.B. (2008, December). Students' perceptions towards Blended Learning in teaching and learning Mathematics: Application of integration. *Proceedings 13th Asian Technology Conference in Mathematic (ATCM08), Suan Sunanda Rajabhat, University Bangkok, Thailand.*

Ahmed, A. (1987). *Better mathematics: A curriculum development study based on the low attainers in mathematics project.* Indiana: HMSO.

AITSL (Australian Institute for Teaching and School Leadership) (2013). *National professional standards for teachers.* Retrieved 1 March 2015, from www.teacherstandards.aitsl.edu.au.

——— (2015). *Australian professional standards for teachers.* Retrieved 12 October 2015 from www.aitsl.edu.au/ australian-professional-standards-for-teachers/standards/list.

Allan, P. (2006). Technology use and changes in teaching mathematics. *ACE papers*, 1(17).

Alson, P. (1992). A qualitative approach to sketch the graph of a function. *School Science and Mathematics*, 92(4), 182–7.

American Association of University Women (1992). *How schools shortchange girls.* Washington, DC: AAUW Education Foundation.

Anderson, J. (2005). Implementing problem solving in mathematics classrooms: What do teachers want? In P. Clarkson, A. Downton, D. Gronn, M. Horne, A. McDonough, R. Pierce & A. Roche (eds) *Building connections: Theory, research and practice. Proceedings of the 28th annual conference of the Mathematical Education Research Group of Australasia.* Melbourne: MERGA.

——— (2009). Mathematics curriculum development and the role of problem solving. *Australian Curriculum Studies Association National Biennial Conference.* Australia: Australian Curriculum Studies Association.

Anderson, J.A. & Bobis, J. (2005). *Reform-oriented teaching practices: A survey of primary school teachers.* In H.L. Chick & J.L. Vincent (eds) *Proceedings of the 29th Conference of the International Group for the Psychology of Mathematics Education* (vol. 2, 65–72). Melbourne: PME.

Anderson, J., Sullivan, P. & White, P. (2004). The influences of perceived constraints on teachers' problem-solving beliefs and practices.

Anthony, G. & Walshaw, M. (2009a). Characteristics of effective teaching of mathematics: A view from the west. *Journal of Mathematics Education*, 2(2), 147–64.

—— (2009b). *Effective pedagogy in mathematics*. Educational Series 19. Brussels: International Academy of Education; Geneva: International Bureau of Education.

Artigue, M. (1991). Analysis. In D. Tall (ed.) *Advanced mathematical thinking* (167–96). Dordrecht: Kluwer Academic Publishers.

Assessment Reform Group (2002). *Assessment for learning: 10 principles*. Cambridge: ARG. Retrieved from www.assessment-reform-group.org.uk.

Attard, C. (2011). 'If I had to pick any subject, it wouldn't be maths': Foundations for engagement with mathematics during the middle years. *Mathematics Education Research Journal, 25*, 569–87.

—— (2013). 'If I had to pick any subject, it wouldn't be maths': Foundations for engagement with mathematics during the middle years. *Mathematics Education Research Journal, 24*(4), 569–87.

Atweh, B. & Goos, M. (2011). The Australian mathematics curriculum: A move forward or back to the future? *Australian Journal of Education, 55*(3), 214–28.

Australian Academy of Sciences (2006). *Mathematics and statistics: Critical skills for Australia's future*. The National Strategic Review of Mathematical Sciences Research in Australia.

Australian Association of Mathematics Teachers (AAMT) (1997). *Numeracy = everybody's business. The Report of the Numeracy Education Strategy Development Conference, May 1997*. Adelaide: AAMT.

—— (2006). *Standards for excellence in teaching mathematics in Australian schools*. Adelaide: AAMT.

—— (2008). *Position paper on the practice of assessing mathematics learning*. Adelaide: AAMT.

—— (2013). *AAMT Make It Count summary*. Retrieved 12 October 2015 from mic.aamt.edu.au/Resources/Make-It-Count-2009–2012/Publications-and-statements.

Australian Bureau of Statistics (ABS) (2011). *2011 Census Quick Stats: People – cultural and language diversity*. Retrieved on 12 May 2015 from www.censusdata.abs.gov.au/census_services/getproduct/census/2011/quickstat/0?opendocument&navpos=220.

Australian Securities & Investments Commission (ASIC) (2014). *National financial literacy strategy 2014–17*.

Averill, R., Anderson, D., Easton, H., Maro, P.T., Smith, D. & Hynds, A. (2009). Culturally responsive teaching of mathematics: Three models from linked studies. *Journal for Research in Mathematics Education, 40*(2), 157–86.

Awang, T. & Zakaria, E. (2013). Enhancing students' understanding in integral calculus through the integration of Maple in learning. *Procedia – Social and Behavioural Sciences, 102*, 204–11.

Ayres, P.L. (2001). Systematic mathematical errors and cognitive load. *Contemporary Educational Psychology*, 26(2), 227–48. doi: 10.1006/ceps.2000.1051.

Ball, D.L. & Bass, H. (2000). Interweaving content and pedagogy in teaching and learning to teach: Knowing and using mathematics. In J. Boaler (ed.) *Multiple perspectives on the teaching and learning of mathematics* (83–104). Westport, CT: Ablex.

Ball, D.L., Hill, H.C. & Bass, H. (2005). Knowing mathematics for teaching: Who knows mathematics well enough to teach third grade, and how can we decide? *American Educator*, 29(1), 14–7, 20–2, 43–6.

Ball, D.L., Hoyles, C., Jahnke, H.N. & Movshovitz-Hadar, N. (2002). The teaching of proof. *Proceedings of the International Congress of Mathematics*, 3, 907–22.

Ballheim. (1999). Readers respond to what's basic. *Mathematics Education Dialogues*, 3, 11.

Baratta, W., Price, B., Stacey, K., Steinle, V. & Gvozdenko, E. (2010). Percentages: The effect of problem structure, number complexity and calculation form. In L. Sparrow, B. Kissane & C. Hurst (eds), *Proceedings of the 33rd annual conference of the Mathematics Education Research Group of Australasia* (61–68). Fremantle, WA: MERGA.

Barnes, M. (1993). *Investigating change: An introduction to calculus for Australian schools* (Unit 5). Carlton: Curriculum Corporation.

Barry, K. & King, L. (1998). *Beginning teaching and beyond* (3rd edn). Katoomba, NSW: Social Science Press.

Barton, B. (2008). The language of mathematics telling mathematical tales. *Mathematics education library*. New York: Springer.

Barton, B., Fairhall, U. & Trinick, T. (1998). Tikanga Reo Ttitai: Issues in the development of a Maori Mathematics Register. *For the Learning of Mathematics*, 18(1), 3–9.

Barton, B. & Neville-Barton, P. (2005). *The relationship between English language and mathematics learning for non-native speakers*.

Bassok, M. (1990). Transfer of domain-specific problem-solving procedures. *Journal of Experimental Psychology: Learning, Memory, and Cognition*, 16(3), 522–33. doi: dx.doi.org/10.1037/0278-7393.16.3.522.

Bassok, M. & Holyoak, K. J. (1989). Interdomain transfer between isomorphic topics in algebra and physics. *Journal of Experimental Psychology: Learning, Memory, and Cognition*, 15(1), 153–66. doi: dx.doi.org/10.1037/0278-7393.15.1.153.

Batanero, C., Green, D. & Serrano, L. (1998). Randomness, its meanings and educational implications. *International Journal of Mathematical Education in Science and Technology*, 29(1), 113–23.

Batanero, C. & Sanchez, E. (2005). What is the nature of high school students' conceptions and misconceptions about probability? In G. Jones (ed.) *Exploring probability in school: Challenges for teaching and learning* (241–66). New York: Springer.

Batterham, R. & Miles, D. (2000). *Australia's innovation future*. Retrieved 20 September 2015 from www.dest.gov.au/science/pmseic/documents/mtg6Item2AAustInnovfuture.doc.

Battista, M.T. (2007). The development of geometric and spatial thinking. In F.K. Lester (ed.) *Second handbook of research on mathematics teaching and learning* (843–908).

Becker, J.R. (1995). Women's ways of knowing in mathematics. In P. Rogers & G. Kaiser (eds) *Equity in mathematics education: Influences of feminism and culture* (163–74). London: Falmer.

Beesey, C., Clarke, B.A., Clarke, D.M., Stephens, M. & Sullivan, P. (1998). *Effective assessment for mathematics*. Carlton, Vic: Board of Studies/Addison Wesley Longman.

Beilock, S.L., Gunderson, E.A., Ramirez, G. & Levine, S.C. (2010). Female teachers' math anxiety affects girls' math achievement. *Proceedings of the National Academy of Sciences*, 107(5), 1860–963.

Belenky, M.F., Clinchy, B.M., Goldberger, N.R. & Tarule, J.M. (1986). *Women's ways of knowing: The development of self, voice, and mind*. New York: Basic Books.

Ben-Hur, M. (2006). *Concept-rich mathematics instruction*. Alexandria, VA: Association for Supervision and Curriculum Development.

Ben-Zvi, D. & Garfield, J. (2004). Statistical literacy, reasoning, and thinking: Goals, definitions, and challenges. In D. Ben-Zvi & J. Garfield (eds) *The challenge of developing statistical literacy, reasoning and thinking* (3–15). Dordrecht: Kluwer Academic Publishers.

Berry, S. & Nyman, M. (2003). Promoting students' graphical understanding of the calculus. *Journal of Mathematical Behaviour*, 22, 481–97.

Betcher, C. & Lee, M. (2009). *The interactive whiteboard revolution: Teaching with IWBs*. ACER Press: Victoria.

Bezuidenhout, J. (2006). First-year university students' understanding of rate of change. *International Journal of Mathematical Education in Science and Technology*, 29(3), 389–99.

Biehler, R., Ben-Zvi, D., Bakker, A. & Makar, K. (2013). Technology for enhancing statistical reasoning at the school level. *Third international handbook of mathematics education* (643–89). New York: Springer.

Bills, C., Bills, L., Watson, A. & Mason, J. (2004). *Thinkers*. Derby, UK: ATM.

Black, P. & William, D. (1998). Inside the black box: Raising standards through classroom assessment. *Phi Delta Kappan*, 80(2), 139–48.

Bloch, I. (2003). Teaching functions in a graphic milieu: What forms of knowledge enable students to conjecture and prove? *Educational Studies in Mathematics*, 52(1), 2–28.

Boaler, J. (1997a). *Experiencing school mathematics: Teaching styles, sex, and setting*. Philadelphia: Open University Press.

——— (1997b). Reclaiming school mathematics: The girls fight back. *Gender and Education*, 9(3), 285–305.

——— (2000). Introduction: Intricacies of knowledge, practice, and theory. In J. Boaler (ed.) *Multiple perspectives on mathematics teaching and learning* (1–18). Westport, CT: Ablex Publishing.

——— (2002a). Exploring the nature of mathematical activity: Using theory, research and 'working hypotheses' to broaden conceptions of mathematics knowing. *Educational Studies in Mathematics*, 51(1), 3–21.

——— (2002b). *Experiencing school mathematics: Traditional and reform approaches to teaching and their impact on student learning*. Mahwah, NJ: Lawrence Erlbaum Associates.

——— (2009). *The elephant in the classroom: Helping children learn and love maths*. London: Souvenir Press.

——— (2013). Ability and mathematics: The mindset revolution that is reshaping education. *FORUM www.wwwords.co.uk/FORUM*, 55(1), 143–52.

——— (2015). *What's math got to do with it? How teachers and parents can transform mathematics learning and inspire success* (revised edn). New York: Penguin Books.

Bobis, J., Mulligan, J. & Lowrie, T. (2013). *Mathematics for children: Challenging children to think mathematically* (4th edn). Frenchs Forest, NSW: Pearson.

Booker, G. (2011). *Building numeracy: Moving from diagnosis to intervention*. Melbourne: Oxford University Press.

Booker, G., Bond, D., Sparrow, L. & Swan, P. (2014). *Teaching primary mathematics* (5th edn). Frenchs Forest, NSW: Pearson.

Borowiski, E.J. & Borwein, J. M. (1989). *Collins Dictionary of Mathematics* (2nd edn). UK: HarperCollins.

Breiner, J., Harkness, S., Johnson, C. & Koehler, C. (2012). What is STEM? A discussion about conceptions of STEM in education and partnerships. *School Science and Mathematics*, 112(1), 3–11.

Brody, L. (2006). Measuring the effectiveness of STEM talent initiatives for middle and high school students. Retrieved 20 September 2015 from dst.sp.maricopa.edu/DWG/STPG/JuniorACE/Shared%20Documents/Policy%20and%20statistics/Measuring%20the%20Effectiveness%20of%20STEM%20Talent%20Initiatives%20for%20M.pdf.

Brown, J. (2004). A difficult function. *Australian Mathematics Teacher*, 60(2), 6–11.

Brown, R. (2003). Blending learning: Rich experiences from a rich picture. *Training and Development in Australia*, 30(3), 14–7.

——— (2010). Does the introduction of the graphics calculator into system-wide examinations lead to change in the types of mathematical skills tested? *Educational Studies in Mathematics, 73*(2), 181–203.

Brown, S.I. & Walter, M.I. (2005). *The art of problem posing* (3rd edn). Mahwah, NJ: Lawrence Erlbaum.

Brualdi, A. (1998). Implementing performance assessment in the classroom. *Practical Assessment, Research & Evaluation, 6*(2). Retrieved from pareonline.net/getvn.asp?v=6&n=2.

Bruder, R. & Prescott, A. (2013). Research evidence on the benefits of IBL. *ZDM Mathematics Education, 45*(6), 811–22.

Brumbaugh, D.K. & Rock, D. (2001). *Teaching secondary mathematics* (2nd edn). Mahwah, NJ: Lawrence Erlbaum.

Bruner, J.S. (2006). *In search of pedagogy: The selected works of Jerome S. Bruner* (vol. 1). Milton Park, UK: Routledge.

Burrill, G., Allison, J., Breaux, G., Kastberg, S., Leatham, K. & Sanchez, W. (2002). *Handheld graphing technology in secondary mathematics: Research findings and implications for classroom practice.* Dallas, TX: Texas Instruments.

Burton, L. (1995). Moving towards a feminist epistemology of mathematics. In P. Rogers & G. Kaiser (eds) *Equity in mathematics education: Influences of feminism and culture* (209–26). London: Falmer.

——— (2001). Fables: The tortoise? The hare? The mathematically underachieving male. In B. Atweh, H. Forgasz & B. Nebres (eds) *Sociocultural research on mathematics education: An international perspective* (379–92). Mahwah, NJ: Lawrence Erlbaum.

Businskas, A.M. (2008). *Conversations about connections: Secondary mathematics teachers conceptualise and contend with mathematical connections* (unpublished doctoral dissertation). Simon Fraser University, Canada.

Bybee, R.W. (2010). Advancing STEM education: A 2020 vision. *Technology and Engineering Teacher, 70*(1), 30–5.

Cai, J. (2005). U.S. and Chinese teachers' constructing, knowing, and evaluating representations to teach mathematics. *Mathematical Thinking and Learning, 7*(2), 135–69. doi: 10.1207/s15327833mtl0702_3.

Cai, J., Lew, H.C., Morris, A., Moyer, J.C., Ng, S.F. & Schmittau, J. (2005). The development of students' algebraic thinking in earlier grades: A cross-cultural comparative perspective. *ZDM – The International Journal on Mathematics Education, 37*, 5–15.

Calder, N., Brown, T., Hanley, U. & Darby, S. (2006). Forming conjectures within a spreadsheet environment. *Mathematics Education Research Journal, 18*(3), 100–16.

Campbell, P.B. (1995). Redefining the 'girl problem in mathematics'. In W.G. Secede & L.B. Adajian (eds) *New directions for equity in mathematics education* (225–41). New York: Cambridge University Press.

Cangelosi, J.S. (2003). *Teaching mathematics in secondary and middle school: An interactive approach* (3rd edn). Upper Saddle River, NJ: Pearson Education.

Carnine, D. (1997). Instructional design in mathematics for students with learning disabilities. *Journal of Learning Disabilities*, 30(2), 130–41.

Carpenter, T.P., Franke, M. & Levi, L. (2003). *Thinking mathematically: Integrating arithmetic and algebra in elementary school.* Portsmouth, NH: Heinemann.

Cavanagh, M. (1999). One secondary teacher's use of problem-solving teaching approaches. In M. Goos, R. Brown & K. Makar (eds) *Navigating currents and charting directions. Proceedings of the 31st annual conference of the Mathematics Education Research Group of Australasia.* Brisbane: MERGA.

Ceci, S.J. & Williams, W.M. (2010). Sex differences in math-intensive fields. *Current Directions in Psychological Science*, 19(5), 275–79.

Chan, C.M.E. (2007). Using open-ended mathematics problems: A classroom experience. In C. Shegar & R.B.B. Rahim (eds) *Redesigning pedagogy: Voices of practitioners* (129–46). Singapore: Pearson Education South Asia.

Chan, K.K. & Leung, S.W. (2014). Dynamic Geometry Software improves mathematical achievement: Systematic review and meta-analysis. *Journal of Educational Computing Research*, 51(3), 311–25.

Chance, B., del Mas, R. & Garfield, J. (2004). Reasoning about sampling distributions. In D. Ben-Zvi & J. Garfield (eds), *The challenges of developing statistical literacy, reasoning and thinking* (295–323). Dordrecht: Kluwer Academic Press.

Charalambous, C.Y. (2008). Mathematical knowledge for teaching and the unfolding of tasks in mathematics lessons: Integrating two lines of research. In O. Figuras, J.L. Cortina, S. Alatorre, T. Rojano & A. Sepulveda (eds) *Proceedings of the 32nd Annual Conference of the International Group for the Psychology of Mathematics Education* (vol. 2, 281–8). Morelia: PME.

Cheeseman, J. (2003). Orchestrating the end of mathematics lessons. In B. Clarke, A. Bishop, H. Forgasz & W.T. Seah (eds) *Making mathematicians* (17–26). Brunswick: Mathematical Association of Victoria.

Chen, C., Crockett, M.D., Namikawa, T., Zilimu, J. & Lee, S.H. (2012). Eighth grade mathematics teachers' formative assessment practices in SES-different classrooms: A Taiwan study. *International Journal of Science and Mathematics Education*, 10, 553–79.

Chen, M. (2001). A potential limitation of embedded-teaching for formal learning. In J. Moore & K. Stenning (eds) *Proceedings of the Twenty-Third Annual Conference of the Cognitive Science Society* (194–99). Edinburgh, Scotland: Lawrence Erlbaum.

Cheung, A.C. & Slavin, R.E. (2013). The effectiveness of educational technology applications for enhancing mathematics achievement in K–12 classrooms: A meta-analysis. *Educational Research Review*, 9, 88–113.

Churchill, R., Ferguson, P., Godinho, S., Johnson, N., Keddie, A., Letts, W., Mackay, J., McGill, M., Moss, J., Nagel, M., Nicholson, P. & Vick, M. (2013). *Teaching: Making a difference* (2nd edn). Milton, Qld: John Wiley & Sons.

Civil, M. (2002). Everyday mathematics, mathematicians' mathematics, and school mathematics: Can we bring them together? In E. Yackel (Series ed.), M.E. Brenner & J.N. Moschkovich (monograph eds) *Everyday and academic mathematics in the classroom* (40–62). Reston, VA: National Council of Teachers of Mathematics.

Clarke, C. & Lovitt, C. (1987). MCTP assessment alternatives in mathematics. *Australian Mathematics Teacher*, 43(3), 11–12.

Clarke, D. (1987). A rationale for assessment alternatives in mathematics. *Australian Mathematics Teacher*, 43(3), 8–10.

———— (1997). *Constructive assessment in mathematics*. Berkeley, CA: Key Curriculum Press.

———— (1998). *Securing their future: Subject-based assessment materials for the school certificate – Mathematics*. Ryde, NSW: Department of Education.

———— (2007). *Constructive assessment in mathematics*. Berkeley, CA: Key Curriculum Press.

Clarke., D., Cheeseman, J., Clarke, B., Gervasoni, A., Gronn, D., Horne, M. & Sullivan, P. (2001). Understanding, assessing and developing young children's mathematical thinking: Research as a powerful tool for professional growth. *Keynote paper presented at the Annual Conference for the Mathematical Education Research Group of Australasia, Sydney, July 2001*.

Clarke, D. & Clarke, B. (2002). Using rich assessment tasks in mathematics to engage students and inform teaching. *Background paper for Seminar for Upper Secondary Teachers, Stockholm, September, 2002*.

———— (2004). Mathematics teaching in Grades K–2: Painting a picture of challenging, supportive, and effective classrooms. In R.N. Rubenstein & G.W. Bright (eds) *Perspectives on the teaching of mathematics* (66th Yearbook of the National Council of Teachers of Mathematics, 67–81). Reston, VA: NCTM.

Clarke, D.M. & Roche, A. (2009). Opportunities and challenges for teachers and students provided by tasks built around 'real' contexts. In R. Hunter, B. Bicknell & T. Burgess (eds) *Proceedings of MERGA 32 conference: Crossing divides* (vol. 2, 722–26). Palmerston North, NZ: MERGA.

Clarke, S. (2001). *Unlocking formative assessment*. London: Hodder & Stoughton.

Clement, J. (1982). Algebra word problem solutions: Thought processes underlying a common misconception. *Journal for Research in Mathematics Education*, 13(1), 16–30. doi: 10.2307/748434.

Clement, J.J.L. & Monk, G. (1981). Translational difficulties in learning mathematics. *American Mathematical Monthly*, 88, 286–90.

Clements, M.K., Bishop, A., Keitel-Kreidt, C., Kilpatrick, J. & Leung, F.K.S. (eds) (2013). *Third international handbook of mathematics education (vol. 27)*. New York: Springer Science + Business Media.

Cobb, P. & McClain, K. (2004). Principles of instructional design for supporting the development of students' statistical reasoning. In D. Ben-Zvi & J. Garfield (eds) *The challenge of developing statistical literacy, reasoning and thinking* (375–95). Dordrecht: Kluwer Academic Publishers.

Commonwealth of Australia (2002). *Style manual for authors, editors and printers* (6th edn). Milton, Qld: John Wiley & Sons.

———— (2008). *National numeracy review report*. Canberra: DEEWR.

Connection (2015). In *Oxford English Dictionary online*. Retrieved from www.oed.com/view/Entry/39356?redirectedFrom=connection#eid.

Cooke, S. & Howard, P. (2009). Can we address the issues surrounding Aboriginal education? Yes we can!!!!! Together! *Keynote address Dare to Lead Conference. Rydges Lakeside Hotel, Canberra, 14 August 2009*.

Cooper, G. & Sweller, J. (1987). Effects of schema acquisition and role automation on mathematical problem-solving transfer. *Journal of Educational Psychology*, 79(4), 347–62.

Cooper, H.M., Robinson, J.C. & Patall, E.A. (2006). *Does homework improve academic achievement? A synthesis of research, 1987–2004*. Review of Educational Research, AERA.

Cornu, B. (1991). Limits. In D. Tall (ed.) *Advanced mathematical thinking* (153–66). Dordrecht: Kluwer Academic Publishers.

Correll, S.J. (2001). Gender and the career choice process: The role of biased self-assessment. *American Journal of Sociology*, 106(6), 1691–730.

Cotrill, J., Dubinsky, E., Nichols, D., Schwingendorf, K., Thomas, K. & Vicakovic, D. (1996). Understanding the limit concept: Beginning with a coordinated process scheme. *Journal of Mathematical Behaviour*, 15, 167–92.

Council of Australian Governments (2008). *National numeracy review report*. Retrieved from archive.coag.gov.au/reports/docs/national_numeracy_review.doc.

Coxford, A.F. (1995). *The case for connection: Connecting mathematics across the curriculum* (1995 NCTM yearbook). Reston, VA: National Council of Teachers of Mathematics.

Cramer, K. & Wyberg, T. (2009). Efficacy of different concrete models for teaching the part-whole construct for fractions. *Mathematical Thinking and Learning*, 11(4), 226–57. doi: 10.1080/10986060903246479.

Cripps Clark, J. (2006). *The role of practical activities in primary school science*. Melbourne: Deakin University.

Crowley, M.L. (1987). The van Hiele model of development of geometric thought. In M.M. Lindquist (ed.) *Learning and teaching geometry: K–12, 1987 Yearbook of the National Council of Teachers of Mathematics* (1–16). Reston, VA: National Council of Teachers of Mathematics.

Cummins, J. (1994). Primary language instruction and the education of language minority students. In C.F. Leyba (ed.) *Schooling and language minority students: A theoretical framework* (3–46). Los Angeles, California State University: National Evaluation, Dissemination and Assessment Center.

Curcio, F. (1989). *Developing graph comprehension*. Reston, VA: National Council of Teachers of Mathematics.

Danielson, C. (1997). *A collection of performance tasks and rubrics: Middle school mathematics*. Larchmont, NT: Eye on Education.

Darby, L. (2008). Negotiating mathematics and science school subject boundaries: The role of aesthetic understanding. In M.V. Thomase (ed.) *Science in Focus* (225–51). Hauppauge, NY: Nova Science Publishers.

David, M.M. & Tomaz, V.S. (2012). The role of visual representations for structuring classroom mathematical activity. *Educational Studies in Mathematics*, 80(3), 413–31.

Davis, G.E. & Tall, D. (2002). What is a scheme? In D. Tall & M. Thomas (eds) *Intelligence, learning and understanding in mathematics: A tribute to Richard Skemp* (131–50). Flaxton, Qld: Post Pressed.

Davis, R. & Vinner, S. (1986). The notion of limit: Some seemingly unavoidable misconception stages. *Journal of Mathematical Behaviour*, 5, 281–303.

De Bortoli, L. & Macaskill, G. (2014). *Thinking it through: Australian students' skills in creative problem solving*. Melbourne: ACER.

de Villiers, M. (1990). The role and function of proof in mathematics. *Pythagoras*, 24, 17–24.

DelliCarpini, M. & Alonso, O.B. (2014). Teacher education that works: Preparing secondary-level math and science teachers for success with English langugage learners through content-based instruction. *Global Education Review*, 1(4), 155–78.

Department for Education and Skills (2007). *Mathematics at key stage 4: Developing your scheme of work*. Norwich, England: Crown.

Department of Education, Employment and Workplace Relations (2009). *What works. The work program: What works in pre-service teacher education*. Retrieved 12 October 2015 from www.whatworks.edu.au/upload/1251164046741_file_SteppingUp.pdf.

Department of Education, Science and Training (DEST) (2006). *Disability standards for education 2005 plus guidance notes*. Retrieved 11 May 2015 from docs.education.gov.au/system/files/doc/other/disability_standards_for_education_2005_plus_guidance_notes.pdf.

Desforges, C. & Cockburn, A. (1987). *Understanding the mathematics teacher: A study of practice in first schools*. London: The Palmer Press.

Dewey, J. (1910). *How we think*. Lexington: D.C. Heath. (Reprinted in 1991 by Prometheus Books, Buffalo).

Diezmann, C., Faragher, R., Lowrie, T., Bicknell & Putt, I. (2004). Exceptional students in mathematics. In R. Perry, G. Anthony & C. Diezmann (eds), *Research in mathematics education in Australasia 2000–2003*. Flaxton: Post Pressed.

Diezmann, C. & Watters, J. (2002). Summing up the education of mathematically gifted students. In B. Barton, K.C. Irwin, M. Pfannkuch & M.O.J. Thomas (eds) *Mathematics education in the South Pacific. Proceedings of the 25th Annual Conference of MERGA* (219–26). Auckland: MERGA.

Doorman, M., Drijvers, P., Dekker, T., Van den Heuvel-Panhuizen, M., de Lange, J. & Vijers, M. (2007). Problem solving as a challenge for mathematics education in The Netherlands. *ZDM Mathematics Education*, 39(5/6), 405–18.

Dorier, J.L. & Maass, K. (2014). Inquiry-based mathematics education. In S. Lerman (ed.) *Encyclopaedia of mathematics education* (300–4). Dordrecht: Springer Reference.

Dreyfus, T. (1991). Advanced mathematical thinking processes. In D. Tall (ed.) *Advanced Mathematical Thinking* (25–36). Dordrecht: Kluwer Academic Publishers.

Drysdale, J.S., Graham, C.R., Spring, K.J. & Halverson, L.R. (2013). An analysis of research trends in dissertations and theses studying blended learning. *The Internet and Higher Education*, 17, 90–100.

Dugger, W.E. (2010). Evolution of STEM in the United States. *Paper presented at the 6th Biennial International Conference on Technology Education Research, Gold Coast, Queensland, Australia.*

Dweck, C.S. (2000). *Self-theories: Their role in motivation, personality, and development.* Philadelphia, PA: Psychology Press.

Ellerton, N.F. & Clements, M.A. (1996). Newman error analysis: A comparative study involving Year 7 students in Malaysia and Australia. In P.C. Clarkson (ed.) *Technology and Mathematics Education: Proceedings of the 19th Annual Conference of the Mathematics Education Research Group of Australasia* (vol. 1, 186–93). Melbourne: MERGA.

Ellington, A. (2003). A meta-analysis of the effects of calculators on students' achievement and attitude levels in precollege mathematics classes. *Journal for Research in Mathematics Education*, 34, 433–63.

Ellison, G. & Swanson, A. (2010). The gender gap in secondary school mathematics at high achievement levels: Evidence from the American mathematics competitions. *Journal of Economic Perspectives*, 24(2), 109–28.

Else-Quest, N.M., Hyde, J.S. & Linn, M. (2010). Cross-national patterns of gender differences in mathematics: A meta-analysis. *Psychological Bulletin*, 136(1), 103–27.

English, L. (2005). Combinatorics and the development of children's combinatorial reasoning. In G. Jones (ed.) *Exploring probability in school: Challenges for teaching and learning* (121–44). New York: Springer.

Ernest, P. (1991). *The philosophy of mathematics education*. London: Falmer Press.

Evans, J. (1987). Investigations: The state of the art. *Mathematics in School*, 16(1), 27–30.

Evitts, T.A. (2004). Investigating the mathematical connections that pre-service teachers use and develop while solving problems from reform curricula (unpublished doctoral dissertation). Pennsylvania State University College of Education.

Fan, S-C. & Ritz, J. (2014). *International views on STEM education*. Retrieved 20 September 2015 from www.iteea.org/Conference/PATT/PATT28/Fan%20 Ritz.pdf.

Fennema, E. & Franke, M.L (1992). Teachers' knowledge and its impact. In D.A. Grouws (ed.) *Handbook of research on Mathematics Teaching and Learning*, (147–64). New York: Macmillan.

Fennema, E. & Romberg, T.A. (1999). *Mathematics classrooms that promote understanding*. Mahwah, NJ: Lawrence Erlbaum.

Fennema, E. & Sherman, J. (1977). Sex-related differences in mathematics achievement, spatial visualization, and affective factors. *American Educational Research Journal*, 14(1), 51–71.

Ferrini-Mundy, J. & Graham, K. (1991). An overview of the calculus reform effort: Issues for learning, teaching, and curriculum development. *The American Mathematical Monthly*, 98(7), 627–35.

Filloy, E., Rojano, T. & Rubio, G. (2001). Propositions concerning the resolution of arithmetical-algebraic problems.

Finger, G., Romeo, G., Lloyd, M., Heck, D., Sweeney, T., Albion, P. & Jamieson-Proctor, R. (2015). Developing Graduate TPACK Capabilities in Initial Teacher Education Programs: Insights from the Teaching Teachers for the Future Project. *The Asia-Pacific Education Researcher*, 1–9.

Forgasz, H. & Leder, G. (2001). A+ for girls, B for boys: Changing perspectives on gender equity and mathematics. In B. Atweh, H. Forgasz & B. Nebres (eds) *Sociocultural research on mathematics education: An international perspective* (347–66). Mahwah, NJ: Lawrence Erlbaum.

Freudenthal, H. (1971). Geometry between the devil and the deep sea. *Educational Studies in Mathematics*, 3, 413–35.

Friel, S., Curcio, F. & Bright, G. (2001). Making sense of graphs: Critical factors influencing comprehension and instructional implications. *Journal for Research in Mathematics Education*, 32(2), 124–58.

Fujita, T. & Jones, K. (2003). *Interpretations of national curricula: The case of geometry in Japan and the United Kingdom*. Proceedings of the British Educational Research Association Annual Conference. Edinburgh, Scotland.

Fulton, K. (2012). The flipped classroom: Transforming education at Byron High School: A Minnesota high school with severe budget constraints enlisted YouTube in its successful effort to boost math competency scores. *The Journal (Technological Horizons In Education)*, 39(3), 18.

Fuson, K. & Kwon, Y. (1991). Chinese based regular and European irregular systems of number words: Disadvantages for English speaking children. In K. Durkin & B. Shire (eds) *Language in Mathematical Education: Research and Practice* (211–26). Milton Keynes: Open University Press.

Fuys, D., Geddes, D. & Tischler, R. (1988). The van Hiele model of thinking in geometry among adolescents. *Journal for Research in Mathematics Education Monograph*, 3. Reston, VA: NCTM.

Gainsburg, J. (2008). Real-world connections in secondary mathematics teaching. *Journal of Mathematics Teacher Education*, 11, 199–219.

Gal, I. (2002). Adults' statistical literacy: Meanings, components, responsibilities. *International Statistical Review*, 70, 1–51.

Galbraith, P. (2011). Models of modelling: Is there a first among equals? In J. Clark, B. Kissane, J. Mousley, T. Spencer & S. Thornton (eds) *Mathematics: Traditions and practices. Proceedings of the 34th Annual Conference of the Mathematics Education Research Group of Australasia.* (279–96). Adelaide: MERGA Inc.

Galbraith, P. & Fitzpatrick, J.B. (1977). *Core Mathematics*. Brisbane: Jackaranda Press.

Gallagher, A.M. & Kaufmann, J.C. (2005). *Gender differences in mathematics: An integrative psychological approach.* Cambridge: Cambridge University Press.

Gallagher, J. & Gallagher, S. (1994). *Teaching the gifted child.* Boston: Allyn & Bacon.

Galligan, L. (1997). Differences in problem processing: A comparison of English and Chinese mathematical word problems. In A. Begg (ed.) *People in mathematics education. Proceedings of the 20th Annual Conference of the Mathematics Education Research Group of Australasia* (vol. 1, 177–83). Rotorua, New Zealand: MERGA.

——— (2001). Possible effects of English–Chinese language differences on processing of mathematical text: A review. *Mathematics Education Research Journal*, 13(2), 112–32.

Galligan, L. & Hobohm, C. (2013). Investigating inking devices to support learning in mathematics. In V. Steinle, L. Ball & C. Bardini (eds) *Mathematics education: Yesterday, today and tomorrow. Proceedings of the 36th annual conference of the Mathematics Education Research Group of Australasia* (322–29). Melbourne: MERGA.

Galton, M., Morrison, I. & Pell, T. (2000). Transfer and transition in English schools: Reviewing the evidence. *International Journal of Educational Research*, 33, 341–63.

Garfield, J. & Ahlgren, A. (1988). Difficulties in learning basic concepts in probability and statistics: Implications for research. *Journal for Research in Mathematics Education*, 19(1), 44–63.

Gavin, M.K. & Sheffield, L.J. (2010). Using curriculum to develop mathematical promise in the middle grades. In M. Saul, S. Assouline & L.J. Sheffield (eds) *The peak in the middle: Developing mathematically gifted students in the middle grades* (51–76). Reston, VA: NCTM, National Association of Gifted Children, & National Middle School Association.

Gentner, D. (1983). Structure-mapping: A theoretical framework for analogy. *Cognitive Science*, 7(2), 155–70.

Gerson, H. (2010). David's understanding of functions and periodicity. *School Science and Mathematics*, 108(1), 28–38.

Gillies, R.M. (2007). *Cooperative learning: Integrating theory and practice*. Los Angeles: Sage Publications.

Glasersfeld, E. von. (1995). *Radical constructivism: A way of knowing and learning*. London: Falmer Press.

Goldenberg, E.P., Cuoco, A.A. & Mark, J. (2012). A role for geometry in general education. In R. Lehrer & D. Chazan (eds) *Designing learning environments for developing understanding of geometry and space*. New Jersey: Lawrence Erlbaum.

Goos, M. (2004). Learning mathematics in a classroom community of inquiry. *Journal for Research in Mathematics Education*, 35(4), 258–91. doi:10.2307/30034810.

——— (2010). *Using Technology to support effective mathematics teaching and learning: What Counts?* Retrieved 2 March 2015 from research.acer.edu.au/cgi/viewcontent.cgi?article=1067&context=research_conference.

Goos, M. & Bennison, A. (2008). Surveying the technology landscape: Teachers' use of technology in secondary mathematics classrooms. *Mathematics Education Research Journal*, 20(3), 102–30.

Goos, M, Galbraith, P, Renshaw, P. & Geiger, V. (2000). Reshaping teacher and student roles in technology-enriched classrooms. *Mathematics Education Research Journal*, 12(3), 303–20.

Goos, M., Stillman, G. & Vale, C. (2007). *Teaching secondary school mathematics: Research and practice for the 21st century*. Crows Nest, NSW: Allen & Unwin.

Gough, J. (2010). Getting wrapped up in trigonometry. *Vinculum*, 47(3), 14–7.

Gray, E.M. (1991). An analysis of diverging approaches to simple arithmetic: Preference and its consequences, *Educational Studies in Mathematics*, 22, 551–74.

Green, J. (2009). Using spreadsheets to make algebra more accessible: Solutions to equations. *Australian Mathematics Teacher*, 65(1), 17–21.

Grimison, L. (1992). Assessment in mathematics – some alternatives. *Mathematics Education Research Group of Australasia 15th Annual Conference, University of Western Sydney*.

Groth, R. & Bergner, J. (2006). Preservice elementary teachers' conceptual and procedural knowledge of the mean, median and mode. *Mathematical Thinking and Learning*, 8(1), 37–63.

Guiso, L., Monte, F., Sapienza, P. & Zingales, L. (2008). Culture, gender, and math. *Science*, 320(5880), 1164–65.

Guven B. (2012). Using dynamic geometry software to improve eighth grade students' understanding of transformation geometry. *Australasian Journal of Educational Technology*, 28(2), 364–82.

Guven, B., Baki, A. & Cekmez, E. (2012). Using dynamic geometry software to develop problem solving skills. *Mathematics and Computer Education*, 46(1), 6.

Hall, J. (2012). Gender issues in mathematics: An Ontario perspective. *Journal of Teaching and Learning*, 8(1), 59–72.

Hall, W. (2010). Student misconceptions of the language of calculus: Definite and indefinite integrals. *Proceedings of the 13th Annual Conference on Research in Undergraduate Mathematics Education*. Raleigh, NC: Mathematical Association of America.

Halliday, M.A.K. (1978). *Language as the social semiotic: The social interpretation of language and meaning*. Baltimore: University Park Press.

Hannula, M.S. (2009). The effect of achievement, gender and classroom context on upper secondary students' mathematical beliefs. *Paper presented at CREME Conference, Lyon, France*. Retrieved 4 April 2015 from ife.ens-lyon.fr/publications/edition-electronique/cerme6/wg1-01-hannula.pdf.

Hannula, M.S., Maijala, H., Pehkonen, E. & Nurmi, A. (2005). Gender comparisons of pupils' self-confidence in mathematics learning. *Nordic Studies in Mathematics Education*, 10(3–4), 29–42.

Hannula, M. & Malmivuori, M.L. (1997). Gender differences and their relation to mathematics classroom context. In. E. Pehkonen (ed.) *Proceedings of the 21st Conference of the International Group for the Psychology of Mathematics Education* (vol. 3, 33–40). Lahti: PME.

Hattie, J. (2009). *Visible learning: A synthesis of over 800 meta-analyses relating to achievement*. New York: Routledge.

Hawera, N. (2006). Maori preservice primary teachers' responses to mathematical investigations. In P. Grootenboer, R. Zevenbergen & M. Chinnappan (eds) *Identities, cultures and learning spaces. Proceedings of the 29th annual conference of the Mathematics Education Research Group of Australasia, Canberra* (vol. 1, 286–92). Sydney: MERGA.

Healy, L., & Hoyles, C. (1998). *Justifying and proving in school mathematics. Technical report on the nationwide survey*. England: Institute of Education, University of London.

Heemskerk, I., Kuiper, E. & Meijer, J. (2014). Interactive whiteboard and virtual learning environment combined: Effects on mathematics education. *Journal of Computer Assisted Learning*, 30(5), 465–78.

Hegarty, M., Mayer, R.E. & Monk, C.A. (1995). Comprehension of arithmetic word problems: A comparison of successful and unsuccessful problem

solvers. *Journal of Educational Psychology*, 87(1), 18–32. doi: dx.doi.
org/10.1037/0022-0663.87.1.18.

Helme, S. & Clarke, D. (2001). Identifying cognitive engagement in the
mathematics classroom. *Mathematics Education Research Journal*, 13,
133–53.

Helme, S. & Teese, R. (2011). How inclusive is Year 12 mathematics? *Paper
presented at the 2011 MERGA Conference*. Retrieved on 4 April 2015 from
www.merga.net.au/node/38?year=2011.

Henke, R.R., Chen, X. & Goldman, G. (1999). What happens in classrooms?
Instructional practices in elementary and secondary schools: 1994–
1995. *Educational Statistics Quarterly*, 1(2), 7–13.

Henningsen, M. & Stein, M.K. (1997). Mathematical tasks and student
cognition: Classroom-based factors that support and inhibit high-
level mathematical thinking and reasoning. *Journal for Research in
Mathematics Education*, 28, 524–49.

Herbel-Eisenmann, B.A. & Breyfogle, M.L. (2005). Questioning our patterns of
questioning. *Mathematics Teaching in the Middle School*, 10(9), 484–9.

Hertzer, K. (2014). *Graphmatica for Windows (Version 2.4): kSoft Inc*. Retrieved
from www.graphmatica.com.

Hoachlander, G. & Yanofsky, D. (2011). Making STEM real. *Educational
Leadership*, 68(6), 1–6.

Hodgson, T.R. (1995). Connections as problem-solving tools. In P.A. House &
A.F. Coxford (eds) *Connecting Mathematics Across the Curriculum* (13–21).
Reston, VA: National Council of Teachers of Mathematics.

Hohenwarter, M., Hohenwarter, J., Kreis, Y. & Lavicza, Z. (2008). Teaching and
learning calculus with free dynamic mathematics software GeoGebra.
11th International Congress on Mathematical Education. Monterrey, Mexico.

Hollebrands, K.F. (2004). High school students' intuitive understandings of
geometric transformations. *The Mathematics Teacher*, 97(3), 207–14.

Hollingsworth, H. (2003). The TIMSS 1999 video study and its relevance to
Australian mathematics education research, innovation, networking
and opportunities. In L. Bragg, C. Campbell, G. Herbert & J. Mousley
(eds) *Mathematics education research: Innovation, networking, opportunity*.
*Proceedings of the 26th Annual Conference of the Mathematics Education
Research Group of Australasia* (7–16). Sydney: MERGA.

Hollingsworth, H., Lokan, J. & McCrae, B., (2003). *Teaching mathematics
in Australia: Results from the TIMSS 1999 Video Study*. Camberwell,
Vic.: Australian Council of Educational Research.

Holyoak, K. & Koh, K. (1987). Surface and structural similarity in analogical
transfer. *Memory & Cognition*, 15(4), 332–40. doi: 10.3758/bf03197035.

Hopper, S. (2009). The effect of technology use on student interest and
understanding in geometry. *Studies in Teaching 2009 Research Digest*,
37–42.

Horn, I.S. (2012). *Strength in numbers: Collaborative learning in secondary mathematics*. Reston, VA: National Council of Teachers of Mathematics.

Hosking, P. & Shield, M. (2001). Feedback practices and the classroom culture. *Paper presented at the 24th Annual MERGA Conference, Sydney*.

House of Representatives Standing Committee on Education and Training (2002). *Boys: Getting it right: Report on the inquiry into the education of boys*. Retrieved 30 April 2015 from www.aph.gov/house/committee/edt/eofb/report/front.pdf.

Hoyles, C. (2010). *Mathematics education and technology: Rethinking the terrain*. Springer.

Huetinck, L. (1990). Gender differences on science exams with respect to item type, format, student interest and experience (unpublished doctoral dissertation). Los Angeles, UCLA.

Huetinck, L. & Munshin, S.N. (2008). *Teaching mathematics for the 21st century: Methods and activities for grades 6–12* (3rd edn). New Jersey: Pearson.

Hughes Hallet, D. (1991). Visualisation and calculus reform. In W. Zimmerman & S. Cunnignham (eds) *Visualisation in teaching and learning mathematics*, MAA Notes no. 19, 121–26.

Hyde, J., Lindberg, S., Linn, M., Ellis, A. & Williams, C. (2008). Gender similarities characterize mathematics performance. *Science*, 321(5888), 494–95.

International Association for K–12 Online Learning (2008). National Standards for Quality Online Teaching.

International GeoGebra Institute (2015). *GeoGebra 5.07*. Linz, Austria: International GeoGebra Institute.

International Society for Technology in Education (2008). *National educational technology standards for teachers*. Retrieved from www.iste.org/standards/nets-for-teachers.aspx.

Irons, R. (2014). Language is the core for the concept of addition [online]. *Educating Young Children: Learning and Teaching in the Early Childhood Years*, 20(1), 38–41.

Jacobs, J.E., Becker, J.R. & Gilmer, G.F. (eds) (2001). *Changing the Faces of Mathematics: Perspectives on gender* (37–41). Reston, VA: NCTM.

Janzen, J. (2008). Teaching English language learners in the content areas. *Review of Educational Research*, 78(4), 1010–38.

Jaworski, B. (1994). *Investigating mathematics teaching: A constructivist enquiry*. London: Falmer.

Jiang, Z., White, A. & Rosenwasser, A. (2011). Randomized control trials on the dynamic geometry approach. *Journal of Mathematics Education at Teachers College*, 2(2).

Johnson, D.R. (1982). *Every minute counts: Making your math class work*. Palo Alto, CA: Dale Seymour Publications.

Jones, G., Langrall, C., Thornton, C. & Mogill, A. (1997). A framework for assessing and nurturing young children's thinking in probability. *Educational Studies in Mathematics*, 32(2), 101–25.

Jones, G., Langrall, C. & Mooney, E. (2007). Research in probability: Responding to classroom realities. In F. Lester (ed.) *Second handbook on mathematics teaching and learning* (909–55). Charlotte, NC: The National Council of Teachers of Mathematics.

Jones, S. (2014). Calculus limits involving infinity: The role of students' informal dynamic reasoning. *International Journal of Mathematics Education in Science and Technology*, 46(1), 105–26.

Jorgensen, R. (2010). Structured failing: Reshaping a mathematical future for marginalised learners. In L. Sparrow, B. Kissane & C. Hurst (eds) *Shaping the future of mathematics education. Proceedings of the 33rd annual conference of the Mathematics Education Research Group of Australasia*, (vol. 1, 26–35). Fremantle: MERGA.

Kahneman, D. & Tversky, A. (1982). Subjective probability: A judgment of representativeness. In D. Kahneman, P. Slovic & A. Tversky (eds) *Judgement under uncertainty: Heuristics and biases* (32–47). Cambridge: Cambridge University Press.

Kalyuga, S. (2007). Expertise reversal effect and its implications for learner-tailored instruction. *Educational Psychology Review*, 19(4), 509–39. doi: 10.1007/s10648-007-9054-3.

———— (2009). Knowledge elaboration: A cognitive load perspective. *Learning and Instruction*, 19(5), 402–10. doi: dx.doi.org/10.1016/j.learninstruc.2009.02.003.

Kalyuga, S. & Renkl, A. (2010). Expertise reversal effect and its instructional implications: Introduction to the special issue. *Instructional Science*, 38(3), 209–15. doi: 10.1007/s11251-009-9102-0.

Kangasniemi, E. (1989). *Curriculum and mathematics achievement*. University of Jyvversit.

Kaplan, A., Ozturk, M. & Ocal, M. (2015). Relieving of misconceptions of derivative concept with derive. *International Journal of Research in Education and Science*, 1(1), 67–74.

Kaput, J. J. (1989). Linking representations in the symbol systems of algebra. In S. Wagner & C. Kieran (eds) *Research issues in the learning and teaching of algebra* (vol. 4, 167–94). Reston, VA: National Council of Teachers of Mathematics.

Kastberg, S. & Leatham, K. (2005). Research on graphing calculators at the secondary level: Implications for mathematics teacher education', *Contemporary Issues in Technology and Teacher Education*, 5(1), 25–37.

Katz, V. (1993). *A history of mathematics: An introduction*. New York: HarperCollins.

Kaur, B. (2001). TIMSS & TIMSS-R – Performance of grade eight Singaporean students. In C. Vale, J. Horwood & J. Roumeliotis (eds) 2001 *A Mathematical Odyssey* (132–144). *Proceedings of the 38th annual conference of the Mathematical Association of Victoria*. Brunswick, Vic: MA.

Kaur, B. & Yeap, B.H. (2009). Mathematical problem solving in Singapore schools. In B. Kaur, B.H. Yeap & Kapur, M. *Mathematical problem solving: Yearbook 2009* (3–13). Singapore: Association of Mathematics Education and World Scientific.

Keene, K., Hall, W. & Duca, A. (2014). Sequence limits in calculus: Using design research and building on intuition to support instruction. *ZDM Mathematics Education*, 46, 561–74.

Kemp, M. & Kissane, B. (2010). A five step framework for interpreting tables and graphs in their contexts. In C. Reading (ed.) *Data and context in statistics education: Towards an evidence-based society. Proceedings of the Eighth International Conference on Teaching Statistics (ICOTS8, July, 2010)* (vol. 8). Ljubljana, Slovenia. Voorburg, The Netherlands: International Statistical Institute.

Kennedy, J., Lyons, T. & Quinn, F. (2014). The continuing decline of science and mathematics enrolments in Australian high schools. *Teaching Science*, 60(2), 34–46.

Khisty, L.L. (1997). Making mathematics accessible to Latino students: Rethinking instructional practice. In J. Trentacosta & M.J. Kenney (eds) *Multicultural and gender equity in the mathematics classroom: The gift of diversity, 1997 yearbook*. Reston, VA: National Council of Teachers of Mathematics.

Kieran, C. (1992). The learning and teaching of school algebra. In D. Grouws (ed.) *Handbook of research on mathematics teaching and learning* (390–419). New York: Macmillan.

——— (2007). Learning and teaching algebra at the middle school through college levels. In F.K. Lester (ed.) *Second Handbook of Research on Mathematics Teaching and Learning* (vol. 2, 707–62). Charlotte, NC: National Council of Teachers of Mathematics.

Killen, R. (2005). *Programming and assessment for quality teaching and learning*. South Melbourne, Vic.: Thomson Social Science Press.

Killen, R. (2013). *Effective teaching strategies: Lessons from research and practice* (6th edn). South Melbourne, Vic.: Cengage Learning.

Kilpatrick, J. (2002). Understanding mathematical literacy: The contribution of research. *Educational Studies in Mathematics*, 47(1), 101–16.

Kilpatrick, J., Swafford, J. & Findell, B. (eds) (2001). *Adding it up: Helping children learn mathematics*. Washington, DC: National Academy Press.

Kincaid, R. (2015). *Weighted dice: A study in applications of probability* (PowerPoint slides). Proceedings of the S.T.E.A.M. & Education Conference. Honolulu, Hawaii.

Kissane, B. (2007a). Hand-held technology in secondary mathematics education. In S. Li, J. Zhang & D. Wang (eds) *Symbolic computation and education* (31–59). Singapore: World Scientific.

———— (2007b). *Exploring the place of hand-held technology in secondary mathematics education.* 12th Asian Technology Conference on Mathematics, *16–20 December*, Taipei, Taiwan.

———— (2010). Using ICT in applications of secondary school mathematics. In B. Kaur & J. Dindyal (eds) *Mathematical applications and modelling*. Singapore: World Scientific.

Klenowski, V. (2009). Assessment for learning revisited: An Asia-Pacific perspective. *Assessment in Education: Principles, Policy and Practice*, 16(3), 263–68.

Kllogjeri, P. & Shyti, B. (2010). GeoGebra: A global platform for teaching and learning math together and using the synergy of mathematicians. *International Journal of Teaching and Case Studies*, 2(3), 225–36.

Koehler, M. & Mishra, P. (2009). What is technological pedagogical content knowledge (TPACK)? *Contemporary issues in technology and teacher education*, 9(1), 60–70.

Konold, C. & Pollatsek, A. (2002). Data analysis as the search for signals in noisy processes. *Journal for Research in Mathematics Education*, 33(4), 259–89.

Koontz, T. (1997). Know thyself, the evolution of an intervention gender-equity program. In J. Trentacosta & M.J. Kenney (eds) *Multicultural and gender equity in the mathematics classroom: The gift of diversity*, 1997 Yearbook. Reston, VA: National Council of Teachers of Mathematics.

Kunimune, S., Egashira, N., Hayakawa, T., Hatta, H., Kondo, H. Matsumoto, S., Kumakura, H. & Fujita, T. (2007). *The teaching of geometry from primary to upper secondary school*. Shizuoka, Japan: Shizuoka University [in Japanese].

Kunimune, S., Fujita, T. & Jones, K. (2009). *Strengthening students' understanding of 'proof' in geometry in lower secondary school*. Proceedings of CERME 6 (756–65). Lyon, France.

Lacey, C. & Lawton, D. (1981). *Issues in evaluation and accountability*. London: Methuen.

Lahann, P. & Lambdin, D.V. (2014). Collaborative learning in mathematics education. In S. Lerman (ed.) *Encyclopaedia of mathematics education* (75–76). Dordrecht: Springer Reference.

Lampert, M. (1990). When the problem is not the question and the solution is not the answer: Mathematical knowing and teaching. *American Educational Research Journal*, 27, 29–63.

Lang, H.R. & Evans, D.N. (2006). *Models, strategies and methods for effective teaching*. Boston, MA: Pearson Education.

Larkin, K, Jamieson-Proctor, R. & Finger, G. (2012). TPACK and pre-service teacher mathematics education: Defining a signature pedagogy for mathematics education using ICT and based on the metaphor 'mathematics is a language'. *Computers in the Schools*, 29(1–2), 207–26. doi: 10.1080/07380569.2012.651424.

Latham, G., Blaise, G., Dole, S., Faulkner, J. & Malone, K. (2011). *Learning to teach: New times, new practices* (2nd edn). South Melbourne, Vic.: Oxford University Press.

Lave, J. & Wenger, E. (1991). *Situated learning: Legitimate peripheral participation.* Cambridge: Cambridge University Press.

Leder, G.C., Brew, C. & Rowley, G. (1999). Gender differences in mathematics achievement – Here today and gone tomorrow? In G. Kaiser, E. Luna & I. Huntley (eds) *International Comparisons in Mathematics Education* (213–24). London: Falmer Press.

Leder, G.C. & Forgasz, H.J. (1992). Perspectives on learning, teaching and assessment. In G.C. Leder (ed.) *Assessment and learning of mathematics* (1–23). Hawthorn, Vic.: ACER.

Leikin, R. & Levav-Waynberg, A. (2007). Exploring mathematics teacher knowledge to explain the gap between theory-based recommendations and school practice in the use of connecting tasks. *Educational Studies in Mathematics*, 66(3), 349–71.

Leinhardt, G., Zaslavsky, O. & Stein, M.K. (1990). Functions, graphs, and graphing: Tasks, learning, and teaching. *Review of Educational Research*, 60(1), 1–64. doi:10.3102/00346543060001001.

Lerman, S. (2000). The social turn in mathematics education research. In J. Boaler (ed.) *Multiple perspectives on mathematics teaching and learning* (19–44). Westport, CT: Ablex Publishing.

——— (2001). Accounting for accounts of learning mathematics: Reading the ZPD in videos and transcripts. In D. Clarke (ed.) *Perspectives on practice and meaning in mathematics and science classrooms* (53–74). Dordrecht: Kluwer.

Lesh, R., Hoover, M., Hole, B., Kelly, A. & Post, T. (2000). Principles for developing thought-revealing activities for students and teachers. In A.E. Kelly & R.A. Lesh (eds) *Handbook of research design in mathematics and science education* (591–646), Mahwah, NJ: Lawrence Erlbaum Associates.

Leung, C. (2005). Mathematical vocabulary: Fixers of knowledge or points of exploration? *Language and Education*, 19(2), 127–35.

Li, C. & Nuttall, R. (2001). Writing Chinese and mathematics achievement: A study with Chinese-American undergraduates. *Mathematics Education Research Journal*, 13(1), 15–27.

Lin, F.L. (2000). An approach for developing well-tested, validated research of mathematics learning and teaching. In T. Nakahara & M. Koyama (eds) *Proceedings of the 24th Conference of the International Group for the Psychology of Mathematics Education* (vol. 1, 84–8).

Linchevski, L. & Williams, J. (1999). Using intuition from everyday life in 'filling' the gap in children's extension of their number concept to include the negative numbers. *Educational Studies in Mathematics*, 39(1/3), 131–47. doi: 10.2307/3483164.

Lipson, K., Kokonis, S. & Francis, G. (2003). Investigation of students' experiences with a web-based computer simulation. *Proceedings of the International Association for Statistical Education Satellite Conference on Statistics and the Internet, Berlin, August, 2003*. [CDRom]. Voorburg, The Netherlands: International Statistical Institute.

Lopez-Real, F. (1997). Effect of the different syntactic structures of English and Chinese in simple algebra problems. In A. Begg (ed.) *People in Mathematics Education. Proceedings of the 20th Annual Conference of the Mathematics Education Research Group of Australasia* (vol. 2, 317–23). Rotorua.

Lotan, R.A. (2002). Group-worthy tasks: Carefully constructed group learning activities can foster students' academic and social growth and help close the achievement gap. *Educational Leadership*, 60(6), 72–5.

Lovitt, C. & Williams, D. (2015). *The licorice factory*. Retrieved from www.maths300.com.

Lowe, I., Kissane, B., Willis, S., Grace, N. & Johnston, J. (1993). *Access to algebra*. Melbourne: Curriculum Corporation.

Lowrie, T., Ramful, A., Logan, T. & Ho, S.Y. (2014). Do students solve graphic tasks with spatial demands differently in digital form? In J. Anderson, M. Cavanagh & A. Prescott (eds) *Curriculum in focus: Research guided practice. Proceedings of the 37th annual conference of the Mathematics Education Research Group of Australasia* (429–36). Sydney: MERGA.

Luke, A., Elkins, J., Weir, K., Land, R., Carrington, V., Dole, S. & Stevens, L. (2003). *Beyond the middle: A report about literacy and numeracy development of target group students in the middle years of schooling*. Australian Department of Education, Employment and Workplace Relations.

Ma, L. (1999). *Knowing and teaching elementary mathematics*. Mahwah, NJ: Erlbaum.

MacGregor, M. (1991). *Making sense of algebra: Cognitive processes influencing comprehension*. Warun Ponds, Vic.: Deakin University press.

———— (1993). Interaction of language competence and mathematics learning. In M. Stephens, A. Waywood, D. Clarke & J. Izard (eds) *Communicating Mathematics: Perspectives from Classroom Practice and Current Research* (51–9). Vic.: ACER.

MacGregor, M. & Moore, R. (1991). *Teaching mathematics in the multilingual classroom*. Melbourne: Institute of Education, The University of Melbourne.

Maddern, S. & Court, R. (1989). *Improving mathematics practice and classroom teaching*. Nottingham, UK: Shell Centre for Mathematics Education.

Maker, K. (2007). Connection levers: Supports for building teachers' confidence and commitment to teach mathematics and statistics through inquiry. *Mathematics Teachers Education and Development*, 8, 48–73.

Malcolm, I., Haig, Y., Königsberg, P., Rochecouste, J., Collard, G., Hill, A. & Cahill, R. (1999). *Two-way English. Towards more user-friendly education for speakers of Aboriginal English*. East Perth: Edith Cowan University.

Malle, G. (1993). Didaktische probleme der elementaren algebra. *Viewteg*, 93–128.

Mamona-Downs, J. (2001). Letting the intuitive bear on the formal: A didactical approach for the understanding of the limit of a sequence. *Educational Studies in Mathematics*, 48(2/3), 259–88.

Manouchehri, A. (2007). Inquiry-discourse: Mathematics instruction. *Mathematics Teacher*, 101(4), 290–300.

Mariotti, M.A. (2007). Proof and proving in mathematics education. In A. Guitierrez & P. Boero (eds) *Handbook of research on the psychology of mathematics education*. Rotterdam: Sense Publishers.

Marsh, C. (2004). *Becoming a teacher: Knowledge, skills and issues* (3rd edn). Sydney: Pearson Education Australia.

—— (2010). *Becoming a teacher: Knowledge, skills and practices* (5th edn). Sydney: Pearson.

Marshall, S. (1984). Sex differences in children's mathematics achievement: Solving, computations and story-problems. *Journal of Educational Psychology*, 76(2), 194–204.

Marton, F. & Tsui, A. (2004). *Classroom discourse and the space of learning*. Mahwah, NJ: Lawrence Erlbaum.

Masingila, J.O., Lester, F.K. & Raymond, A.M. (2002). *Mathematics for elementary teachers via problem solving*. Upper Saddle River, NJ: Prentice Hall.

Mason, M. (2014). The van Hiele levels of geometric understanding. In *Professional Handbook for Teachers* (4–8). McDougal Little Inc.

Matsuda, N., Yarzebinski, E., Keiser, V., Raizada, R., Cohen, W.W., Stylianides, G.J. & Koedinger, K.R. (2013). Cognitive anatomy of tutor learning: Lessons learned with SimStudent. *Journal of Educational Psychology*, 105(4), 1152–63. doi: dx.doi.org/10.1037/a0031955.

Matthews, R.S., Cooper, J.L., Davidson, N. & Hawkes, P. (1995). Building bridges between cooperative and collaborative learning. *Change*, 29(4), 35–40.

Mau, T.S. & Leitze, A.R. (2001). Powerless gender or gender-less power: The promise of constructivism for females in the mathematics classroom. In J.E. Jacobs, J.R. Martin, A.J. (2003). *How to motivate your child for school and beyond*. Sydney: Bantam.

Mayer, R. (1981). Frequency norms and structural analysis of algebra story problems into families, categories, and templates. *Instructional Science*, 10(2), 135–75. doi: 10.1007/bf00132515.

Mayer, R.E. (1985). Mathematical ability. In R.J. Sternberg (ed.) *Human abilities: An information-processing approach* (127–50). New York: Freeman.

McDermott, L.C., Rosenquist, M.L. & van Zee, E.H. (1987). Student difficulties in connecting graphs and physics: Examples from kinematics. *American Journal of Physics*, 55(6), 503–13.

McPhan, G., Morony, W., Pegg, J., Cooksey, R. & Lynch, T. (2008). *Maths? Why not? Final report prepared for the Department of Education, Employment and Workplace Relations (DEEWR)*. Canberra: Department of Education, Employment and Workplace Relations.

McSeveny, A., Conway, R. & Wilkes, S. (2004). *New signpost mathematics 8*. Melbourne: Pearson Education Australia.

Meiers, M. (2010). Language in the mathematics classroom. *QCT Research Digest, QCT*, 7.

Mercer, N., Wegerif, R. & Dawes, L. (1999). Children's talk and the development of reasoning in the classroom. *British Educational Research Journal*, 25, 95–111.

Meyer, J. & Land, R. (2003). Threshold concepts and troublesome knowledge: Linkages to ways of thinking and practising within the disciplines. In C. Rust (ed.) *Improving Student Learning – Ten years on*. Oxford: OCSLD.

Miller, D. & Glover, D. (2010). Presentation or mediation: Is there a need for 'interactive whiteboard technology-proficient' teachers in secondary mathematics? *Technology, Pedagogy and Education*, 19(2), 253–59.

Miller, G.A. (1956). The magical number seven, plus or minus two: Some limits on our capacity for processing information. *Psychological Review*, 63(2), 81–97. doi: dx.doi.org/10.1037/h0043158.

——— (2003). The cognitive revolution: A historical perspective. *TRENDS in Cognitive Sciences*, 7, 141–4.

Miller, S.P. & Mercer, C.D. (1997). Educational aspects of mathematics disabilities. *Journal of Learning Disabilities*, 30(1), 47–56.

Ministerial Council on Education, Employment, Training and Youth Affairs (MCEETYA) (2008a). *The Melbourne declaration on educational goals for young Australians*. Melbourne: MCEETYA.

——— (2008b). *National Assessment Program Literacy and Numeracy: Year 7 numeracy test (calculator allowed)*. Melbourne: Curriculum Corporation.

Minstrell, J. (2001). Facets of students' thinking: Designing to cross the gap from research to standards-based practice. In K. Crowley, C. Schunn & T. Okada (eds) *Designing for science: Implications from professional, instructional, and everyday science*. Mawah, NJ: Lawrence Erlbaum Associates.

——— (2015). *Facets of thinking: Diagnostic assessment of speed versus time graph*. Retrieved on 6 March 2015 from www.diagnoser.com.

Mokros, J. & Russell, S. (1995). Children's concepts of average and representativeness. *Journal for Research in Mathematics Education*, 26(1), 20–39.

Moore, D. (1990). Uncertainty. In L. Steen (ed.) *On the shoulders of giants: New approaches to numeracy* (95–137). Washington, DC: National Academy Press.

Moore, K.C. (2014). Quantitative reasoning and the sine function: The case of Zac. *Journal for Research in Mathematics Education*, 45(1), 102–38.

Morgan, C. (2005). Words, definitions and concepts in discourses of mathematics, teaching and learning. *Language and Education*, 19(2), 103–16.

Morrison, J. (2006). *STEM education monograph series: Attributes of STEM education*. Baltimore, MD: Teaching Institute for Essential Science.

Morrison, J. & Bartlett, R. (2009). STEM as curriculum. *Education Week*, 23, 28–31.

Morrow, C. & Morrow, J. (1995). Connecting women with mathematics. In P. Rogers & G. Kaiser (eds) *Equity in mathematics education: Influences of feminism and culture* (13–26). London: Falmer Press.

Motz, L. & Weaver, J. (1993). *The story of mathematics*. New York: Plenum Press.

Nathan, G., Trinick, T., Tobin, E. & Barton, B. (1993). Tahi, Rua, Toru, Wha: Mathematical counts in Maori renaissance. In M. Stephens, A. Waywood, D. Clarke & J. Izard (eds) *Communicating mathematically: Perspectives from classroom practice and current research*. Melbourne: ACER.

Nathan, M.J., Kintsch, W. & Young, E. (1992). A theory of algebra-word-problem comprehension and its implications for the design of learning environments. *Cognition and Instruction*, 9(4), 329–89.

National Academy of Sciences (2003). *Engaging schools: Fostering high school students' motivation to learn*. Washington, DC: National Academy Press.

National Association for Gifted Children (2007). What is giftedness? Retrieved 30 April 2015 from www.nag.org.

National Consumer and Financial Literacy Framework (2011, September). Retrieved 6 May 2015 from scseec.edu.au/site/DefaultSite/filesystem/documents/Reports%20and%20publications/Publications/Miscellaneous/National%20Consumer%20and%20Financial%20Literacy%20Framework-2011.pdf.

National Council of Teachers of Mathematics (2000). *Principles and standards for school mathematics*. Reston, VA: NCTM.

———— (2009). *Guiding principles for mathematics curriculum and assessment*. Retrieved from old.nctm.org/uploadedFiles/Math_Standards/NCTM%20Guiding%20Principles%206209.pdf.

———— (2015). *Executive summary: Principles and standards for teaching mathematics*. Retrieved from www.nctm.org/uploadedFiles/Standards_and_Positions/PSSM_ExecutiveSummary.pdf.

National Curriculum Board (2009). *Shape of the Australian Curriculum: Mathematics*. Canberra: Commonwealth of Australia.

National Research Council (NRC) (1998). *High school mathematics at work: Essays and examples for the education of all students*. Washington, DC: National Academy Press.

———— (2001). *Adding it up: Helping children learn mathematics*. Washington, DC: National Academy Press.

NCTM (National Council of Teachers of Mathematics) (1995). *Assessment standards for school mathematics*. Reston, VA: Author.

———— (2014). *Principles to actions: Ensuring mathematical success for all*. Retrieved 1 March, 2015 from www.nctm.org/PtA.

Neal, D., Muir, T., Manuel, K., Livy, S. & Chick, H. (2014). Desperately seeking birthday mates!: Or what maths teachers get up to on Saturday nights! [online]. *Australian Mathematics Teacher*, 70 (1), 36–40. Retrieved from search.informit.com.au/documentSummary;dn=234505759890312;res=IELHSS.

Newman, M.A. (1977). An analysis of sixth-grade pupils' errors on written mathematical tasks. *Victorian Institute for Educational Research Bulletin*, 39, 31–43.

———— (1983). *Strategies for diagnosis and remediation*. Sydney: Harcourt, Brace Jovanovich.

Ngu, B.H., Chung, S.F. & Yeung, A.S. (2015). Cognitive load in algebra: element interactivity in solving equations. *Educational Psychology*, 35(3), 271–93. doi: 10.1080/01443410.2013.878019.

Ngu, B.H. & Phan, H. (2016a). Comparing balance and inverse methods on learning conceptual and procedural knowledge in equation solving: A cognitive load perspective. *Pedagogies: An International Journal*, 11(1), 63–83. doi: 10.1080/1554480X.2015.1047836.

———— (2016b). Unpacking the complexity of linear equations from a cognitive load theory perspective. *Educational Psychology Review*. doi: 10.1007/s10648-015-9298-2.

Ngu, B.H. & Yeung, A.S. (2012). Fostering analogical transfer: The multiple components approach to algebra word problem solving in a chemistry context. *Contemporary Educational Psychology*, 37(1), 14–32. doi: dx.doi.org/10.1016/j.cedpsych.2011.09.001.

———— (2013). Algebra word problem solving approaches in a chemistry context: Equation worked examples versus text editing. *The Journal of Mathematical Behavior*, 32(2), 197–208. doi: dx.doi.org/10.1016/j.jmathb.2013.02.003.

Ngu, B.H., Yeung, A.S. & Phan, H.P. (2015). Constructing a coherent problem model to facilitate algebra problem solving in a chemistry context. *International Journal of Mathematical Education in Science & Technology*, 46(3), 388–403. doi: 10.1080/0020739x.2014.979899.

Ngu, B.H., Yeung, A. & Tobias, S. (2014). Cognitive load in percentage change problems: Unitary, pictorial, and equation approaches to instruction. *Instructional Science*, 42(5), 685–713. doi: 10.1007/s11251-014-9309-6.

Niss, M. (1999). Aspects of the nature and state of research in mathematics education. *Educational Studies in Mathematics*, 40, 1–24.

Nosek, B., Banaji, M. & Greenwald, A. (2002). Math = male, me = female, therefore me = math. *Journal of Personality and Social Psychology*, 83(1), 44–59.

Noss, R., Healy, L. & Hoyles, C. (1997). The construction of mathematical meanings: Connecting the visual with the symbolic. *Educational Studies in Mathematics*, 33(2), 203–233.

O'Connor, M.C. & Michaels, S. (1996). Shifting participant frameworks: Orchestrating thinking practices in group discussion. In D. Hicks (ed.) *Discourse, learning and schooling* (63–103). New York: Cambridge University Press.

OECD (2006). *Evolution of student interest in science and technology studies: Policy report*. Paris: OECD.

Office of the Chief Scientist (2014). *Science, technology, engineering and mathematics: Australia's future*. Canberra: Commonwealth of Australia.

Ohlsen, M.T. (2007). Classroom assessment practices of secondary school members of NCTM. *American Secondary Education*, 36(1), 4–14.

Okolica, S. & Macrina, G. (1992). Integrating transformational geometry into traditional high school geometry. *Mathematics Teacher*, 85(9), 716–9.

Organisation for Economic Co-operation and Development (OECD) (2006). *Encouraging student interest in science and technology studies: Policy report*. Paris: OECD and Development Global Science Forum.

Orton, A. (1983a). Students' understanding of differentiation. *Educational Studies in Mathematics*, 14, 235–50.

——— (1983b). Students' understanding of integration. *Educational Studies in Mathematics*, 14, 1–18.

Özgün-Koca, A., Edwards, M. & Meagher, M. (2013). Spagehetti sine curves. *Mathematics Teacher*, 107(3), 180–7.

Panizzon, D. & Pegg, J. (2007). Assessment practices: Empowering mathematics and science teachers in rural secondary schools to enhance student learning. *International Journal of Science and Mathematics Education*, 6, 417–36.

Parameswaran, R. (2006). On understanding the notion of limits and infinitesimal quantities. *International Journal of Science and Mathematics Education*, 5, 193–216.

Parker, M. (2014). *Things to make and do in the fourth dimension*. New York: Farrar, Straus and Giroux.

Parker, M. & Leinhardt, G. (1995). Percent: A privileged proportion. *Review of Educational Research*, 65(4), 421–481. doi: 10.3102/00346543065004421.

Parkin, B. & Hayes, J. (2006). Scaffolding the language of maths. *Literacy Learning: The Middle Years*, 14(1), 23–35.

Pawley, D., Ayres, P., Cooper, M. & Sweller, J. (2005). Translating words into equations: A cognitive load theory approach. *Educational Psychology*, 25(1), 75–97. doi: 10.1080/0144341042000294903.

Peled, I. & Suzan, A. (2011). Pedagogical, mathematical, and epistemological goals in designing cognitive conflict tasks for teacher education. In O. Zaslavsky & P. Sullivan (eds) *Constructing knowledge for teaching secondary mathematics: Tasks to enhance prospective and practicing teacher learning* (73–88). New York: Springer.

Perry, B. & Howard, P. (2008). Mathematics in Indigenous contexts. *Australian Primary Mathematics Classroom*, 13(4), 4–9.

Perso, T. (2011). Assessing numeracy and NAPLAN. *The Australian Mathematics Teacher*, 67(4), 32–5.

Peterson, I. (1998). *The jungles of randomness: Mathematics at the edge of uncertainty*. London: Penguin Books.

Pfannkuch, M. & Wild, C. (2004). Towards an understanding of statistical thinking. In D. Ben-Zvi & J. Garfield (eds) *The challenge of developing statistical literacy, reasoning and thinking* (3–15). Dordrecht: Kluwer Academic Publishers.

Phet Interactive Simulations (2015). *Projectile simulator*. Retrieved from phet.colorado.edu/en/simulation/legacy/projectile-motion.

Piaget, J. (1972). *The principles of genetic epistemology* (W. Mays trans.). London: Routledge & Kegan Paul.

Pierce, R., & Bardini, C. (2015). Computer algebra systems: Permitted but are they used? *Australian Senior Mathematics Journal*, 29(1), 32.

Pierce, R. & Stacey, K. (2010). Mapping pedagogical opportunities provided by mathematics analysis software. *International Journal of Computers for Mathematical Learning*, 15(1), 1–20.

——— (2011). Using dynamic geometry to bring the real world into the classroom. In *Model-centered learning* (41–55). Sense Publishers.

Pierce, R., Stacey, K., Wander, R. & Ball, L. (2011). The design of lessons using mathematics analysis software to support multiple representations in secondary school mathematics. *Technology, Pedagogy and Education*, 20(1), 95–112.

Piggott, J. (2015). *Rich tasks and contexts*. Retrieved from nrich.maths.org/5662.

Pimm, D. (1991). Communicating mathematically. In K. Durkin & B. Shire (eds) *Language in mathematical education: Research and practice* (17–23). Milton Keynes: Open University Press.

Pirie, S. (1987). *Mathematical investigations in your classroom: A guide for teachers.* Basingstoke, UK: Macmillan.

Polya, G. (1957). *How to solve it.* Princeton, NJ: Princeton University Press.

Ponte, J.P. (2007). Investigations and explorations in the mathematics classroom. *ZDM Mathematics Education*, 39(5–6), 419–30.

Ponte, J.P. & Matos, J.F. (1992). Cognitive processes and social interaction in mathematical investigations. In J.P. Ponte, J.F. Matos, J.M. Matos & D. Fernandes (eds) *Mathematical problem solving and new information technologies: Research in contexts of practice* (239–54). Berlin: Springer.

Popham, W.J. (2008). *Transformative assessment.* Alexandria, VA: Association for Supervision and Curriculum Development.

Przenioslo, M. (2004). Images of the limit of function formed in the course of mathematical studies at the university. *Educational Studies in Mathematics*, 55(1/3), 103–32.

Rakes, C.R., Valentine, J.C., McGatha, M.B. & Ronau, R.N. (2010). Methods of instructional improvement in algebra: A systematic review and meta-analysis. *Review of Educational Research*, 80(3), 372–400.

Ramprasad, A. (1983). On the definition of feedback. *Behavioral Science*, 28(1), 4–13.

Rasslan, S. & Tall, D. (1997). Definitions and images for the definite integral concept. In A. Cockburn & E. Nardi (eds) *Proceedings of the 26th Conference PME* (89–96). Norwich: PME.

Reaburn, R. (2012, July). Strategies used by students to compare two data sets. In J. Dindyal, L.P. Cheng & S.F. Ng (eds) *Mathematics education: Expanding Horizons.* Paper presented at the 35th annual conference of the Mathematics Education Research Group of Australasia. Singapore: MERGA.

———— (2013). Students' understanding of conditional probability of entering university. In V. Steinle, L. Ball & C. Bardini (eds) *Mathematics education: Yesterday, today and tomorrow. Proceedings of the 36th annual conference of the Mathematics Education Research Group of Australasia.* Melbourne: MERGA.

Reid, D.A. (2011). *Understanding proof and transforming teaching.* In L.R. Wiest & T. Lamberg (eds) *Proceedings of the 33rd Annual Meeting of the North American Chapter of the International Group for the Psychology of Mathematics Education.* Reno, NV.

Richland, L.E., Zur, O. & Holyoak, K.J. (2007). Cognitive supports for analogies in the mathematics classroom. *Science*, 316(5828), 1128–29. doi: 10.2307/20036317.

Riegle-Crumb, C. (2006). The path through math: Course-taking trajectories and student performance at the intersection of gender and race/ethnicity. *American Journal of Education*, 113(1), 101–22.

Rittle-Johnson, B. & Star, J.R. (2007). Does comparing solution methods facilitate conceptual and procedural knowledge? An experimental study on learning to solve equations. *Journal of Educational Psychology*, 99(3), 561–74. doi: 10.1037//1082-989x.7.2.147.

———— (2009). Compared with what? The effects of different comparisons on conceptual knowledge and procedural flexibility for equation solving. *Journal of Educational Psychology*, 101(3), 529–44. doi: dx.doi.org/10.1037/a0014224.

Roberts, A. & Cantu, D. (2015). *Applying STEM instructional strategies to design and curriculum curriculum*. Retrieved 20 September 2015 from www.ep.liu.se/ecp/073/013/ecp12073013.pdf.

Robert, A. & Schwarzenberger, R. (1991). Research in teaching and learning mathematics at an advanced level. In D. Tall (ed.) *Advanced Mathematical Thinking* (127–39). Dordrecht: Kluwer Academic Publishers.

Ronda, E. (2010). *What is mathematical investigation?* Retrieved from math4teaching.com/2010/03/09/what-is-mathematical-investigation.

Rossouw, A., Hacker, M. & de Vries, M. (2010). Concepts and contexts in engineering and technology education: An international and interdisciplinary Delphi study. *International Journal of Technology and Design Education*, 21(4), 409–24.

Rotigel, J. & Fello, S. (2005). Mathematically gifted students: How can we meet their needs? *Gifted Child Today*, 27(4), 46–65.

Ruthven, K., Laborde, C., Leach, J. & Tiberghien, A. (2009). Design tools in didactical research: Instrumenting the epistemological and cognitive aspects of the design of teaching sequences. *Educational Researcher*, 38(5), 329–42.

Ryan, J. & Williams, B. (2007). *Children's mathematics 4–15: Learning from errors and misconceptions*. Maidenhead: McGraw-Hill Education.

Sadler, D.R. (1989). Formative assessment and the design of instructional systems. *Instructional Science*, 18, 119–44.

Sanders, M. (2009). STEM, STEM education, STEMmania. *The Technology Teacher*, December/January, 20–6.

Saranen, E. (1992). Lukion yleisen oppimäärän opiskelijoiden matematiikan taidot ja käsitykset matematiikasta. [Upper secondary school general course mathematics students' skills in and conceptions about mathematics]. Kasvatustieteiden tutkimuslaitoksen julkaisusarja A, Tutkimuksia; 38. Jyväskylä: Kasvatusteiteen tutkimuslaitos.

Sattler, J. (2008). *Assessment of children: Cognitive foundations*. La Mesa, CA: Jerome's Sattler.

Schoenfeld, A.H. (1988). When good teaching leads to bad results: The disasters of 'well-taught' mathematics courses. *Educational Psychologist*, 23, 145–66.

Schultz, K.A. & Austin, J.D. (1983). Directional effects in transformational tasks. *Journal for Research in Mathematics Education*, 14(2), 95–101.

Schwalbach, E. & Dosemagen, D. (2000). Developing students' understanding: Contextualising calculus concepts. *School Science and Mathematics*, 100(2), 90–8.

Schwartzman, S. (1994). *The words of mathematics: An etymological dictionary of mathematical terms used in English.* Washington: Mathematical Association of America.

Sealey, V. (2006). Definite integrals, Riemann sums, and area under a curve: What is necessary and sufficient? In S. Alatorre, J. Cortina, M. Sáiz & A. Méndez (eds) *Proceedings of the 28th annual meeting of the North American Chapter of the International Group for the Psychology of Mathematics Education.* México: Universidad Pedagógica Nacional.

Senk, S.L. (1985). How well do students write geometry proofs? *Mathematics Teacher*, 78(6), 449–56.

Senk, S.L., Beckmann, C.E. & Thompson, D.R. (1997). Assessment and grading in high school mathematics classrooms. *Journal for Research in Mathematics Education*, 28(2), 187–215.

Sfard, A. (1991). On the dual nature of mathematical conceptions: reflections on processes and objects as different sides of the same coin. *Educational Studies in Mathematics*, 22, 1–36.

Shaughnessy, J. (2007). Research on statistics learning and reasoning. In F. Lester (ed.) *Second handbook on mathematics teaching and learning* (957–1009). Charlotte, NC: The National Council of Teachers of Mathematics.

Shaughnessy, J. & Chance, B. (2005). *Statistical questions from the classroom.* Reston, VA: National Council of Teachers of Mathematics.

Sheffield, L.J. (1997). From Doogie Howser to dweebs – or how we went in search of Bobby Fischer and found that we are dumb and dumber. *Mathematics Teaching in the Middle School*, 2(6), 376–9.

——— (1999). *Developing mathematically promising students.* Reston, VA: National Council of Teachers of Mathematics.

Shell Centre for Mathematics Education (n.d.). *Problems with patterns and numbers.* Retrieved from: www.mathshell.com/publications/tss/ppn/ppn_teacher.pdf.

Shield, M. (2008). The function concept in middle-years mathematics. *Australian Mathematics Teacher*, 64(2), 36–40.

Shinba, S., Sonoda, H. & Kunimune, S. (2004). *Teaching similarity of geometrical figures: An emphasised learning process from constructing to proving.* Memoirs of Center for Educational Research and Teacher Development, Shizuoka University, 10, 11–22 [in Japanese].

Shulman, L.S. (1986). Those who understand: Knowledge growth in teaching. *Educational Researcher*, 15(2), 4–14.

Siemon, D., Beswick, K., Brady, K., Clark, J., Faragher, R. & Warren, E. (2011). *Teaching mathematics: Foundations to middle years*. South Melbourne, Vic.: Oxford University Press.

——— (2015). *Teaching mathematics* (2nd edn). South Melbourne, Vic.: Oxford University Press.

Sinclair, M.P. (2003). Some implications of the results of a case study for the design of pre-constructed, dynamic geometry sketches and accompanying materials. *Educational Studies in Mathematics*, 52(3), 289–317.

Sinclair, N. & Robutti, O. (2013). Technology and the role of proof: The case of dynamic geometry. In M.A. Clements, A.J. Bishop, Ch. Keitel, J. Kilpatrick, & F.K.S. Leung (eds) *Third International Handbook of Mathematics Education* (571–96). New York: Springer.

Singh, S. (2005). *Fermat's last theorem*. London: Fourth Estate.

Singhal, R., Henz, M. & McGee, K. (2014). Automated generation of geometry questions for high school mathematics (14–25). In S. Zvacek, M.T. Restivo, J. Uhomoibhi & M. Helfert (eds) *Proceedings 6th International Conference on Computer Supported Education*. Barcelona, Spain.

Siskin, L.S. (1994). *Realms of knowledge: Academic departments in secondary schools*. London: Falmer Press.

Skemp, R. (1976). Relational understanding and instrumental understanding. *Mathematics Teaching*, 77, 20–26.

——— (1987). *The psychology of learning mathematics*. Hillsdale, NJ: Lawrence Erlbaum Associates.

Skilling, K. (2014). Teacher practices: How they promote or hinder student engagement in mathematics. In J. Anderson, M. Cavanagh & A. Prescott (eds) *Curriculum in focus: Research guided practice. Proceedings of the 37th Annual Conference of the Mathematics Education Research Group of Australasia* (589–96). Sydney: MERGA.

Slavin, R.E., Lake, C. & Groff, C. (2009). Effective programs in middle and high school mathematics: A best-evidence synthesis. *Review of Educational Research*.

Sokolowski, A. (2013). Modelling transformations of quadratic functions: A proposal of inductive inquiry. *Australian Senior Mathematics Journal*, 27(2), 45–54.

Soro, R. (2002). Teachers' beliefs about gender differences in mathematics: 'Girls or boys?' scale. In A.D. Cockburn & E. Nardi (eds) *Proceedings of the 26th Conference of the International Group for the Psychology of Mathematics Education* (vol. 4, 225–32). Norwich: PME.

Sousa, D.A. (2008). *How the brain learns mathematics*. Thousand Oaks, CA: Sage Publications.

Spanos, G., Rhodes, N., Dale, T. & Crandall, J.A. (1988). Linguistic features of mathematical problem solving: Insights and applications. In R. Cocking & J. Mestre (eds) *Linguistic and cultural Influences on learning Mathematics*. New Jersey: Erlbaum.

Spelke, E.S. (2005). Sex differences in intrinsic aptitude for mathematics and science? A critical review. *American Psychologist*, 60, 950–8.

Spencer, S.J., Steele, C.M. & Quinn, D.M. (1999). Stereotype threat and women's math performance. *Journal of Experimental Social Psychology*, 35, 4–28.

SPSS (1989, 2013). *IBM SPSS Statistics*, version 22.0. IBM Corporation and others, 1989, 2013.

Stacey, K. (2003). The need to increase attention to mathematical reasoning. In H. Hollingsworth, J. Lokan & B. McCrae *Teaching Mathematics in Australia: Results from the TIMSS 1999 Video Study* (119–22), Camberwell, Vic.: ACER.

——— (2005). The place of problem solving in contemporary mathematics curriculum documents. *Journal of Mathematical Behaviour*, 24, 341–50.

Stacey, K. & MacGregor, M. (1999). Learning the algebraic method of solving problems. *The Journal of Mathematical Behavior*, 18(2), 149–67. doi: dx.doi.org/10.1016/S0732-3123(99)00026-7.

Stacey, K., Price, B., Gvozdenko, E. & Steinle, V. (2013). *Specific mathematics assessments that reveal thinking*. Retrieved 27 October 2015 from www.smartvic.com/teacher.

Stagg, P. (2007). *Careers from science: An investigation for the Science Education Forum*. Warwick University, England: Centre for Education and Industry (CEI).

Steen, L.A. & Forman, S.L. (2000). Beyond eighth grade: Functional mathematics for life and work. In M.J. Burke & F.R. Curcio (eds) *Learning mathematics for a new century*. NCTM Yearbook (127–57). Reston, VA: National Council of Teachers of Mathematics.

Stein, M.K., Boaler, J. & Silver, E.A. (2003). Teaching mathematics through problem solving: Research perspectives. In H.L. Schoen, *Teaching mathematics through problem solving: Grades 6–12*. Reston, VA: National Council of Teachers of Mathematics.

Stenmark, J.K. (1989). *Assessment alternatives in mathematics: An overview of assessment techniques that promote learning*. Berkeley, CA: University of California.

——— (1991). *Mathematics assessment: Myths, models, good questions and practical suggestions*. Reston, VA: NCTM.

Stephan, M. (2014a). Learner-centered teaching in mathematics education. In S. Lerman (ed.) *Encyclopaedia of mathematics education* (338–43). Dordrecht: Springer Reference.

——— (2014b). Teacher-centered teaching in mathematics education. In S. Lerman (ed.) *Encyclopaedia of mathematics education* (593–8). Dordrecht: Springer Reference.

Stephens, M. (1987). Towards an AAMT policy on assessment and reporting in school mathematics. *Australian Mathematics Teacher*, 43(3), 2–3.

——— (1988). AAMT discussion on assessment and reporting in school mathematics. *Australian Mathematics Teacher*, 44(1), 16a–16c.

——— (2009). *Numeracy in practice: Teaching, learning and using mathematics.* Melbourne: Department of Education and Early Childhood Development.

Stevens, T., Wang, K., Olivarez, A., Jr. & Hamman, D. (2007). Use of self-perceptions and their sources to predict the mathematics enrolment intentions of girls and boys. *Sex Roles*, 56(3), 51–63.

Stiggins, R.J. (1994). *Student-centred classroom assessment.* New York: Macmillan.

Stigler, J.W. & Hiebert, J. (1999). *The teaching gap: Best ideas from the world's teachers for improving education in the classroom.* New York: Free Press.

Stillman, G. (2002a). Assessing higher order mathematical thinking through applications (unpublished Doctor of Philosophy thesis). University of Queensland, Brisbane.

——— (2002b). The role of extra-mathematical knowledge in application and modelling activity. *Teaching Mathematics*, 27(2), 18–31.

——— (2012). *Applications and modelling research in secondary classrooms: What have we learnt? Proceedings of the 12th International Congress on Mathematical Education.* South Korea: ICME.

Stillman, G. & Galbraith, P.L. (1998). Applying mathematics with real world connections: Metacognitive characteristics of secondary students. *Educational Studies in Mathematics*, 36(2), 157–95.

Stipek, D.J. (1996). Motivation and instruction. In D.C. Berliner & R.C. Calfee (eds) *Handbook of educational psychology* (85–113). New York: Macmillan.

Subotnik, R.F., Tai, R.H., Rickoff, R. & Almarode, J. (2009). Specialized public high schools of science, mathematics, and technology and the STEM pipeline: What do we know now and what will we know in 5 years? *Roeper Review*, 32(1), 7–16.

Sullivan, P. (2011). *Teaching mathematics: Using research-informed strategies.* Canberra: ACER.

Sullivan, P., Clarke, D. & Clarke, B. (2013). *Teaching with tasks for effective mathematics learning.* New York: Springer.

Sullivan, P., Tobias, S. & McDonough, A. (2006). Perhaps the decision of some students not to engage in learning mathematics in school is deliberate. *Educational Studies in Mathematics*, 62, 81–99.

Swan, M. (2005). *Improving learning in mathematics: Challenges and strategies.* Sheffield, England: Department of Education and Skills Standards Unity.

———— (2006). *Collaborative learning in mathematics: A Challenge to our beliefs and practices.* London: National Research and Development Centre for Adult Literacy and Numeracy.

Sweller, J. (2012). Human cognitive architecture: Why some instructional procedures work and others do not. In K. Harris, S. Graham & T. Urdan (eds) *APA Educational Psychology Handbook,* (vol. 1, 295–325). Washington, DC: American Psychological Association.

Sweller, J., Ayres, P. & Kalyuga, S. (2011). *Cognitive load theory.* New York: Springer. doi:10.1007/978-1-4419-8126-4.

Sweller, J., van Merrienboer, J.G. & Paas, F.W.C. (1998). Cognitive architecture and instructional design. *Educational Psychology Review,* 10(3), 251–96. doi: 10.1023/a:1022193728205.

Swetz, F. & Hartzler, J.S. (1991). *Mathematical modeling in the secondary school curriculum.* Reston, VA: National Council of Teachers of Mathematics.

Szydlik, J. (2000). Mathematical beliefs and conceptual understanding of the limit of a function. *Journal for Research in Mathematics Education,* 31(3), 258–76.

Taggart, G.L., Adams, P.E., Eltze, E., Heinrichs, J., Hohman, J. & Hickman, K. (2007). Fermi questions. *Mathematics Teaching in the Middle School,* 13(3), 164–7.

Tall, D. (1992a). Students' difficulties in calculus. Plenary presentation in Working Group 3. *Proceedings of Working Group 3 on Students' Difficulties in Calculus, ICME-7.* Québec: ICME.

———— (1992b). The transition to advanced mathematical thinking: Functions, limits, infinity, and proof. In D. Grouws (ed.) *Handbook of research on mathematics teaching and learning* (495–511). New York: Macmillan.

———— (2000). Cognitive development in advanced mathematics using technology. *Mathematics Education Research Journal,* 12(3), 196–218.

Tall, D. & Vinner, S. (1981). Concept image and concept definition in mathematics with particular reference to limits and continuity. *Educational Studies in Mathematics,* 12(2), 151–69.

Tannock, R. & Martinussen, R. (2001). Reconceptualising ADHD. *Educational Leadership,* 59(3), 20–5.

Tarvainen, K. (2006). How to make sure that the error in the f(x) dx term is insignificant when setting up definite integrals. *International Journal of Mathematical Education in Science and Technology,* 29(3), 359–70.

Teacher Education Ministerial Advisory Group (2014). *Action now: Classroom ready teachers.* Retrieved 20 February 2015 from www.studentsfirst.gov.au/teacher-education-ministerial-advisory-group.

TESOL (2008). *Position statement on teacher preparation for Content Based Instruction (CBI)*. Alexandria, VA: Author.

Thickett, G. (2000). *Macmillan chemistry pathways 1*. Melbourne, Vic.: Macmillan.

Thomas, G. & Finney, R. (1996). *Calculus and analytical geometry* (9th edn). Massachusetts: Addison-Wesley Publishing Co.

Thompson, A.G. (1992). Teachers' beliefs and conceptions: A synthesis of the research. In D.A. Grouws (ed.) *NCTM handbook of research on mathematics teaching and learning* (127–46). New York: Macmillan.

Thompson, A. & Mishra, P. (2008). Breaking news: TPCK becomes TPACK! *Journal of Computing for Teacher Educators*, 24(2), 38.

Thompson, P. & Silverman, J. (2007). The concept of accumulation in calculus. In M. Carlson & C. Rassmussen (eds) *Making the connection: Research in teaching in undergraduate mathematics* (117–31). Washington, DC: Mathematical Association of America.

Thompson, S. & Fleming, N. (2004). *Summing it up: Mathematics achievement in Australian schools in TIMSS 2002*. Melbourne: ACER.

Thompson, T.D. & Preston, R.V. (2004). Measurement in the middle grades: Insights from NAEP and TIMSS. *Mathematics Teaching in the Middle School*, 9(9), 514–9.

Thomson, S. (2014). *Gender and mathematics: Quality and equity*. Retrieved 4 April 2015 from research.acer.edu.au/cgi/viewcontent. cgiarticle=1226&context=research_conference.

Thomson, S., De Bortoli, L. & Buckley, S. (2013). *PISA 2012: How Australia measures up*. Melbourne: ACER.

Thomson, S., Hillman, K. & De Bortoli, L. (2013). *A teacher's guide to PISA mathematical literacy*. Melbourne: ACER.

Thomson, S., Hilman, K. & Wernert, N. (2012). *Monitoring Australian Year 8 student achievement internationally: TIMSS 2011*. Melbourne: ACER.

Thomson, S., Hillman, K., Wernert, N., Schmid, M., Buckley, S & Munene, A. (2012). *Highlights from TIMSS & PIRLS 2011 from Australia's perspective*. Melbourne: ACER.

Tinkerplots (2012). Massachusetts: Key Curriculum Press.

Toohey, P. (1995). Adolescent perceptions of the concept of randomness (unpublished Master's thesis). University of Adelaide, Adelaide, Australia.

Tortolani, M. (2007). Presentation given at the 2007 National Association of Multicultural Engineering Program Advocates National Conference. Baltimore, MD.

Treacy, K. & Frid, S. (2008). Recognising different starting points in Aboriginal students' learning of number. Paper presented to the 31st Annual Conference of the Mathematics Education Research Group of Australasia (MERGA). Brisbane: MERGA.

Tytler, R., Osborne, J., Williams, G., Tytler, K., Cripps Clark, J., Tomei, A. & Forgasz, H. (2008). *Opening up pathways: Engagement in STEM across the primary-secondary school transition, A review of the literature concerning supports and barriers to Science, Technology, Engineering and Mathematics engagement at primary-secondary transition.* Australian Department of Education, Employment and Workplace Relations.

Tytler, R., Symington, D., Smith, C. & Rodrigues, S. (2007). *An innovation framework based on best practice exemplars from the Australian School Innovation in Science, Technology and Mathematics (ASISTM) Project.* Canberra: Commonwealth of Australia.

Tzur, R. (2008). A researcher perplexity: Why do mathematical tasks undergo metamorphosis in teacher hands? In O. Figuras, J.L. Cortina, S. Alatorre, T. Rojano & A Sepulveda (eds) *Proceedings of the 32nd annual conference of the International Group for the Psychology of Mathematics Education* (vol. 1, 139–47). Morelia: PME.

Ubuz, B. (2007). Interpreting a graph and constructing its derivative graph: Stability and change in students' conceptions. *International Journal of Mathematical Education in Science and Technology*, 38(5), 609–37.

Vale, C. & Bartholomew, H. (2008). Gender and mathematics: Theoretical frameworks and findings. In H. Forgasz, A. Barkatssas, A. Bishop, B. Clarke, S. Keast, W.T. Seah & P. Sullivan (eds) *Research in Mathematics Education in Australasia 2004–2007* (271–90). Rotterdam: Sense Publishers.

Van de Walle, J.A., Karp, K.S. & Bay-Williams, J.M. (2014). *Elementary and middle school mathematics: Teaching developmentally* (8th international edition). Essex, England: Pearson Education.

van Hiele, P.M. (1959). A child's thought and geometry. In D. Fuys, D. Geddes & R. Tischler (eds) *English translation of selected writings of Dina van Miele-Geldof and Pierre M. van Hiele* (1984). Washington, DC: Research in Science Education Program.

van Merrienboer, J.J.G., Kirschner, P.A. & Kester, L. (2003). Taking the load off a learner's mind: Instructional design for complex learning. *Educational Psychologist*, 38(1), 5–13. doi: 10.1207/s15326985ep3801_2.

Van Tassel-Baska, J. & Brown, E.F. (2007). Toward best practice: An analysis of the efficacy of curriculum models in gifted education. *Gifted Child Quarterly*, 51(4), 342–58.

Vincent, J., Price, B., Caruso, N., McNamara, A. & Tynan, D. (2011). *MathsWorld 7 Australian curriculum edition.* Melbourne: Macmillan.

Vincent, J. & Stacey, K. (2008). Do mathematics textbooks cultivate shallow teaching? Applying the TIMSS Video Study criteria to Australian eighth-grade mathematics textbooks. *Mathematics Education Research Journal*, 20(1), 82–107.

Vygotsky, L.S. (1962). *Thought and Language.* Cambridge, MA: MIT Press.

——— (1978). *Mind in society: The development of higher psychological processes.* Cambridge, MA: Harvard University Press.

Wai, J., Cacchio, M., Putallaz, M.C. & Mackel, M.C. (2010). Sex differences in the right tail of cognitive abilities: A 20-year examination. *Intelligence,* 38(4), 412–23.

Walshaw, M. & Anthony, G. (2008). The teacher's role in classroom discourse: A review of recent research into mathematics classrooms. *Review of Educational Research,* 78(3), 516–51. doi:10.3102/0034654308320292.

Warren, E. & Cooper, T. (2005). Young children's ability to use the balance strategy to solve for unknowns. *Mathematics Education Research Journal,* 17(1), 58–72. doi: 10.1007/bf03217409.

Warren, E. & Devries, E. (2010). Young Australian indigenous students. *Australian Primary Mathematics Classroom,* 15(1), 4–9.

Watson, A. & Sullivan, P. (2008). Teachers learning about tasks and lessons. In D. Tirosh & T. Wood (eds) *Tools and resources in mathematics teacher education* (109–135). Rotterdam: Sense Publishers.

Watson, J. (2006). *Statistical literacy at school: Growth and goals.* Mahwah, NJ: Lawrence Erlbaum and Associates.

Watson, J. & Kelly, B. (2007). The development of conditional probability reasoning. *International Journal of Mathematics Education in Science and Technology,* 38(2), 213–35.

Watt, H.M.G. (2005). Attitudes to the use of alternative assessment methods in mathematics: A study with secondary mathematics teachers in Sydney, Australia. *Educational Studies in Mathematics,* 58, 21–44.

Webb, D.C. (2001). Instructionally embedded assessment practices of two middle grades mathematics teachers. Unpublished doctoral dissertation, University of Wisconsin-Madison.

Weinberg, S.L. (2001). Is there a connection between fractions and division? Students' inconsistent responses. Paper presented at the Annual Meeting of the American Educational Research Association, 10–14 April 2001, Seattle, WA.

White, B., Barnes, A & Lawson, M. (2012). Student views on the value and use of interactive whiteboards in a secondary school. ACEC2012 Conference paper, Perth.

White, B. & Geer, R. (2010, October). Learner practice and satisfaction in a blended learning environment. *World Conference on E-Learning in Corporate, Government, Healthcare, and Higher Education* (vol. 2010, no. 1, 569–78).

White, P. & Sullivan, P. (2005). Using a schematic model to represent influences on, and relationships between, teachers' problem-solving beliefs and practices. *Mathematics Education Research Journal,* 17(2), 9–38.

Wiggins, G. & McTighe, J. (1998). *Understanding by design*. Alexandria, VA: Association for Supervision and Curriculum Development.

Wiliam, D. (2007). Keeping learning on track: Classroom assessment and the regulation of learning. In F. Lester (ed.) *Second Handbook of Research on Mathematics Teaching and Learning* (1053–98). Charlotte, NC: Information Age Publishing.

—— (2008). Changing classroom practice. *Educational Leadership*, 65(4), 36–41.

—— (2011). What is assessment for learning? *Studies in Educational Evaluation*, 37, 3–14.

Wilkerson-Jerde, M.H. & Wilensky, U.J. (2011). How do mathematicians learn math? Resources and acts for constructing and understanding mathematics. *Journal of Mathematical Behavior*, 78, 21–43.

Wilkie, K. & Clarke, D. (2014). Developing students' functional thinking in algebra through different visualisations of a growing pattern's structure. In J. Anderson, M. Cavanagh & A. Prescott (eds) *Proceedings of the 37th Annual Conference of the Mathematics Education Research Group of Australasia* (637–44). Sydney: MERGA.

William, S. (1991). Model of limit held by college calculus students. *Journal for Research in Mathematics Education*, 22(3), 219–36.

Williams, J. (2011). STEM education: Proceed with caution. *Design and Technology Education: An International Journal*, 16(1), 26–35.

Wood, E. (2003). Sunrise, sunset and the sine function. *Australian Senior Mathematics Journal*, 17(2), 58–63.

Yackel, E. & Cobb, P. (1996). Sociomathematical norms, argumentation, and autonomy in mathematics. *Journal for Research in Mathematics Education*, 27, 458–77.

Yeo, J. (2008). Secondary school students investigating mathematics. *Proceedings of the 31st Annual Conference of the Mathematical Education Research Group of Australasia* (613–8).

Yerushalmy, M. (2000). Problem solving strategies and mathematical resources: A longitudinal view on problem solving in a function based approach to algebra. *Educational Studies in Mathematics*, 43(1), 125–47.

Yilmaz, M. (1996). The challenge of teaching statistics to non-specialists. *Journal of Statistics Education*, 4(1). Retrieved from www.amstat.org/publications/jse.

Yopp, D. (2011). How some research mathematicians and statisticians use proof in undergraduate mathematics. *Journal of Mathematical Behavior*, 31, 115–30.

Zawojewski, J. (2007). Problem solving versus modelling. Paper presented at the 13th biennial conference of the International Community of Teachers of Mathematical Modelling and Applications, Indiana University, 22–26 July 2007.

Zawojewski, J., Lesh, R. & English, L. (2003). A models and modeling perspective on the role of small group learning activities. In R. Lesh & H.M. Doerr (eds) *Beyond constructivism: Models and modeling perspectives on mathematics problem solving, learning, and teaching* (337–58). Mahwah, NJ: Lawrence Erlbaum.

Zbiek, R. & Hollebrands, K. (2008). A research-informed view of the process of mathematics technology into classroom practice by in-service and prospective teachers In M.K. Heid & G.W. Blume (eds) *Research on technology and the teaching and learning of mathematics* (vol. 1). Charlotte, NC: Information Age Publishing.

Zembylas, M. (2005). Three perspectives on linking the cognitive and the emotional in science learning: Conceptual change, socio-constructivism and poststructualism. *Studies in Science Education*, 41, 91–116.

Index

pedagogy (*cont.*)
  pedagogical content knowledge, 62
  pedagogical strategies, 172–4
  positive pedagogical relationships, 24–5
  TPACK, 98
peer-assisted learning, 174
'per', 55
percentage change problems
  challenging, 240–2
  equation approach 1, 238–9, 240–1
  equation approach 2, 239, 241–2
  errors, 240
  teaching and learning, 237, 240–2
percentage quantity, 238, 239
percentages, 9
performance goals, 23, 24
periodicy, 349–52
Piaget, Jean, 17–18
  empirical abstraction–reflective
    abstraction, distinction, 19
  encapsulation (of a process), 19
Plan–Teach–Evaluate model, 198
planar objects, 270
plane mirrors, 258–60
plotting, 301, 305, 306–8, 335
plural, 54
point of enlargement, 267
Poisson distributions, 318
policy, technology use requirements, 97–8
polynomial graphs, 359
polynomials, 165, 206, 270, 373–4, 376–7
  exploring simple polynomials,
  polynomial graphs, 72
population, 289
  conclusion made regarding samples
    from *see* statistical inference
  population parameters, 300
  population size (*N*), 299–300
  representative samples of, 6, 289–90
  samples varying from parent, 290–1
position, 383
positive appropriate behaviour
  interventions, 172

practice *see* professional practice
pre-service teachers
  making connections, 69
  technology use, 100–1
prefixes, names of, 55
previous learning experiences
  algebra introduction using, 329–30
  with patterns, 331
prime numbers, 87
prior knowledge, 63, 22063, 220
  assessment, 242
  before beginning calculus, 358–65
  of percentage quantity, 238
probability, 107, 311–20
  approaches, 312
  in Australian Curriculum, 311–12
  conditional, 316–17, 322–3
  formal components, 313–17
  rules, 314–17
  statistics and, 288
  teaching implications, 317
  theoretical and experimental, 25, 89
  weighted dice activity, 87–9
  of zero, 320
probability distributions, 317–20
  example, 317
  probabilities all add to one, 317
problem-based learning 4–5, 130
problem-solving, 10–11, 254
  algebraic expression problems, 219
  algebraic problem-solving, 217, 236
  appropriate problem choice, 132
  computer program tools for, 102
  formalisation of tasks, 11
  General Capabilities on importance
    of, 84
  as instructional strategy, 73–4
  vs. investigation, 192
  language problem? 36, 42–5
  for mathematical connections
    activities, 83
  mathematical word problems, 42–3
  molarity chemistry problems, 245